Modeling and Simulation in Science, Engineering and Technology

Series Editor

Nicola Bellomo
Politecnico di Torino
Italy

Advisory Editorial Board

K.J. Bathe
Massachusetts Institute of Technology
USA

W. Kliemann
Iowa State University
USA

S. Nikitin
Arizona State University
USA

V. Protopopescu
CSMD
Oak Ridge National Laboratory
USA

P. Degond
Université P. Sabatier Toulouse 3
France

P. Le Tallec
INRIA, BP 105
France

K.R. Rajagopal
Texas A&M University
USA

Y. Sone
Kyoto University
Japan

E.S. Subuhi
Istanbul Technical University
Turkey

M. Avellaneda
Courant Institute of Mathematical Sciences
New York University
USA

J. Douglas, Jr.
Purdue University
USA

H.G. Othmer
University of Utah
USA

Vincent Giovangigli

Multicomponent Flow Modeling

Birkhäuser
Boston • Basel • Berlin

Vincent Giovangigli
Centre de Mathématiques Appliquées
École Polytechnique
91128 Palaiseau Cedex
France

Library of Congress Cataloging-in-Publication Data
Giovangigli, Vincent, 1957–
Multicomponent flow modeling / Vincent Giovangigli.
p. cm.—(Modeling and simulation in science, engineering and technology)
Includes bibliographical references and index.
ISBN 0-8176-4048-7
1. Multiphase flow—Mathematical models. I. Title. II. Series: Modeling and simulation in science, engineering & technology.
QA922.G56 1999
532′.051—dc21 99-27050
CIP

AMS Subject Classifications: 76A, 80A, 82B, 35B, 35Q

Printed on acid-free paper.

Birkhäuser

ISBN 0-8176-4048-7
ISBN 3-7643-4048-7 SPIN 19901762

Typeset by the author in TEX.
Printed and bound by Maple-Vail Book Manufacturing Group, York, PA.
Printed in the United States of America.

9 8 7 6 5 4 3 2 1

Contents

Preface

The goal of this is book to give a detailed presentation of multicomponent flow models and to investigate the mathematical structure and properties of the resulting system of partial differential equations. These developments are also illustrated by simulating numerically a typical laminar flame. Our aim in the chapters is to treat the general situation of multicomponent flows, taking into account complex chemistry and detailed transport phenomena.

In this book, we have adopted an interdisciplinary approach that encompasses a physical, mathematical, and numerical point of view. In particular, the links between molecular models, macroscopic models, mathematical structure, and mathematical properties are emphasized. We also often mention flame models since combustion is an excellent prototype of multicomponent flow.

This book still does not pretend to be a complete survey of existing models and related mathematical results. In particular, many subjects like multiphase–flows, turbulence modeling, specific applications, porous media, biological models, or magneto–hydrodynamics are not covered. We rather emphasize the fundamental modeling of multicomponent gaseous flows and the qualitative properties of the resulting systems of partial differential equations.

Part of this book was taught at the post-graduate level at the University of Paris, the University of Versailles, and at École Polytechnique in 1998–1999 to students of applied mathematics.

The author wishes to thank many of his colleagues for stimulating discussions, especially to Professor Alexandre Ern of École Nationale des Ponts et Chaussées, Doctor Marc Massot of Laboratoire d'Analyse Numérique de l'Université de Lyon, and Professor Mitchell D. Smooke of Yale University

Mechanical Engineering Department.

The author wishes to thank the series editor, Professor Nicola Bellomo, for his encouragement and interest, the executive editor Wayne Yuhasz, and the senior production editor Louise Farkas for their precious help in editing the book.

I also wish to thank my wife Professor Hélène Dumontet of Université de Cergy, and my children Laure, Antoine, and Marine for their love and support.

Paris, France
July 1999

Vincent Giovangigli

1

Introduction

Multicomponent reactive flows arise in various engineering applications, such as combustion, crystal growth, atmospheric reentry, or chemical reactors. Modeling pollutant formation, chemical vapor deposition reactors, laminar flame extinction limits, or gas dissociation behind bow shocks around space vehicles, for instance, requires us to take into account complex chemistry mechanisms and detailed transport phenomena. There is, thus, a strong motivation for investigating the equations governing multicomponent flows and analyzing their mathematical structure and properties. The present book is an attempt to fill this need.

In Chapters 2 to 5 we first give a detailed presentation of multicomponent flow models, and, in Chapters 6 to 11, we then analyze some of their mathematical structures and properties. These developments are also illustrated by simulating numerically a typical laminar flame in Chapter 12. Our aim, in these chapters, will always be to take into account multicomponent aspects, such as complex chemistry and detailed transport.

In Chapter 2 we give a detailed presentation of the governing equations for multicomponent reactive flows, as obtained from the kinetic theory of gases. We present the fundamental conservation equations for mass, momentum, and energy and discuss various alternative formulations. We investigate thermodynamic properties, chemical production rates, as well as transport fluxes and several possible extensions of the model. We also present the entropy governing equation, which will play a fundamental role.

In Chapter 3 we discuss various ideas that can be used to simplify the general equations in special situations. These simplifications can be in the chemistry aspects, fluid aspects, or coupling between them. This chapter is devoted to filling the gap between the complete equations presented in the previous chapter and the simplified models often used in the literature. We specifically discuss the simplifications associated with one-reaction chemistry flows, small Mach number flows, strained flows, and the uncouplings resulting from the dilution limit and the incompressible limit.

In Chapter 4 we summarize the derivation from the kinetic theory of gases. This chapter treats the general situation of mixtures of polyatomic gases with chemical reactions and reveals the kinetic origin of the model presented in Chapter 2. It is the only chapter that specifically considers the molecular aspects of multicomponent flows and goes deeper into the physical understanding of the model. We present the kinetic framework and consider the Enskog expansion in a regime in which chemical characteristic times are larger than the mean free time of the molecules. We obtain, in particular, the macroscopic conservation equations, thermodynamic properties, transport fluxes, and chemical production rates. We further obtain the macroscopic entropy conservation equation and discuss the link between the kinetic entropy and the macroscopic entropy. We also introduce the transport linear systems investigated in the next chapter.

In Chapter 5 we investigate the evaluation of the transport coefficients in multicomponent mixtures, which is an important task in practical applications. These coefficients are not explicitly given by the kinetic theory, but their evaluation requires solving transport linear systems. Iterative solution of transport linear systems is investigated as well as direct inversions. Various practical transport linear systems of reduced dimension are also presented.

Chapters 2 to 5 then constitute a detailed description of multicomponent reactive gas models from the molecular underlying up to practical approximations. In the following chapters, we analyze the mathematical quality of these models, that is, we investigate the structure and properties of the resulting set of partial differential equations.

In Chapter 6 we investigate various mathematical aspects of thermochemistry. We investigate, in particular, chemical equilibrium states, differential and convexity properties of various thermodynamic functions, and stability inequalities. The existence of equilibrium points and various stability inequalities will be used in Chapters 9, 10, and 11.

In Chapter 7 we investigate the mathematical structure of multicomponent transport fluxes and transport coefficients. We obtain, in particular, the fundamental diffusion inequality, which reveals that the natural entropy production norm is a nonlinear solution-weighted norm, and we elucidate the diagonal diffusion problem. Chapters 6 and 7 are necessary preliminaries before investigating mathematically multicomponent flow models.

In Chapter 8 we investigate the general structure of the set of partial differential equations governing multicomponent flows. We first restate the link between the existence of an entropy function and the symmetrizability of hyperbolic–parabolic composite systems and then discuss normal forms. These results are then applied to multicomponent flows, and we obtain, in particular, symmetric formulations of the governing equations. These symmetric formulations are important for existence results as well as for streamline upwind Petrov–Galerkin finite element simulations.

In Chapter 9 we obtain a global existence theorem around equilibrium states as well as asymptotic stability and decay estimates. Our method of proof relies on the normal form of the governing equations, on entropic estimates, and on the local dissipativity properties of linearized equations. We use, in particular, the results of Chapters 6, 7, and 8.

In Chapter 10 we then investigate the mathematical structure of the equations governing chemical equilibrium flows. These flows are a limiting model that is of interest for various applications, such as chemical vapor deposition reactors, flows around space vehicles, or diverging nozzle rocket flows. These simplified models provide reasonable predictions when the characteristic chemical times are small in comparison with the flow time. The associated computational costs are significantly reduced in comparison with nonequilibrium flows. Chemical equilibrium flows are governed by equations expressing the conservation of atoms, momentum, and energy. We first present these governing equations and then investigate their symmetrization. Upon using a normal form, we finally obtain global existence results around constant states.

In Chapter 11 we investigate the anchored wave problem with complex chemistry and detailed transport. Traveling waves in inert or reactive flows can indeed be classified into deflagration and detonation waves. In the context of combustion—which does not decrease the generality of the problem but makes things more explicit—weak deflagration waves correspond to plane laminar flames. We investigate the anchored flame model using, in particular, the formalism of Chapters 6 and 7. Our method of proof mainly relies on entropic estimates and on Leray–Schauder topological degree theory.

The results obtained in Chapters 6 to 11 reveal the mathematical quality of the models derived from the kinetic theory. Although these mathematical results are far from being complete, they still rigorously justify the quality of the model.

Finally, in Chapter 12 we illustrate the preceding developments by simulating numerically a typical laminar flame model. We also indicate some of the difficulties typically associated with the multicomponent aspects, such as the presence of multiple time scales and multiple space scales. Our computational model is that of an axisymmetric hydrogen–air Bunsen laminar flame with complex chemistry and detailed transport.

Every chapter ends with notes in which we discuss, in particular, various model generalizations. These notes are intented to supplement the text and place it in a better perspective.

The main logical connections between the chapters are indicated in the following diagram.

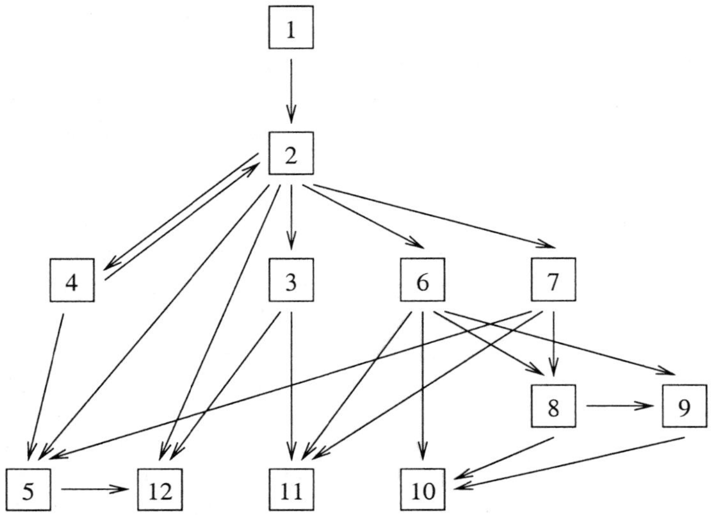

The reader may want to take different reading paths, depending on one's center of interest.

2

Fundamental Equations

2.1. Introduction

In this chapter we present in detail the equations governing multicomponent reactive flows. The derivation of these equations from the kinetic theory of gases can be found in classical textbooks, usually for nonreactive and/or monatomic mixtures [CC70] [FK72] [WT62], and is summarized in Chapter 4 in the general situation of polyatomic reactive gas mixtures [EG94].

We first present the primitive conservation equations in Section 2.2 and further obtain various alternative formulations. The thermodynamic properties are then expressed in Section 2.3, and the chemical production rates are expressed in Section 2.4. The transport fluxes and transport coefficients are presented in Section 2.5. Note, however, that the transport coefficients are not given explicitly by the kinetic theory of gases. Their evaluation requires solving linear systems, as will be discussed in Chapters 4 and 5. These results are then combined to derive the entropy governing equation in Sections 2.6. The concept of entropy will be of fundamental interest in the modeling as well as in the mathematical analysis of multicomponent reactive gas mixtures. Finally we present typical boundary conditions associated with multicomponent flows, such as solid–gas interfaces, in Section 2.7.

The equations presented in Sections 2.2–2.7 can be used to model very different phenomena, such as laminar flames [Smo82] [Dix84] [Wil85] [EG98b], supersonic flames [DH87] [DM89] [KOB89] [Mas96], epitaxial growth [RJ87] [KEC87] [EGS96], or reentry problems into the earth's atmosphere [And89] [LGC92]. Some applications require the full set of equations, without any simplification, but simplifications are sometimes feasible under various approximations, as will be discussed in Chapter 3.

2.2. Conservation equations

In this section we present the primitive conservation equations governing multicomponent flows, which express the conservation of species mass, momentum, and energy. These primitive equations are then combined to derive various alternative formulations. We obtain, in particular, governing equations for total mass, species mass fractions, kinetic energy, and internal energy, and discuss the simplified situation of species independent specific forces.

2.2.1. Species, momentum, and energy

The balance equations associated with multicomponent reactive flows express the conservation of species mass, momentum, and energy. Species mass conservation can be written in the form

$$\partial_t \rho_k + \boldsymbol{\partial_x}\cdot(\rho_k \boldsymbol{v}) + \boldsymbol{\partial_x}\cdot\boldsymbol{\mathcal{F}}_k = m_k \omega_k, \qquad k \in S, \tag{2.2.1}$$

where ρ_k is the mass density of the kth species, $\boldsymbol{v}$ is the mass average flow velocity, $\boldsymbol{\mathcal{F}}_k$ is the diffusion flux of the kth species, m_k is the molar mass of the kth species—also termed the molecular weight of the kth species,—ω_k is the molar production rate of the kth species, S is the set of species indices $S = \{1,\ldots,n\}$, and n is the number of species. In these equations, t denotes the time, $\boldsymbol{x}$ denotes the space variable, ∂_t denotes the time derivative operator, and $\boldsymbol{\partial_x}$ denotes the space derivative operator, that is, $\boldsymbol{\partial_x} = (\partial_x, \partial_y, \partial_z)^t$ if $\boldsymbol{x} = (x, y, z)^t$ denotes the spatial coordinates and ^{t} denotes the transposition. Note that vectors belonging to the physical space $\mathbb{R}^3$ are written in boldface style as are tensors acting on these vectors in this book. We denote by the symbol $\cdot$ the scalar product in $\mathbb{R}^3$.

The momentum conservation equation can be written

$$\partial_t(\rho \boldsymbol{v}) + \boldsymbol{\partial_x}\cdot(\rho \boldsymbol{v}\otimes\boldsymbol{v} + p\boldsymbol{I}) + \boldsymbol{\partial_x}\cdot\boldsymbol{\Pi} = \sum_{k\in S} \rho_k \boldsymbol{b}_k, \tag{2.2.2}$$

where ρ is the total mass density, $\boldsymbol{v}\otimes\boldsymbol{v}$ is the tensor product of velocity vectors, p is the pressure, $\boldsymbol{I}$ is the unit tensor, $\boldsymbol{\Pi}$ is the viscous tensor, and $\boldsymbol{b}_k$ is the external specific force acting on the kth species.

Finally, the energy conservation equation reads

$$\partial_t\big(\rho e^{\text{tot}}\big) + \boldsymbol{\partial_x}\cdot\big((\rho e^{\text{tot}} + p)\boldsymbol{v}\big) + \boldsymbol{\partial_x}\cdot(\boldsymbol{\mathcal{Q}} + \boldsymbol{\Pi}\cdot\boldsymbol{v}) = \sum_{k\in S}(\rho_k \boldsymbol{v} + \boldsymbol{\mathcal{F}}_k)\cdot\boldsymbol{b}_k, \tag{2.2.3}$$

where e^{tot} is the specific total energy and $\boldsymbol{\mathcal{Q}}$ is the heat flux.

A nonlocal interpretation of these equations is easily obtained by integrating over an arbitrary control volume, making use of classical vector analysis, as shown in classical textbooks on fluid mechanics [Bat67] [Can90].

2.2.2. Total mass conservation

By summing the n species conservation equations, we can derive the total mass conservation equation associated with the total density

$$\rho = \sum_{k\in S} \rho_k. \tag{2.2.4}$$

Anticipating the mass constraint

$$\sum_{k\in S} \boldsymbol{\mathcal{F}}_k = 0, \tag{2.2.5}$$

between the diffusion fluxes $\boldsymbol{\mathcal{F}}_k$, $k \in S$, and the mass constraint

$$\sum_{k\in S} m_k\omega_k = 0, \tag{2.2.6}$$

between the chemical source terms ω_k, $k \in S$—which will later be established—we obtain the total mass conservation equation

$$\partial_t\rho + \boldsymbol{\partial_x}\cdot(\rho\boldsymbol{v}) = 0. \tag{2.2.7}$$

Using the total mass conservation equation (2.2.7), one may then easily establish that, for any function ζ of $(t, \boldsymbol{x})$, we have the identity

$$\partial_t(\rho\zeta) + \boldsymbol{\partial_x}\cdot(\rho\boldsymbol{v}\zeta) = \rho\partial_t\zeta + \rho\boldsymbol{v}\cdot\boldsymbol{\partial_x}\zeta, \tag{2.2.8}$$

in such a way that all previous conservation equations written in conservative form can also be rewritten in nonconservative form. As a typical example, the momentum conservation equation can be rewritten as

$$\rho\partial_t\boldsymbol{v} + \rho\boldsymbol{v}\cdot\boldsymbol{\partial_x}\boldsymbol{v} + \boldsymbol{\partial_x}p + \boldsymbol{\partial_x}\cdot\boldsymbol{\Pi} = \sum_{k\in S}\rho_k\boldsymbol{b}_k. \tag{2.2.9}$$

The equivalence between conservative and nonconservative forms for smooth solutions will be used freely in this book.

2.2.3. Species mass fractions

The species mass fractions Y_k, $k \in S$, are defined by

$$Y_k = \frac{\rho_k}{\rho}, \qquad k \in S, \tag{2.2.10}$$

and satisfy—by definition—the relation

$$\sum_{k\in S} Y_k = 1. \tag{2.2.11}$$

The mass fraction Y_k of the kth species locally represents the mass of the kth species with respect to the total mass of the mixture. The associated conservation equations now read as

$$\partial_t(\rho Y_k) + \boldsymbol{\partial_x}\cdot(\rho Y_k \boldsymbol{v}) + \boldsymbol{\partial_x}\cdot\boldsymbol{\mathcal{F}}_k = m_k\omega_k, \qquad k \in S, \tag{2.2.12}$$

and constitute an alternative for describing species conservation.

Note that the n species conservation equations (2.2.1) or (2.2.12) and the total mass conservation equations (2.2.7) are a priori linearly dependent. Therefore, it is possible to only solve the n species equations (2.2.1) with the species unknowns $(\rho_1, \dots, \rho_n)$. An alternative is to solve the total mass conservation equations and $n-1$ species equations, using the species unknowns $(\rho, \rho_1, \dots, \rho_{n-1})$ or, equivalently, $(\rho, Y_1, \dots, Y_{n-1})$. In this situation, one evaluates ρ_n from $\rho_n = \rho - \sum_{k\neq n} \rho_k$ or from $Y_n = 1 - \sum_{k\neq n} Y_k$.

A third important possibility is to consider *all* mass fractions as formally independent unknowns. In this situation, the total mass conservation equation (2.2.7) and the n species conservation equations (2.2.12) are solved with the species unknowns $(\rho, Y_1, \dots, Y_n)$. The relation $\sum_{k\in S} Y_k = 1$, which was a consequence from definition (2.2.10), must now be *deduced* from the governing equations. More specifically, the quantity $\mathsf{Y} = \sum_{k\in S} Y_k$ then satisfies a conservation equation easily obtained from (2.2.7) and (2.2.12):

$$\rho\partial_t \mathsf{Y} + \rho\boldsymbol{v}\cdot\boldsymbol{\partial_x}\mathsf{Y} = 0, \tag{2.2.13}$$

from which we ultimately recover that $\mathsf{Y} = \sum_{k\in S} Y_k = 1$, making use of initial and boundary conditions. This procedure is often used for complex chemistry flows, in particular, when the first-order equation (2.2.7) and the second-order equations (2.2.12) are discretized differently and a symmetric role is given to all species. In order to encompass all possible situations, we will sometimes keep factors in the form $\sum_{k\in S} Y_k$ in various expressions, e.g., by writing $\sum_{k\in S} \rho_k$ instead of ρ, understanding that, when the species unknowns are $(\rho_1, \dots, \rho_n)$, these factors $\sum_{k\in S} Y_k$ are a priori unity, whereas, when the species unknowns are $(\rho, Y_1, \dots, Y_n)$, these factors are a posteriori unity.

2.2.4. Kinetic and internal energy

The total specific energy of the mixture e^{tot} can be decomposed into

$$e^{\text{tot}} = e + \tfrac{1}{2}\boldsymbol{v}\cdot\boldsymbol{v}, \tag{2.2.14}$$

where e is the specific internal energy and $\frac{1}{2}\boldsymbol{v}\cdot\boldsymbol{v}$ is the specific kinetic energy. Taking the scalar product of the momentum conservation equation

(2.2.2) by the velocity vector $\boldsymbol{v}$ yields—after some algebra—the conservation equation for kinetic energy

$$\partial_t(\rho\tfrac{1}{2}\boldsymbol{v}\cdot\boldsymbol{v})+\boldsymbol{\partial_x}\cdot\big((\rho\tfrac{1}{2}\boldsymbol{v}\cdot\boldsymbol{v}+p)\boldsymbol{v}\big)+\boldsymbol{\partial_x}\cdot\big(\boldsymbol{\Pi}\cdot\boldsymbol{v}\big)=p\,\boldsymbol{\partial_x}\cdot\boldsymbol{v} + \boldsymbol{\Pi}:\boldsymbol{\partial_x}\boldsymbol{v}+\sum_{k\in S}\rho_k\boldsymbol{b}_k\cdot\boldsymbol{v}, \tag{2.2.15}$$

where the symbol : denotes the full contraction between two tensors in $\mathbb{R}^3$, so that the term $-\boldsymbol{\Pi}:\boldsymbol{\partial_x}\boldsymbol{v}$ represents the viscous dissipation rate. Subtracting this equation from the total energy conservation equation (2.2.3), we obtain an equation for the internal energy e

$$\partial_t(\rho e)+\boldsymbol{\partial_x}\cdot(\rho e\boldsymbol{v})+\boldsymbol{\partial_x}\cdot\boldsymbol{\mathcal{Q}}=-p\,\boldsymbol{\partial_x}\cdot\boldsymbol{v}-\boldsymbol{\Pi}:\boldsymbol{\partial_x}\boldsymbol{v}+\sum_{k\in S}\boldsymbol{\mathcal{F}}_k\cdot\boldsymbol{b}_k. \tag{2.2.16}$$

A temperature equation will be derived from (2.2.16) in Section 2.3, once the expression of thermodynamic functions is available.

Finally, note that a total pressure tensor $\boldsymbol{\mathcal{P}}$ can also be defined as

$$\boldsymbol{\mathcal{P}}=p\boldsymbol{I}+\boldsymbol{\Pi}, \tag{2.2.17}$$

and the momentum and energy conservation equations are easily rewritten in terms of $\boldsymbol{\mathcal{P}}$. It has been preferred, however, to clearly separate dissipative effects associated with $\boldsymbol{\Pi}$ from nondissipative effects associated with $p\boldsymbol{I}$.

2.2.5. Species independent specific forces

For most practical applications, the only external force acting on the species is gravity, so that

$$\boldsymbol{b}_k=\boldsymbol{g}, \tag{2.2.18}$$

where $\boldsymbol{g}$ denotes the gravity vector. In this situation, the momentum and energy conservation equations can be simplified by using

$$\sum_{k\in S}\rho_k\boldsymbol{b}_k=\rho\boldsymbol{g} \tag{2.2.19}$$

and

$$\sum_{k\in S}\boldsymbol{\mathcal{F}}_k\cdot\boldsymbol{b}_k=0,$$

anticipating the mass conservation constraint $\sum_{k\in S}\boldsymbol{\mathcal{F}}_k=0$.

As a consequence, in the particular situation where $\boldsymbol{b}_k=\boldsymbol{g}$, $k\in S$, we obtain the simplified equations

$$\partial_t(\rho\boldsymbol{v})+\boldsymbol{\partial_x}\cdot(\rho\boldsymbol{v}\otimes\boldsymbol{v})+\boldsymbol{\partial_x}p+\boldsymbol{\partial_x}\cdot\boldsymbol{\Pi}=\rho\boldsymbol{g}, \tag{2.2.20}$$

$$\partial_t(\rho e) + \boldsymbol{\partial_x}\cdot(\rho e \boldsymbol{v}) + \boldsymbol{\partial_x}\cdot \boldsymbol{\mathcal{Q}} = -p\boldsymbol{\partial_x}\cdot\boldsymbol{v} - \boldsymbol{\Pi} : \boldsymbol{\partial_x} \boldsymbol{v}. \qquad (2.2.21)$$

2.3. Thermodynamics

In this section, we introduce the fundamental thermodynamic relations. It is worthwhile to point out that thermodynamics obtained in the framework of the kinetic theory of gases is valid out of equilibrium and has, therefore, a wider range of validity than classical thermodynamics introduced for stationary homogeneous equilibrium states. The formalism obtained from the kinetic theory still coincides with the Gibbs formalism applied to intensive variables.

We first consider the state law and the internal energy and then introduce the specific enthalpy and derive the corresponding balance equation. We next obtain the temperature balance equation from the enthalpy conservation equation. Finally, we consider the specific entropy and Gibbs function. The entropy will play an important role as a modeling concept and as a mathematical tool. The entropy conservation equation will be established in Section 2.6. The mathematical properties of thermodynamic functions are investigated in Chapter 6.

2.3.1. Density and internal energy

The kinetic theory of dilute gas mixtures yields the perfect gas law

$$\rho = \frac{p\overline{m}}{RT}, \qquad (2.3.1)$$

where $\overline{m}$ is the mean molar mass of the mixture—also termed the mean molecular weight,—R is the perfect gas constant, and T is the absolute temperature. The mean molar mass of the mixture $\overline{m}$ appearing in (2.3.1) is given by

$$\frac{\sum_{k\in S}\rho_k}{\overline{m}} = \sum_{k\in S}\frac{\rho_k}{m_k}. \qquad (2.3.2)$$

The specific internal energy—the internal energy per unit mass—can be decomposed into

$$\rho e = \sum_{k\in S}\rho_k e_k, \qquad (2.3.3)$$

where e_k is the specific internal energy of the kth species. The internal energy per unit mass of the kth species e_k is given by

$$e_k = e_k^{\text{st}} + \int_{T^{\text{st}}}^{T} c_{vk}(T')\,dT', \qquad (2.3.4)$$

where $e_k^{\mathrm{st}} = e_k(T^{\mathrm{st}})$ is the formation energy of the kth species at the positive standard temperature T^{st} and c_{vk} is the constant-volume specific heat of the kth species. The mixture specific heat at constant volume c_v is then obtained from

$$\rho c_v = \sum_{k \in S} \rho_k c_{vk}. \tag{2.3.5}$$

2.3.2. Enthalpy

The specific enthalpy of the kth species h_k is given by

$$h_k = e_k + \frac{RT}{m_k} = e_k + r_k T, \tag{2.3.6}$$

where $r_k = R/m_k$. The mixture enthalpy h is expressed in terms of the species enthalpies per unit mass h_k, $k \in S$, from the relation

$$\rho h = \sum_{k \in S} \rho_k h_k. \tag{2.3.7}$$

These relations imply—after some algebra—that

$$h = e + \frac{p}{\rho}. \tag{2.3.8}$$

We further define the constant-pressure specific heat of the kth species c_{pk} from

$$c_{pk} = c_{vk} + \frac{R}{m_k} = c_{vk} + r_k, \tag{2.3.9}$$

in such a way that

$$h_k = h_k^{\mathrm{st}} + \int_{T^{\mathrm{st}}}^{T} c_{pk}(T')\, dT', \tag{2.3.10}$$

where $h_k^{\mathrm{st}} = e_k^{\mathrm{st}} + RT^{\mathrm{st}}/m_k$ is the formation enthalpy of the kth species at the positive standard temperature T^{st}. The constant-pressure specific heat of the mixture c_p is then given by

$$\rho c_p = \sum_{k \in S} \rho_k c_{pk}, \tag{2.3.11}$$

in such a way that

$$\rho c_p = \rho c_v + \sum_{k \in S} R \frac{\rho_k}{m_k}. \tag{2.3.12}$$

2.3.3. Mole fractions and molar concentrations

Modeling multicomponent flows further requires introducing molar quantities. Indeed, although mass-weighted quantities naturally appear in conservation equations—momentum being conserved in molecular collisions—transport processes and chemical production rates depend on collision frequencies between molecules, which are intrinsically molar quantities. In this section, we introduce various molar quantities that will be used in the modeling.

The mole fraction of the kth species X_k is defined by

$$X_k = \frac{\overline{m}}{m_k} Y_k, \qquad k \in S, \tag{2.3.13}$$

and locally represents the number of moles of the kth species with respect to the number of moles of the mixture.

Other molar quantities are the species partial pressures

$$p_k = \frac{\rho_k RT}{m_k}, \qquad k \in S, \tag{2.3.14}$$

and the species molar concentrations, that is, the species number of moles per unit volume

$$\gamma_k = \frac{p_k}{RT} = \frac{\rho_k}{m_k}, \qquad k \in S, \tag{2.3.15}$$

that often appears in chemical production rates.

2.3.4. Enthalpy and temperature equations

From the relation $\rho h = \rho e + p$ and the internal energy conservation equation (2.2.16), we obtain the enthalpy balance equation

$$\partial_t(\rho h) + \boldsymbol{\partial_x}\cdot(\rho h \boldsymbol{v}) + \boldsymbol{\partial_x}\cdot\boldsymbol{\mathcal{Q}} = -\boldsymbol{\Pi}:\boldsymbol{\partial_x}\boldsymbol{v} + \sum_{k\in S} \boldsymbol{\mathcal{F}}_k\cdot\boldsymbol{b}_k + \partial_t p + \boldsymbol{v}\cdot\boldsymbol{\partial_x} p. \tag{2.3.16}$$

On the other hand, expanding the derivatives of ρh, we obtain that

$$\begin{aligned}\partial_t(\rho h) + \boldsymbol{\partial_x}\cdot(\rho h \boldsymbol{v}) &= \rho\partial_t h + \rho\boldsymbol{v}\cdot\boldsymbol{\partial_x} h \\ &= \rho c_p \partial_t T + \rho c_p \boldsymbol{v}\cdot\boldsymbol{\partial_x} T + \sum_{k\in S} h_k(\rho\partial_t Y_k + \rho\boldsymbol{v}\cdot\boldsymbol{\partial_x} Y_k),\end{aligned}$$

and we observe that

$$\sum_{k\in S} h_k \boldsymbol{\partial_x}\cdot\boldsymbol{\mathcal{F}}_k. = \boldsymbol{\partial_x}\cdot\Big(\sum_{k\in S} h_k\boldsymbol{\mathcal{F}}_k\Big) - \Big(\sum_{k\in S}\boldsymbol{\mathcal{F}}_k c_{pk}\Big)\cdot\boldsymbol{\partial_x} T.$$

Upon using these relations and the species conservation equations, we finally obtain the balance equation for the absolute temperature T

$$\rho c_p \partial_t T + \rho c_p \boldsymbol{v}\cdot\boldsymbol{\partial_x} T + \boldsymbol{\partial_x}\cdot\Big(\boldsymbol{\mathcal{Q}} - \sum_{k\in S} h_k \boldsymbol{\mathcal{F}}_k\Big) = -\boldsymbol{\Pi}:\boldsymbol{\partial_x}\boldsymbol{v} + \partial_t p + \boldsymbol{v}\cdot\boldsymbol{\partial_x} p$$
$$+ \sum_{k\in S} \boldsymbol{\mathcal{F}}_k\cdot\boldsymbol{b}_k - \Big(\sum_{k\in S} \boldsymbol{\mathcal{F}}_k c_{pk}\Big)\cdot\boldsymbol{\partial_x} T - \sum_{k\in S} h_k m_k \omega_k. \tag{2.3.17}$$

Note, in particular, the term $-\sum_{k\in S} h_k m_k \omega_k$, which represents the heat release rate due to chemical reactions.

2.3.5. Entropy and Gibbs function

The specific entropy of the mixture s can be expressed in terms of the species specific entropies s_k, $k \in S$, from the relation

$$\rho s = \sum_{k\in S} \rho_k s_k, \tag{2.3.18}$$

and the specific entropy of the kth species s_k is given by

$$s_k = s_k^{\mathrm{st}} + \int_{T^{\mathrm{st}}}^{T} \frac{c_{pk}(T')}{T'}\, dT' - \frac{R}{m_k}\log\frac{p_k}{p^{\mathrm{st}}}, \tag{2.3.19}$$

where $s_k^{\mathrm{st}} = s_k(T^{\mathrm{st}}, p^{\mathrm{st}})$ is the formation entropy of the kth species at the standard positive temperature T^{st} and the standard pressure $p^{\mathrm{st}} = p^{\mathrm{atm}}$. We can also express the entropy per unit mass s_k in terms of the mass density ρ_k of the kth species

$$s_k = s_k^{\mathrm{st}} + \int_{T^{\mathrm{st}}}^{T} \frac{c_{vk}(T')}{T'}\, dT' - \frac{R}{m_k}\log\left(\frac{\rho_k}{\gamma^{\mathrm{st}} m_k}\right), \tag{2.3.20}$$

where $\gamma^{\mathrm{st}} = p^{\mathrm{st}}/RT^{\mathrm{st}}$ is the standard concentration, that is, the concentration at the standard state T^{st}, p^{st}.

For future use, we also denote by $s_k^{\mathrm{atm}} = s_k^{\mathrm{atm}}(T)$ the entropy of the kth species at atmospheric pressure $p_k = p^{\mathrm{st}} = p^{\mathrm{atm}}$, that is,

$$s_k^{\mathrm{atm}} = s_k^{\mathrm{st}} + \int_{T^{\mathrm{st}}}^{T} \frac{c_{pk}(T')}{T'}\, dT'. \tag{2.3.21}$$

Similarly, we can express the specific Gibbs function g in terms of the species specific Gibbs functions g_k, $k \in S$, from the relation

$$\rho g = \sum_{k\in S} \rho_k g_k, \tag{2.3.22}$$

where g_k is given by

$$g_k = h_k - Ts_k, \tag{2.3.23}$$

in such a way that

$$g = h - Ts. \tag{2.3.24}$$

We can also introduce the Gibbs function of the kth species at atmospheric pressure

$$g_k^{\text{atm}} = h_k - Ts_k^{\text{atm}}, \tag{2.3.25}$$

which is easily expressed in terms of the temperature. The species Gibbs functions naturally appear in entropy differentials as well as in the study of chemical equilibrium points.

2.3.6. Alternative formulations

In the previous sections, the energy conservation equation has been written in terms of the mass densities e and h. Similarly, we have also written the species conservation equations in terms of the mass densities Y_k, $k \in S$. It is also possible, however, to introduce volumetric as well as molar densities. These various quantities provide alternative descriptions of the mixture and are briefly discussed in this section.

For the species, we have already written the equations in terms of the volumetric densities ρ_k, $k \in S$, and molar densities have also been introduced. Similarly, the energy conservation equation can be rewritten in terms of the energy per unit volume $\mathcal{E}$ given by

$$\mathcal{E} = \rho e, \tag{2.3.26}$$

or the enthalpy per unit volume $\mathcal{H}$, which reads

$$\mathcal{H} = \mathcal{E} + p = \rho h. \tag{2.3.27}$$

Similarly, we can define the entropy $\mathcal{S}$ per unit volume

$$\mathcal{S} = \rho s, \tag{2.3.28}$$

and the Gibbs function $\mathcal{G}$ per unit volume

$$\mathcal{G} = \mathcal{H} - T\mathcal{S} = \rho g. \tag{2.3.29}$$

These quantities will be used, in particular, in Chapters 8, 9, and 10. The corresponding equations are easily obtained after simple substitutions.

We can further introduce the enthalpy, energy, and specific heats per unit mole of the species or per unit mole of the mixture. Denoting these molar quantities with uppercase letters, we obtain the following relations between the molar constant-volume specific heats C_{vk} and energy E_k and the corresponding mass properties c_{vk} and e_k :

$$c_{vk} = \frac{C_{vk}}{m_k}, \qquad e_k = \frac{E_k}{m_k}, \tag{2.3.30}$$

in such a way that

$$c_v = \frac{C_v}{\overline{m}}, \qquad e = \frac{E}{\overline{m}}, \tag{2.3.31}$$

where C_v is the heat capacity at constant volume per unit mole of the mixture and E is the energy per unit mole of the mixture. Correspondingly, we have

$$c_{pk} = \frac{C_{pk}}{m_k}, \qquad h_k = \frac{H_k}{m_k}, \tag{2.3.32}$$

in such a way that

$$c_p = \frac{C_p}{\overline{m}}, \qquad h = \frac{H}{\overline{m}}, \tag{2.3.33}$$

where C_p is the heat capacity at constant pressure per unit mole of the mixture and H is the enthalpy per unit mole of the mixture. We then have obvious relations, such as

$$\mathcal{E} = \rho e = \gamma E = \sum_{k \in S} \rho_k e_k = \sum_{k \in S} \gamma_k E_k,$$

where $\gamma = \sum_{k \in S} \gamma_k$ is the total concentration of the mixture.

Similarly, the molar quantities S_k and G_k are easily obtained from the corresponding mass quantities s_k and g_k and the relations

$$s_k = \frac{S_k}{m_k}, \qquad g_k = \frac{G_k}{m_k}, \tag{2.3.34}$$

in such a way that

$$s = \frac{S}{\overline{m}}, \qquad g = \frac{G}{\overline{m}}. \tag{2.3.35}$$

As for the energy per unit volume, we also have the relations

$$\mathcal{S} = \rho s = \gamma S = \sum_{k \in S} \rho_k s_k = \sum_{k \in S} \gamma_k S_k.$$

All of these descriptions are useful and may be found in the literature. Volumetric densities are often used in fluid dynamics when conservation is under concern and tabulated data are often in the form of molar densities. Mass fractions and mass densities, on the other hand, are often used in combustion studies.

2.3.7. Thermodynamic data

Thermodynamic data needed for modeling multicomponent reactive flows are the species constant-pressure specific heats $c_{pk}(T)$, $k \in S$, or, equivalently, the species constant-volume specific heats $c_{vk}(T)$, $k \in S$, which depend on temperature, species formation enthalpies h_k^{st}, $k \in S$, and species formation entropies s_k^{st}, $k \in S$, at the standard temperature T^{st} and standard pressure $p^{\mathrm{st}} = p^{\mathrm{atm}}$.

These quantities are usually evaluated from their molar counterparts $C_{pk}(T)$, $k \in S$, H_k^{st}, $k \in S$, and S_k^{st}, $k \in S$. The molar specific heats $C_{pk}(T)$ are generally evaluated from polynomial approximations. Absolute data can be found in the JANAF tables [StPr71] [Cal85] and addenda. The formation enthalpy and entropy can also be found in the JANAF tables.

Note, however, that one may lack thermodynamic data when developing new applications that are off the beaten track.

2.4. Chemistry

We first give a general description of chemical reaction mechanisms under consideration. These mechanisms—sometimes termed chemical networks—are constituted by an arbitrary number of reversible elementary reactions. We then express the rates of progress of the chemical reactions and discuss chemical data. We also present notational shortcuts often used for describing complex chemical mechanisms. The mathematical properties of chemical production rates are investigated in Chapter 6.

2.4.1. Elementary reactions

We consider a system of n^{r} elementary reactions for n species, which can be formally written as

$$\sum_{k \in S} \nu_{ki}^{\mathrm{f}} \, \mathfrak{M}_k \quad \rightleftharpoons \quad \sum_{k \in S} \nu_{ki}^{\mathrm{b}} \, \mathfrak{M}_k, \qquad i \in R, \tag{2.4.1}$$

where $\mathfrak{M}_k$ is the chemical symbol of the kth species, ν_{ki}^{f} and ν_{ki}^{b} are the forward and backward stoichiometric coefficients of the kth species in the ith reaction, respectively, and $R = \{1, \ldots, n^{\mathrm{r}}\}$ is the set of reaction indices. Chemical reactions usually involve less than three reactants and three products, in such a way that the matrices of stoichiometric coefficients ν_{ki}^{f} and ν_{ki}^{b} $i \in R$, $k \in S$, are sparse matrices.

As a typical example, the following reaction is an important step in combustion chemistry:

$$\mathrm{H} + \mathrm{O_2} \rightleftharpoons \mathrm{OH} + \mathrm{O},$$

where H is the symbol of atomic hydrogen, O_2 is the symbol of oxygen, O is the symbol of atomic oxygen, and OH is the symbol of the hydroxyl radical. Typical examples of elementary reaction mechanisms can be found in [B*al*92] for combustion and atmospheric chemistry, in [EGS96] for metalorganic chemical vapor deposition, or in [And89] for reentry chemistry into the earth's atmosphere. The size of chemical reaction networks is typically $n = 9$ and $n^{\mathrm{r}} = 19$ for hydrogen combustion [GS87], $n = 33$ and $n^{\mathrm{r}} = 126$ for propane combustion [D*al*88], $n \geq 140$ and $n^{\mathrm{r}} \geq 500$ for acetylene combustion with soot formation [F*al*86], $n \geq 450$ and $n^{\mathrm{r}} \geq 2800$ for iso–octane oxidation [C*al*96], and thousands of species are involved in atmospheric chemistry [SePa98].

Note that elementary chemical reactions—which are real life molecular events—are always reversible [Gar84] [Z*al*85] and the number of reactions n^{r} is arbitrary. This situation differs from that of global or semiglobal chemical reactions, which are often taken to be irreversible, but do not represent real molecular events, that is, real life detailed chemical processes. In other words, it is important to distinguish elementary reactions from artificial global transformations going directly from initial reactants to final products.

2.4.2. Maxwellian production rates

The molar production rates that we consider are the Maxwellian production rates obtained from the kinetic theory, as discussed in Chapter 4 [EG94]. These rates are obtained in a reactive kinetic framework when the chemistry characteristic times are larger than the mean free times of the molecules and the characteristic times of internal energy relaxation. These production rates ω_k, $k \in S$, are compatible with the law of mass action and written in the form

$$\omega_k = \sum_{i \in R} \nu_{ki}\, \tau_i, \tag{2.4.2}$$

where

$$\nu_{ki} = \nu_{ki}^{\mathrm{b}} - \nu_{ki}^{\mathrm{f}} \tag{2.4.3}$$

and τ_i is the rate of progress of the ith reaction. This rate of progress τ_i is given by

$$\tau_i = K_i^{\mathrm{f}} \prod_{l \in S} \gamma_l^{\nu_{li}^{\mathrm{f}}} - K_i^{\mathrm{b}} \prod_{l \in S} \gamma_l^{\nu_{li}^{\mathrm{b}}}, \tag{2.4.4}$$

where $\gamma_k = \rho_k/m_k$ is the molar concentration of the kth species and K_i^{f} and K_i^{b} are the forward and backward rate constants of the ith reaction, respectively.

The quantities K_i^{f} and K_i^{b} are functions of the temperature, and their ratio is the equilibrium constant K_i^{e} of the ith reaction

$$K_i^{\mathrm{e}}(T) = \frac{K_i^{\mathrm{f}}(T)}{K_i^{\mathrm{b}}(T)}, \tag{2.4.5}$$

given by

$$\log K_i^{\mathrm{e}}(T) = -\sum_{k\in S} \frac{\nu_{ki} m_k}{RT}\Big(g_k^{\mathrm{atm}}(T) - \frac{RT}{m_k}\log\big(\frac{p^{\mathrm{atm}}}{RT}\big)\Big). \tag{2.4.6}$$

The quantity $g_k^{\mathrm{atm}} - (RT/m_k)\log(p^{\mathrm{atm}}/RT)$ involved in the equilibrium constant simply represents the Gibbs function of the kth species at unit concentration.

Another expression for the rates of progress τ_i, $i \in R$, will be obtained in Chapter 6. This expression involves a single reaction constant and can also be derived from the kinetic theory of gases, as shown in Chapter 4.

We already pointed out that each elementary reaction considered in the chemical network is reversible [Z*al*85]. Indeed, the macroscopic constants K_i^{f} and K_i^{b} are Maxwellian averaged values of molecular chemical transition probabilities appearing in reactive collisional source terms of the species Boltzmann equations [EG94]. However, forward and backward chemical transition probabilities are always proportional—as are nonreactive cross sections in any Boltzmann equation—as can be shown from quantum mechanics. A direct consequence is the fundamental proportionality relation between the macroscopic forward and backward rate constants $K_i^{\mathrm{b}}(T) = K_i^{\mathrm{f}}(T)/K_i^{\mathrm{e}}(T)$ [EG94] [Z*al*85], which are also established in Chapter 4. Consequently, we cannot assume that one elementary reaction constant vanishes without assuming that both vanish. A second reason for assuming full reversibility is that the model must also be able to describe equilibrium mixtures. Statistical mechanics for uniform mixtures at equilibrium, however, also shows that detailed balance must prevail, so that irreversibility again cannot be assumed [Kub65]. Last, but not least, hundreds of experimental measurements have shown the validity of the fundamental relation $K_i^{\mathrm{b}}(T) = K_i^{\mathrm{f}}(T)/K_i^{\mathrm{e}}(T)$.

Remark 2.4.1. The forward rate constant is usually approximated by using a generalized Arrhenius empirical relation

$$K_i^{\mathrm{f}}(T) = \mathfrak{A}_i T^{\mathfrak{b}_i} \exp\Big(-\frac{\mathfrak{E}_i}{RT}\Big), \tag{2.4.7}$$

where $\mathfrak{A}_i > 0$ is the preexponential factor, $\mathfrak{b}_i$ is the preexponential exponent, and $\mathfrak{E}_i \geq 0$ is the activation energy of the ith reaction, but the exact expression of $K_i^{\mathrm{f}}(T)$ will not be needed in the following. Of course, the empirical Arrhenius law may either be used for the forward or backward reaction constant, but certainly not for *both*. Otherwise, the fundamental relation $K_i^{\mathrm{b}} = K_i^{\mathrm{f}}/K_i^{\mathrm{e}}$—which is a direct consequence from the kinetic theory of reactive gases—would be violated. In this situation, positiveness of entropy production due to chemical reactions is not even guaranteed [GM98a]. ■

2.4.3. Total mass conservation

The stoichiometric coefficients satisfy atom conservation relations in the form

$$\sum_{k\in S} \nu_{ki}^{\mathrm{f}} \mathfrak{a}_{kl} = \sum_{k\in S} \nu_{ki}^{\mathrm{b}} \mathfrak{a}_{kl}, \qquad i \in R, \quad l \in \mathfrak{A}, \tag{2.4.8}$$

where $\mathfrak{a}_{kl}$ denotes the number of lth atom in the kth species, $\mathfrak{A}$ denotes the set of atom indices $\mathfrak{A} = \{1, \ldots, n^{\mathrm{a}}\}$, and n^{a} denotes the number of atoms, or elements, in the mixture. A direct consequence of these atom conservation relations is the total mass conservation relations

$$\sum_{k\in S} \nu_{ki}^{\mathrm{f}} m_k = \sum_{k\in S} \nu_{ki}^{\mathrm{b}} m_k, \qquad i \in R. \tag{2.4.9}$$

After a little algebra, it is easily checked that these mass conservation relations for stoichiometric coefficients (2.4.9) imply the mass conservation relation for the production rates

$$\sum_{k\in S} m_k \omega_k = 0,$$

already used in Section 2.2.2. More details about the chemical source terms will be given in Chapter 6.

2.4.4. Notation for three-body reactions

In this section, we describe a compact notation often used for three-body reactions. This notation simplifies considerably the description of complex reaction mechanisms.

More specifically, for some reactions, it is required that a third body interact with the reactants. The third body can be any species from the mixture and appears simultaneously as a reactant and as a product in the chemical reaction. In this situation, it is often the case that the temperature dependent part of the reaction constants are proportional, that is to say, only the preexponential factors depend on the type of third body. Since the equilibrium constant of the corresponding reactions does not depend on the third body, it is readily seen that one may sum the contribution of all of these reactions and consider that they are equivalent to a fictitious total reaction involving an abstract third body often denoted by M. In this situation, the rate of progress of the symbolic total reaction—which corresponds to the sum of all elementary reaction rates obtained when the third body ranges in the set of species—becomes

$$q_i = \gamma_{\mathrm{M}} \left(\mathcal{K}_i^{\mathrm{f}} \prod_{l\in S} \gamma_l^{\nu_{li}^{\mathrm{f}}} - \mathcal{K}_i^{\mathrm{b}} \prod_{l\in S} \gamma_l^{\nu_{li}^{\mathrm{b}}}, \right),$$

where third bodies are no longer taken into account in the stoichiometric coefficients. The concentration γ_{M} of the equivalent third body M is now given by

$$\gamma_{\mathrm{M}} = \sum_{l \in S} \alpha_{li} \gamma_l,$$

where the factors α_{li} are termed the third-body efficiencies. These efficiencies are unity when all species contribute similarly as third bodies and the concentration of the third body is then the total concentration $\gamma = p/RT$. Note that three-body reactions may also have pressure dependent rate constants. The corresponding expressions can be fairly complex and are out of the scope of the present book. For more details on pressure dependent reaction constants, we refer the reader to [B*al*92].

As a typical example, in the reactive mixture constituted by the $n = 9$ species H_2, O_2, H_2O, N_2, H, O, OH, HO_2, and H_2O_2, associated with the combustion of hydrogen in air, the symbolic notation

$$\mathrm{H} + \mathrm{O}_2 + \mathrm{M} \rightleftharpoons \mathrm{HO}_2 + \mathrm{M},$$

corresponds to the $n = 9$ different reversible reactions obtained when M is taken to be any of the species H_2, O_2, H_2O, N_2, H, O, OH, HO_2, and H_2O_2. This notation for three-body reactions thus appears to be very practical, especially for reactions mechanisms involving a large number of interacting species.

Let us emphasize again that this notational problem for three-body reactions is only a practical shortcut that is described here for completeness. From a theoretical point of view, the reaction mechanisms described in (2.4.1) already involve the most general complexity, since they already consider an arbitrary number of species and an arbitrary number of elementary chemical reactions.

2.4.5. Chemistry data

The chemistry data needed for modeling complex chemistry flows are the forward and backward stoichiometric coefficients, ν_{ki}^{f} and ν_{ki}^{b}, $k \in S$, $i \in R$, respectively, the Arrhenius parameters $\mathfrak{A}_i$, $\mathfrak{b}_i$ et $\mathfrak{E}_i$, $i \in R$, when (2.4.7) is used, the third-body efficiencies α_{ki} and sometimes the pressure dependencies of three-body reaction constants. We also need thermodynamic properties, in particular, for evaluating equilibrium constants of elementary reactions.

Note, however, that reaction constants are generally difficult to measure over wide temperature intervals. As a result, large uncertainties may arise, even by several orders of magnitude in reaction constants. Nevertheless, the size of the reaction mechanisms used for various applications is steadily increasing.

2.5. Transport fluxes

In this section, we present the general form of transport fluxes as given by the kinetic theory of gases. These fluxes are the viscous tensor $\boldsymbol{\Pi}$, the species mass fluxes $\boldsymbol{\mathcal{F}}_k$, $k \in S$, and the heat flux $\boldsymbol{\mathcal{Q}}$. They are written in terms of macroscopic variable gradients and transport coefficients. We further specify the main properties of transport coefficients and study the diffusion velocities $\boldsymbol{\mathcal{V}}_k$, $k \in S$, which are only defined when the species mass fractions are positive. We also present alternative flux formulations in terms of the thermal diffusion ratios. The evaluation of transport coefficients from the kinetic theory of gases will be developed in Chapters 4 and 5.

2.5.1. Viscous tensor, species mass fluxes, and heat flux

The transport fluxes $\boldsymbol{\Pi}$, $\boldsymbol{\mathcal{F}}_k$, $k \in S$, and $\boldsymbol{\mathcal{Q}}$ obtained from the kinetic theory of gases can be written in the form [Wal58] [WT62] [CC70] [FK72] [EG94]

$$\boldsymbol{\Pi} = -(\kappa - \tfrac{2}{3}\eta)(\boldsymbol{\partial_x}\cdot\boldsymbol{v})\boldsymbol{I} - \eta(\boldsymbol{\partial_x}\boldsymbol{v} + \boldsymbol{\partial_x}\boldsymbol{v}^t), \tag{2.5.1}$$

$$\boldsymbol{\mathcal{F}}_k = -\sum_{l\in S} C_{kl}\boldsymbol{d}_l - \rho Y_k \theta_k \boldsymbol{\partial_x} \log T, \qquad k \in S, \tag{2.5.2}$$

$$\boldsymbol{\mathcal{Q}} = \sum_{k\in S} h_k \boldsymbol{\mathcal{F}}_k - \widehat{\lambda}\,\boldsymbol{\partial_x} T - p\sum_{k\in S} \theta_k \boldsymbol{d}_k, \tag{2.5.3}$$

where κ is the volume viscosity, η is the shear viscosity, ^{t} is the transposition operator, C_{kl}, $k,l \in S$, are the multicomponent flux diffusion coefficients, $\boldsymbol{d}_k$ is the diffusion driving force of the kth species, θ_k is the thermal diffusion coefficient of the kth species, and $\widehat{\lambda}$ is the partial thermal conductivity. The species diffusion driving forces incorporate the effects of various state variable gradients and external forces

$$\boldsymbol{d}_k = \boldsymbol{\partial_x}\Big(\frac{p_k}{p}\Big) + \Big(\frac{p_k}{p} - \frac{\rho_k}{\sum_{l\in S}\rho_l}\Big)\boldsymbol{\partial_x}\log p + \frac{\rho_k}{p}\Big(\frac{\sum_{l\in S}\rho_l \boldsymbol{b}_l}{\sum_{l\in S}\rho_l} - \boldsymbol{b}_k\Big), \quad k \in S. \tag{2.5.4}$$

The first term in the expression (2.5.2) of the diffusion flux $\boldsymbol{\mathcal{F}}_k$ yields diffusion effects due to mole fraction gradients, pressure gradients, and differences between specific forces acting on the species. The second term represents diffusion arising from temperature gradients and is termed the Soret effect. The first term in the expression (2.5.3) of the heat flux $\boldsymbol{\mathcal{Q}}$ represents the transfer of energy due to species molecular diffusion, whereas the second term represents Fourier's law. The third term corresponds to the Dufour effect, that is, heat diffusion due to concentration gradients, which is the symmetric of the Soret effect.

The main properties of the flux diffusion matrix $C = (C_{kl})_{k,l \in S}$ are

$$C\mathcal{Y} = \mathcal{Y}C^t, \tag{2.5.5}$$

$$N(C) = \mathbb{R}Y, \tag{2.5.6}$$

$$R(C) = \mathcal{U}^\perp, \tag{2.5.7}$$

where $\mathcal{Y} = \mathrm{diag}(Y_1, \ldots, Y_n)$ is the diagonal matrix of species mass fractions, $Y = (Y_1, \ldots, Y_n)^t$ is the mass fraction vector, and $\mathcal{U} = (1, \ldots, 1)^t$ is the unit vector. In addition, $\mathbb{R}Y$ is the linear space spanned by Y, $\mathcal{U}^\perp$ is the orthogonal complement of $\mathbb{R}\mathcal{U}$, and, for any matrix A, we denote by $N(A)$ its nullspace and by $R(A)$ its range. The thermal diffusion vector $\theta = (\theta_1, \ldots, \theta_n)^t$ also satisfies the relation

$$\langle \theta, Y \rangle = 0, \tag{2.5.8}$$

where $\langle x, y \rangle$ denotes the scalar product between two vectors x and y.

An important consequence of (2.5.7) and (2.5.8) is the mass conservation constraint

$$\sum_{k \in S} \mathcal{F}_k = 0, \tag{2.5.9}$$

between the species mass fluxes, already presented in Section 2.2.2 and used in Section 2.2.5.

The kinetic theory of gases actually yields the relations (2.5.2) and (2.5.3) with $\widehat{\boldsymbol{d}}_k$, $k \in S$, in place of $\boldsymbol{d}_k$, $k \in S$, where

$$\widehat{\boldsymbol{d}}_k = \frac{\boldsymbol{\partial}_{\boldsymbol{x}} p_k - \rho_k \boldsymbol{b}_k}{p}, \qquad k \in S. \tag{2.5.10}$$

However, using the relations

$$\boldsymbol{d}_k = \widehat{\boldsymbol{d}}_k - \rho_k \frac{\sum_{l \in S} \widehat{\boldsymbol{d}}_l}{\sum_{l \in S} \rho_l} = \widehat{\boldsymbol{d}}_k - Y_k \frac{\sum_{l \in S} \widehat{\boldsymbol{d}}_l}{\sum_{l \in S} Y_l}, \tag{2.5.11}$$

and properties (2.5.5)–(2.5.8), it is easily checked that both formulations are equivalent.

The formulation (2.5.2)–(2.5.4) is often more practical, since the diffusion driving forces $\boldsymbol{d}_k$, $k \in S$, are linearly dependent, as are the diffusion fluxes $\mathcal{F}_k$, $k \in S$. This linear dependence is needed, for instance, when manipulating generalized inverses of the flux diffusion matrix C. It is also needed in the fundamental diffusion inequality investigated in Chapter 7.

The mathematical properties of the transport coefficients κ, η, C, θ, and $\widehat{\lambda}$ will be investigated in Chapter 7.

2.5.2. Diffusion velocities and diffusion matrix

For nonzero mass fractions, we can further introduce the species diffusion velocity $\mathcal{V}_k$, $k \in S$, as

$$\mathcal{V}_k = \frac{\mathcal{F}_k}{\rho Y_k}, \qquad k \in S. \tag{2.5.12}$$

From the relation (2.5.2), we now obtain that

$$\mathcal{V}_k = -\sum_{l \in S} D_{kl} \boldsymbol{d}_l - \theta_k \boldsymbol{\partial_x} \log T, \qquad k \in S, \tag{2.5.13}$$

where the multicomponent diffusion coefficients D_{kl}, $k, l \in S$, are

$$D_{kl} = \frac{C_{kl}}{\rho Y_k}, \qquad k, l \in S. \tag{2.5.14}$$

The main properties of the diffusion matrix $D = (D_{kl})_{k,l \in S}$ are

$$D = D^t, \tag{2.5.15}$$

$$N(D) = \mathbb{R}Y, \tag{2.5.16}$$

$$R(D) = Y^{\perp}, \tag{2.5.17}$$

$$D \text{ is positive definite over } \mathcal{U}^{\perp}. \tag{2.5.18}$$

It is important to note that the diffusion matrix D is symmetric. Symmetric diffusion matrices have been introduced by Waldmann [Wal58] [WT62] and used by Chapmann and Cowling [CC70], Curtiss [Cur68], and Ferziger and Kaper [FK72], and this symmetric formulation is compatible with Onsager reciprocal relations. Another definition of diffusion matrices due to Hirschfelder et al. [HCB54] imposes the constraint $D_{kk} = 0$, for $k \in S$, instead of (2.5.16) and artificially destroys the natural symmetry of the diffusion process [Van67] [FK72] [Gio91] [EG94].

As for the diffusion fluxes, the kinetic theory of gases actually yields expressions (2.5.13) with $\widehat{\boldsymbol{d}}_k$ $k \in S$, in place of $\boldsymbol{d}_k$ $k \in S$. Both formulations are equivalent because of the mass constraint relations (2.5.16). The properties (2.5.9) and (2.5.12) further imply the total mass conservation constraint

$$\sum_{k \in S} Y_k \mathcal{V}_k = 0, \tag{2.5.19}$$

which is also a consequence of (2.5.17) and (2.5.8). The mathematical properties of the matrix D will be investigated in Chapter 7. In particular, the property (2.5.18) corresponds to the positivity of the entropy production quadratic form $(p/T)\,\langle Dx, x\rangle$ associated with diffusive processes over the physical hyperplane $\mathcal{U}^\perp$, keeping in mind the constraint $\sum_{k\in S} \boldsymbol{d}_k = 0$ between the diffusion driving forces [Gio91] [EG94].

2.5.3. Alternative formulations

We define the species thermal diffusion ratios $\chi = (\chi_1, \ldots, \chi_n)^t$ from the relations [WT62] [EG94]

$$\begin{cases} C\chi = \rho\mathcal{Y}\theta, \\ \sum_{k\in S} \chi_k = 0, \end{cases} \tag{2.5.20}$$

where $\mathcal{Y} = \mathrm{diag}(Y_1, \ldots, Y_n)$ and the thermal conductivity λ from

$$\lambda = \widehat{\lambda} - (p/T) \sum_{k\in S} \theta_k \chi_k. \tag{2.5.21}$$

One may then easily establish the following alternative expressions for the species diffusion velocities and the heat flux:

$$\boldsymbol{\mathcal{V}}_k = -\sum_{l\in S} D_{kl}(\boldsymbol{d}_l + \chi_l \boldsymbol{\partial_x} \log T), \qquad k \in S, \tag{2.5.22}$$

$$\boldsymbol{\mathcal{Q}} = \sum_{k\in S} h_k \boldsymbol{\mathcal{F}}_k - \lambda \boldsymbol{\partial_x} T + p \sum_{k\in S} \chi_k \boldsymbol{\mathcal{V}}_k. \tag{2.5.23}$$

These alternative formulations are important in the theory [GM98a] and for numerical modeling, since evaluating λ and χ is computationally more efficient than evaluating $\widehat{\lambda}$ and θ [EG95b]. In addition, the thermal conductivity λ is accessible to experimental measurements but not the partial thermal conductivity $\widehat{\lambda}$.

For the mathematical treatment of transport coefficients, it will also be convenient to introduce the rescaled thermal diffusion ratios $\widetilde{\chi}_k$, $k \in S$, defined by

$$\chi_k = \frac{p_k}{p} \widetilde{\chi}_k, \qquad k \in S,$$

in such a way that

$$\boldsymbol{\mathcal{F}}_k = -\sum_{l\in S} C_{kl}(\boldsymbol{d}_l + \frac{p_l}{p} \widetilde{\chi}_l \boldsymbol{\partial_x} \log T), \qquad k \in S, \tag{2.5.24}$$

and

$$\boldsymbol{\mathcal{Q}} = \sum_{k\in S} h_k \boldsymbol{\mathcal{F}}_k - \lambda \boldsymbol{\partial_x} T + RT \sum_{k\in S} \frac{\widetilde{\chi}_k}{m_k} \boldsymbol{\mathcal{F}}_k. \tag{2.5.25}$$

The rescaled thermal diffusion ratio vector defined by $\widetilde{\chi} = (\widetilde{\chi}_1, \ldots, \widetilde{\chi}_n)^t$ is such that $\widetilde{\chi} \in X^\perp$, where $X = (X_1, \ldots, X_n)^t$.

2.5.4. Transport coefficients

The preceding expressions for transport fluxes show that accurate modeling of transport coefficients is a must before any theoretical, mathematical, or numerical investigation of multicomponent reactive flows.

These multicomponent transport coefficients, however, are *not* explicitly given by the kinetic theory of gases. Evaluating these coefficients—which are functions of state variables—requires solving linear systems termed "transport linear systems." These transport linear systems arise from Galerkin approximation procedures applied to linearized Boltzmann integral equations.

The derivation of the transport linear systems will be sketched in Chapter 4. Transport linear systems of reduced dimensions associated with each transport coefficient are also presented in Chapter 5. These transport linear systems involve complex analytic expressions depending on species molecular parameters and collision integrals. The mathematical structure of transport linear systems is also investigated in Chapters 4 and 5. These structure properties allow the solution of the transport linear systems either by direct methods or by iterative techniques. Note also that the coefficients κ, η, C, ρD, $\widehat{\lambda}$, λ, $\rho\theta$, and χ only depend on (T, Y). On the other hand, the coefficients D and θ also depend on pressure, but are simply inversely proportional to the pressure p.

Finally, the mathematical properties of transport coefficients are investigated in Chapter 7. Among the properties of transport coefficients are (2.5.5)–(2.5.7) and (2.5.15)–(2.5.18).

2.6. Entropy

In this section, we derive an equation governing entropy. In particular, we explicit the entropy flux and the entropy production rate, which is also termed the dissipation rate.

The concept of entropy is of fundamental importance in the modeling of multicomponent flows. Entropy will be used when investigating thermochemistry as well as in the study of multicomponent transport. It will also be used as a fundamental mathematical tool for a priori estimates, since entropy Hessians form a metrics that correlates the fluxes with the macroscopic variable gradients. Entropy also plays an important role in various numerical techniques, such as the streamline diffusion method.

2.6.1. Entropy differential

In order to derive an equation governing entropy, we need to relate the differential of entropy to the differentials of energy, total density, and species mass fractions, which are available from the governing equations. From thermodynamics, making use of the state variables $(T, p, Y_1, \dots, Y_n)$, one can easily establish the Gibbs relation

$$T\,\mathbb{D}s = \mathbb{D}e + p\mathbb{D}(\frac{1}{\rho}) - \sum_{k\in S} g_k\,\mathbb{D}Y_k\,, \tag{2.6.1}$$

where $\mathbb{D}$ denotes the total derivative. These differential relations between thermodynamic functions will be investigated in detail in Chapter 6.

2.6.2. Entropy equation

Upon using the expression of the entropy differential (2.6.1) and the species mass, momentum, and energy equations, we obtain that

$$\begin{aligned} T\rho(\partial_t s + \boldsymbol{v}\cdot\boldsymbol{\partial_x} s) = &- \boldsymbol{\Pi}:\boldsymbol{\partial_x}\boldsymbol{v} - \boldsymbol{\partial_x}\cdot\boldsymbol{\mathcal{Q}} \\ &+ \sum_{k\in S}\boldsymbol{\mathcal{F}}_k\cdot\boldsymbol{b}_k + \sum_{k\in S} g_k\boldsymbol{\partial_x}\cdot\boldsymbol{\mathcal{F}}_k - \sum_{k\in S} g_k m_k\omega_k. \end{aligned}$$

Dividing this relation by T and integrating the $1/T$ factor into the divergence operators $\boldsymbol{\partial_x}\cdot$ then yields

$$\partial_t(\rho s) + \boldsymbol{\partial_x}\cdot(\rho\boldsymbol{v}s) + \boldsymbol{\partial_x}\cdot\Big(\frac{\boldsymbol{\mathcal{Q}}}{T} - \sum_{k\in S}\frac{g_k}{T}\boldsymbol{\mathcal{F}}_k\Big) = \mathfrak{v}, \tag{2.6.2}$$

where

$$\begin{aligned} \mathfrak{v} = &-\frac{\boldsymbol{\Pi}:\boldsymbol{\partial_x}\boldsymbol{v}}{T} - \frac{\boldsymbol{\mathcal{Q}}\cdot\boldsymbol{\partial_x}T}{T^2} \\ &+ \sum_{k\in S}\frac{\boldsymbol{\mathcal{F}}_k\cdot\boldsymbol{b}_k}{T} - \sum_{k\in S}\boldsymbol{\mathcal{F}}_k\cdot\boldsymbol{\partial_x}\big(\frac{g_k}{T}\big) - \sum_{k\in S}\frac{g_k m_k\omega_k}{T}. \end{aligned}$$

From (2.6.2), we deduce that $(\boldsymbol{\mathcal{Q}} - \sum_{k\in S} g_k\boldsymbol{\mathcal{F}}_k)/T$ is the entropy flux and $\mathfrak{v}$ is the entropy production. In order to reformulate $\mathfrak{v}$, we can use

$$\boldsymbol{\partial_x}\Big(\frac{g_k}{T}\Big) = -\frac{h_k}{T^2}\boldsymbol{\partial_x}T + \frac{R}{m_k p_k}\boldsymbol{\partial_x}p_k,$$

in such a way that

$$\begin{aligned} \mathfrak{v} = &-\frac{1}{T^2}\Big(\boldsymbol{\mathcal{Q}} - \sum_{k\in S} h_k\boldsymbol{\mathcal{F}}_k\Big)\cdot\boldsymbol{\partial_x}T - \frac{p}{T}\sum_{k\in S}\boldsymbol{\mathcal{V}}_k\cdot\Big(\frac{\boldsymbol{\partial_x}p_k - \rho_k\boldsymbol{b}_k}{p}\Big) \\ &-\frac{1}{T}\boldsymbol{\Pi}:\boldsymbol{\partial_x}\boldsymbol{v} - \sum_{k\in S}\frac{g_k m_k\omega_k}{T}. \end{aligned} \tag{2.6.3}$$

It is important to note that (2.6.2) and (2.6.3) are valid only when the mass fractions Y_k, $k \in S$, are positive.

2.6.3. Entropy production

We now use the expression of transport fluxes obtained in Section 2.5 to rewrite the entropy production term $\mathfrak{v}$. The term $-\boldsymbol{\Pi}:\boldsymbol{\partial_x} \boldsymbol{v}$ in (2.6.3) can first be rewritten as

$$- \boldsymbol{\Pi}:\boldsymbol{\partial_x} \boldsymbol{v} = \kappa\,(\boldsymbol{\partial_x}\cdot\boldsymbol{v})^2$$
$$+ \frac{\eta}{2}\Big(\boldsymbol{\partial_x}\boldsymbol{v} + \boldsymbol{\partial_x}\boldsymbol{v}^t - \tfrac{2}{3}(\boldsymbol{\partial_x}\cdot\boldsymbol{v})\,\boldsymbol{I}\Big):\Big(\boldsymbol{\partial_x}\boldsymbol{v} + \boldsymbol{\partial_x}\boldsymbol{v}^t - \tfrac{2}{3}(\boldsymbol{\partial_x}\cdot\boldsymbol{v})\,\boldsymbol{I}\Big).$$

On the other hand, (2.5.23) yields

$$\mathcal{Q} - \sum_{k\in S} h_k \boldsymbol{\mathcal{F}}_k = -\lambda \boldsymbol{\partial_x} T + p \sum_{k\in S} \chi_k \boldsymbol{\mathcal{V}}_k.$$

Substituting this expression into (2.6.3) and regrouping in $\mathfrak{v}$ the two sums involving the diffusion velocities $\boldsymbol{\mathcal{V}}_k$, $k \in S$, we obtain a term in the form

$$-\frac{p}{T}\sum_{k\in S}\boldsymbol{\mathcal{V}}_k\cdot\Big(\frac{\boldsymbol{\partial_x} p_k - \rho_k \boldsymbol{b}_k}{p} + \chi_k\frac{\boldsymbol{\partial_x} T}{T}\Big) = -\frac{p}{T}\sum_{k\in S}\boldsymbol{\mathcal{V}}_k\cdot\Big(\boldsymbol{d}_k + \chi_k\frac{\boldsymbol{\partial_x} T}{T}\Big),$$

making use of $\sum_{k\in S} Y_k \boldsymbol{\mathcal{V}}_k = 0$. We can now use the expression of diffusion velocities $\boldsymbol{\mathcal{V}}_k$, $k \in S$, and rewrite this term as

$$\frac{p}{T}\sum_{k,l\in S} D_{kl}\Big(\boldsymbol{d}_k + \chi_k\frac{\boldsymbol{\partial_x} T}{T}\Big)\cdot\Big(\boldsymbol{d}_l + \chi_l\frac{\boldsymbol{\partial_x} T}{T}\Big).$$

Combining the previous results, we can finally rewrite the entropy production rate $\mathfrak{v}$ in the form

$$\mathfrak{v} = \frac{\lambda}{T^2}\boldsymbol{\partial_x} T\cdot\boldsymbol{\partial_x} T + \frac{p}{T}\sum_{k,l\in S} D_{kl}\Big(\boldsymbol{d}_k + \chi_k\frac{\boldsymbol{\partial_x} T}{T}\Big)\cdot\Big(\boldsymbol{d}_l + \chi_l\frac{\boldsymbol{\partial_x} T}{T}\Big)$$
$$+ \frac{\eta}{2T}\Big(\boldsymbol{\partial_x}\boldsymbol{v} + \boldsymbol{\partial_x}\boldsymbol{v}^t - \tfrac{2}{3}(\boldsymbol{\partial_x}\cdot\boldsymbol{v})\,\boldsymbol{I}\Big):\Big(\boldsymbol{\partial_x}\boldsymbol{v} + \boldsymbol{\partial_x}\boldsymbol{v}^t - \tfrac{2}{3}(\boldsymbol{\partial_x}\cdot\boldsymbol{v})\,\boldsymbol{I}\Big)$$
$$+ \frac{\kappa}{T}(\boldsymbol{\partial_x}\cdot\boldsymbol{v})^2 - \sum_{k\in S}\frac{g_k m_k \omega_k}{T}. \tag{2.6.4}$$

Note that the term associated with the diffusion matrix can also be written as

$$\frac{p}{T}\sum_{k,l\in S} D_{kl}\Big(\widehat{\boldsymbol{d}}_k + \chi_k\frac{\boldsymbol{\partial_x} T}{T}\Big)\cdot\Big(\widehat{\boldsymbol{d}}_l + \chi_l\frac{\boldsymbol{\partial_x} T}{T}\Big),$$

where $\widehat{\boldsymbol{d}}_k$, $k \in S$, are the unconstrained diffusion driving forces (2.5.10). However, the coercivity of the diffusion matrix D only holds on hyperplanes which do not contain the vector Y. This is why the linearly dependent diffusion driving forces $\boldsymbol{d}_k$, $k \in S$, are more convenient to use in the expression of entropy production $\mathfrak{v}$.

2.7. Boundary conditions

In this section we briefly examine classical boundary conditions associated with multicomponent flows and investigate solid–gas interfaces [Wil85]. When a gaseous mixture is in contact with a solid body, the interfacial equations are indeed the boundary conditions of the gas phase equations. We only examine here the simplified situations of porous plates and catalytic surfaces. We just give an insight into solid wall/gas interactions and do not pretend to present a complete study.

2.7.1. Dirichlet and Neumann boundary conditions

Dirichlet and Neumann boundary conditions are universal boundary conditions that can be used to model various phenomena associated with single component or multicomponent flows [Bat67] [Wil85] [Can90].

Dirichlet boundary conditions are typically associated with inflow phenomena in infinite length domains, isothermal walls, or classical velocity adherence conditions.

On the other hand, Neumann boundary conditions are often associated with symmetry boundaries, adiabatic walls, or nonreactive walls.

2.7.2. Porous walls

We consider a porous wall separating a mixing chamber filled with a homogeneous uniform mixture from a "reacting" chamber, where the mixture evolves. We denote by "fresh gases" the uniform mixture in the mixing chamber. This situation is typical of cooled porous plug burners in combustion chambers.

We assume that the porous wall is an ideal wall that does not allow diffusion of reacting gases into the mixing chamber, and we neglect catalytic effects. Boundary conditions are classically obtained by integrating the conservation equations into a thin layer around the interface and by letting the layer thickness go to zero.

Denoting by $\boldsymbol{n}$ the unit normal vector oriented towards the gas, the conservation of chemical species can be written in the form

$$(\rho Y_k \boldsymbol{v} + \mathcal{F}_k)\cdot\boldsymbol{n} = \rho^{\mathrm{fr}} Y_k^{\mathrm{fr}} \boldsymbol{v}^{\mathrm{fr}}\cdot\boldsymbol{n}, \qquad k \in S, \tag{2.7.1}$$

where ρ^{fr} is the density of fresh gases, $\boldsymbol{v}^{\text{fr}} = v^{\text{fr}}\boldsymbol{n}$ is the fresh gas velocity, and Y_k^{fr} is the mass fraction of the kth species in the fresh gases. By summing these relations, we obtain the mass flux conservation

$$\rho\boldsymbol{v}\cdot\boldsymbol{n} = \rho^{\text{fr}}\boldsymbol{v}^{\text{fr}}\cdot\boldsymbol{n}. \tag{2.7.2}$$

The tangential adherence condition is also

$$\boldsymbol{v}\cdot\boldsymbol{\tau} = 0, \tag{2.7.3}$$

where $\boldsymbol{\tau}$ denotes any tangent vector to the porous wall. For an isothermal wall, we have

$$T = T^{\text{fr}}, \tag{2.7.4}$$

where T^{fr} is the fresh mixture temperature and the wall temperature. When the wall is adiabatic, on the contrary, we have

$$(\rho h\boldsymbol{v} + \boldsymbol{\mathcal{Q}})\cdot\boldsymbol{n} = \rho^{\text{fr}}h^{\text{fr}}\boldsymbol{v}^{\text{fr}}\cdot\boldsymbol{n}, \tag{2.7.5}$$

where $\boldsymbol{\mathcal{Q}}$ is the heat flux. These relations can be used, for instance, to model porous injectors, such as porous plug burners often used in laboratories.

2.7.3. Catalytic plates

We now consider a solid–gas interface with normal vector $\boldsymbol{n}$ oriented towards the gas. We assume that the wall is catalytic, that is to say, we assume that heterogeneous reactions take place on the surface without modifying the wall, or without ablation or deposition. The wall is only used as a catalyst through adsorption and desorption of gaseous species and surface reactions between adsorbed species without surface regression or growth. We assume, for simplicity, that the plate is at rest, and we neglect surface diffusion and diffusion into the wall.

We denote symbolically by $\mathfrak{M}_{n+k}$, $k \in \{1, \dots, \hat{n}\}$, the species adsorbed on the catalytic surface, where $\hat{n}$ is the number of surface species. Note that, when there are several types of adsorption sites, it is necessary to introduce different symbols for identical molecules adsorbed on different types of sites. We then have the n gaseous species $\mathfrak{M}_k$, $k \in S$, and the $\hat{n}$ surface adsorbed species $\mathfrak{M}_k$, $k \in \{n+1, \dots, n+\hat{n}\}$. The surface reaction mechanism can then be written in a form similar to that of the homogeneous reaction mechanism

$$\sum_{k\in\{1,\dots,n+\hat{n}\}} \widehat{\nu}_{ki}^{\text{f}}\,\mathfrak{M}_k \rightleftharpoons \sum_{k\in\{1,\dots,n+\hat{n}\}} \widehat{\nu}_{ki}^{\text{b}}\,\mathfrak{M}_k, \qquad i \in \{1, \dots, \widehat{n}^{\text{r}}\}, \tag{2.7.6}$$

where $\widehat{n}^{\text{r}}$ is the number of chemical surface reactions and $\widehat{\nu}_{ki}^{\text{f}}$ and $\widehat{\nu}_{ki}^{\text{b}}$ are the corresponding surface stoichiometric coefficients. Since the wall is a catalyst, only the species $\mathfrak{M}_k$, $k \in \{1,\dots,n+\hat{n}\}$, are concerned, that is, the species present *in* the wall play no role.

The interfacial gaseous species conservation equations then read

$$(\rho Y_k \boldsymbol{v} + \boldsymbol{\mathcal{F}}_k)\boldsymbol{\cdot n} = m_k \widehat{\omega}_k, \qquad k \in S, \tag{2.7.7}$$

where $\widehat{\omega}_k$ is the surface molar production rate of the kth species. The velocity $\boldsymbol{v\cdot n}$ induced by gaseous mass consumption due to surface reaction is then

$$\rho \boldsymbol{v\cdot n} = \sum_{k\in S} m_k \widehat{\omega}_k. \tag{2.7.8}$$

The surface species conservation equations read

$$\partial_t(\widehat{\rho}_k) = m_k \widehat{\omega}_k, \qquad k \in \{n+1,\dots,n+\hat{n}\}, \tag{2.7.9}$$

where $\widehat{\rho}_k$ denotes the mass surface concentration of the surface molecule $\mathfrak{M}_k$, $k \in \{n+1,\dots,n+\hat{n}\}$, keeping in mind that surface diffusion is neglected. The tangential adherence condition can be written in the form

$$\boldsymbol{v\cdot \tau} = 0, \tag{2.7.10}$$

where $\boldsymbol{\tau}$ denotes any tangent vector to the catalytic surface. Note that, at steady state, we have $\partial_t(\widehat{\rho}_k) = 0$ and thus $\boldsymbol{v\cdot n} = 0$, since there is neither regression nor deposition so that $\boldsymbol{v} = 0$ in this situation. Finally, the conservation of energy reads as

$$(\rho h \boldsymbol{v} + \boldsymbol{\mathcal{Q}})\boldsymbol{\cdot n} = \boldsymbol{\mathcal{Q}}_{\mathrm{S}}\boldsymbol{\cdot n} - \sum_{k\in\{n+1,\dots,n+\hat{n}\}} h_k m_k \widehat{\omega}_k, \tag{2.7.11}$$

where $\boldsymbol{\mathcal{Q}}_{\mathrm{S}}$ is the heat flux towards the wall. On the contrary, if the wall is maintained at constant temperature, say, by cooling, one has to replace (2.7.11) by the isothermal condition

$$T = T_{\mathrm{S}}, \tag{2.7.12}$$

where T_{S} is the surface temperature. The relation (2.7.11) is then used to determine the interfacial heat flux $\boldsymbol{\mathcal{Q}}_{\mathrm{S}}.\boldsymbol{n}$.

Finally, the surface production rates $\widehat{\omega}_k$ can be expressed in terms of surfacic concentrations $\widehat{\rho}_k$ [W*al*94] [EGS96]. As for gaseous reactions, it is necessary to obtain kinetic data, although such data are even more subject to uncertainties than for homogeneous reactions.

2.8. Notes

2.1. As opposed to a molecular derivation from the kinetic theory of gases, a continuum derivation of the conservation equations is also feasible. This derivation involves the idea of a multicomponent continuum composed of coexistent continua, each obeying the laws of dynamics and thermodynamics. Conservation equations are then derived by expressing the conservation of various species quantities over arbitrary control volumes and by summing over the species. The expressions for transport fluxes are then obtained from Onsager reciprocal relations after an appropriate formulation of entropy production as a quadratics. This procedure, however, suffers from three drawbacks. First, the approach appears to be somewhat artificial. Indeed, it requires introducing balance equations for quantities that *cannot* be governed by partial differential equations, since they are not associated with any collisional invariant, such as species momentum. More fundamentally, this approach does not allow the determination of transport coefficients or even their *qualitative* properties, such as the asymptotic behavior for small concentrations. In addition, the Onsager linear formalism is fundamentally wrong for chemical production rates, which depend exponentially—and not linearly, as suggested by Onsager formalism—on chemical affinities.

2.2. The system of partial differential equations governing multicomponent reactive flows appears to have a complex structure. We note, in particular, that all species second derivatives are coupled in the conservation equations through the flux diffusion matrix C and to the temperature derivatives through the thermal diffusion coefficients θ. Multicomponent fluxes also naturally involve mole fractions and pressure derivatives, although mass fractions naturally appear in the governing equations. Species derivatives also appear in the energy and temperature equations. Chapter 8 will be devoted to the mathematical structure of the resulting system of partial differential equations.

2.3. When radiation effects are taken into account, there is an extra term in the heat flux $\mathcal{Q}$ in the form of the radiation heat flux [Wil85]. Correspondingly, there are heat loss terms in the boundary condition (2.7.11). Radiation modeling in multicomponent mixtures is out of the scope of this book, and we refer the reader to [Wil85] [And89] for more details.

2.4. The model presented in this chapter can also be used to describe vibrational nonequilibrium. More specifically, when each vibrational quantum level is treated as a separate "chemical species," allowing detailed state-to-state relaxation models, the same *formal* equations are obtained as those presented in this chapter [COB97] [CAG97] [GM98b]. When the vibrational quantum levels are partially at equilibrium between them but not at equilibrium with the translational/rotational states—allowing the definition of a vibrational temperature—a different structure is obtained, which is beyond the scope of this book.

2.5. The case of infinitely fast chemistry, that is, the case of equilibrium flows, will be considered in Chapters 6 and 10. In this situation, one has to solve the momentum and energy equations together with equations expressing the conservation of atomic elements [EG98a].

2.6. The fluid dynamic part of the governing equations are usually termed the compressible Navier-Stokes equations. Note, however, that the expression of the stress tensor and the general form of the momentum conservation equations were derived first by Navier (1822) and Poisson (1829) by using specific assumptions concerning the molecular mechanism of internal friction and later by Saint-Venant (1843) and Stokes (1845) in a very general way [Bat67].

2.7. Multiphase flows will not be considered in this book. Modeling multiphase flows requires introducing a Boltzmann-type equation for each family of particles present in the mixture, either liquid for sprays or solid for clouds, and coupling these condensed species kinetic equations to the gas phase equations [Wil58] [Wil85] [AOB89] [Sai95]. Similarly, typical applications will not be considered in this book, and we refer the reader, in particular, to Williams [Wil85] for combustion applications, Rosner [Ros86] for chemical engineering applications, and Poinsot et al. [P*al*96] for direct numerical simulation of turbulent reactive flows.

2.9. References

[AOB89] A. Amsden, P. J. O'Rourke, and T. D. Butler, *KIVA-II: A Computer Program for Chemically Reactive Flows with Sprays,* Los Alamos National Laboratory Report, LA-11560-MS, May, (1989).

[And89] J. D. Anderson, Jr., *Hypersonics and High Temperature Gas Dynamics,* McGraw-Hill Book Company, New York, (1989).

[Bat67] G. K. Batchelor, *An Introduction to Fluid Dynamics,* Cambridge University Press, Cambridge, (1967).

[B*al*92] D. L. Baulch, C. J. Cobos, R. A. Cox, C. Esser, P. Frank, T. Just, J. A. Kerr, M. J. Pilling, J. Troe, R. J. Walker, and J. Warnatz, *Evaluated Kinetic Data for Combustion Modelling,* J. Phys. Chem. Ref. Data, **21**, (1992), pp. 411–734.

[CC70] S. Chapman and T. G. Cowling, *The Mathematical Theory of Non-Uniform Gases,* Cambridge University Press, Cambridge, (1970).

[Can90] S. M. Candel, *Mécanique des Fluides,* Dunod, Paris, (1990).

[COB97] G. V. Candler, J. Olejniczak, and B. Harrold, *Detailed Simulation of Nitrogen Dissociation in Stagnation Regions,* Phys. Fluids, **9**, (1997), pp. 2108–2117.

[CAG97] M. Capitelli, I. Armenise, and C. Gorse, *State-to-State Approach in the Kinetics of Air Components Under Re-Entry Conditions,* J. Thermophys. Heat Transfer, **11**, (1997), pp. 570–578.

[Cal85] M. W. Chase, Jr., C. A. Davies, J. R. Downey, Jr., D. J. Frurip, R. A. McDonald, and A. N. Syverud, *JANAF Thermochemical Tables, Third Edition,* J. Phys. Chem. Ref. Data, **14**, Suppl. 1, (1985), pp. 1–1856.

[Cal96] G. M. Côme, V. Warth, P. A. Glaude, R. Fournet, F. Battin–Leclerc, and G. Sacchi, *Computer–Aided Design of Gas–Phase Oxidation Mechanisms, Application to the Modeling of n-Heptane and Iso–Octane Oxidation,* 26th Symposium International on Combustion, The Combustion Institute, Pittsburgh, (1996), pp. 755–762.

[Cur68] C. F. Curtiss, *Symmetric Gaseous Diffusion Coefficients,* J. Chem. Phys., **49**, (1968), pp. 2917–2919.

[Dal88] N. Darabiha, S. Candel, V. Giovangigli, and Smooke M., *Extinction of Strained Premixed Propane-Air Flames with Complex Chemistry,* Comb. Sci. Tech., **60**, (1988), pp. 267–284.

[Dix84] G. Dixon-Lewis, *Computer Modeling of Combustion Reactions in Flowing Systems with Transport,* in W. C. Gardiner, Ed., *Combustion Chemistry.* Springer, New York, (1984), pp. 21–125.

[DH87] J. P. Drumond and M. Y. Hussaini, *Numerical Simulation of a Supersonic Reacting Mixing Layer,* AIAA 19th Fluid Dynamics, Plasma Physics and Laser Conference, Honolulu, Hawaii, Paper AIAA-87-1325, (1987).

[DM89] J. P. Drumond and H. S. Mukunda, *A Numerical Study of Mixing Enhancement in Supersonic Reacting Flow Fields,* 3rd International Conference on Numerical Combustion, Sophia-Antipolis (Antibes), A. Dervieux and B. Larouturrou, Eds., Lecture Notes in Physics, Springer-Verlag, **351**, (1989), pp. 36–64.

[EG94] A. Ern and V. Giovangigli, *Multicomponent Transport Algorithms,* Lecture Notes in Physics, New Series "Monographs", **m 24**, Springer-Verlag, Berlin, (1994).

[EG95b] A. Ern and V. Giovangigli, *Fast and Accurate Multicomponent Property Evaluations,* J. Comp. Phys., **120**, (1995), pp. 105–116.

[EG98a] A. Ern and V. Giovangigli, *The Kinetic Equilibrium Regime,* Physica-A, **260**, (1998), pp. 49–72.

[EG98b] A. Ern and V. Giovangigli *Thermal Diffusion Effects in Hydrogen/Air and Methane/Air Flames,* Comb. Theor. Mod., **2**, (1998), pp. 349–372.

[EGS96] A. Ern, V. Giovangigli, and M. Smooke, *Numerical Study of a Three-Dimensional Chemical Vapor Deposition Reactor with Detailed Chemistry,* J. Comp. Phys., **126**, (1996), pp. 21–39.

[Fei95] M. Feinberg, *The Existence and Uniqueness of Steady States for a Class of Chemical Reaction Networks,* Arkiv Rat. Mech. Anal., **132**, (1995), pp. 311–370.

[FK72] J. H. Ferziger and H. G. Kaper, *Mathematical Theory of Transport Processes in Gases,* North Holland Publishing Company, Amsterdam, (1972).

[F*al*86] M. Frenklach, D. W. Clary, T. Yuan, W. C. Gardiner, and S. E. Stein, *Mechanism of Soot Formation in Acetylene-Oxygen Mixtures,* Comb. Sci. Tech., **50**, (1986), pp. 79–115.

[Gar84] W. C. Gardiner, Jr., *Combustion Chemistry,* Springer-Verlag, New York, (1984).

[Gio91] V. Giovangigli, *Convergent Iterative Methods for Multicomponent Diffusion,* IMPACT Comput. Sci. Eng., **3**, (1991), pp. 244–276.

[GM98a] V. Giovangigli and M. Massot, *Asymptotic Stability of Equilibrium States for Multicomponent Reactive Flows,* Math. Mod. Meth. Appl. Sci., **8**, (1998), pp. 251–297.

[GM98b] V. Giovangigli and M. Massot, *The Local Cauchy Problem for Multicomponent Reactive Flows in Full Vibrational Nonequilibrium,* Math. Meth. Appl. Sci., **21**, (1998), pp. 1415–1439.

[GS87] V. Giovangigli and M. Smooke, *Extinction Limits of Strained Premixed Laminar Flames with Complex Chemistry,* Comb. Sci. Tech., **53**, (1987), pp. 23–49.

[HCB54] J. O. Hirschfelder, C. F. Curtiss, and R. B. Bird, *Molecular Theory of Gases and Liquids,* John Wiley & Sons, Inc., New York, (1954).

[KOB89] K. Kailasanath, E. S. Oran, and J. P. Boris, *Numerical Simulation of Flames and Detonations,* 3rd International Conference on Numerical

Combustion, Sophia-Antipolis (Antibes), A. Dervieux and B. Larouturrou, Eds., Lecture Notes in Physics, Springer-Verlag, **351**, (1989), pp. 82–97.

[KEC87] R. J. Kee, G. H. Evans, and M. E. Coltrin, *Application of Supercomputers to Model Fluid Transport and Chemical Kinetics in Chemical Vapor Deposition Reactors,* Supercomputer Research in Chemistry and Chemical Engineering, K. F. Jensen and D. G. Truhlar, Eds., ACS Symposium Series, **353**, (1987), pp. 334–352.

[Kub65] R. Kubo, *Statistical Mechanics,* North Holland Publishing Company, Amsterdam, (1965).

[LGC92] B. Laboudigue, V. Giovangigli, and S. Candel, *Numerical Solution of a Free-Boundary Problem in Hypersonic Flow Theory: Nonequilibrium Viscous Shock Layers,* J. Comp. Phys., **102**, (1992), pp. 297–309.

[Mas96] M. Massot, *Modélisation Mathématique et Numérique de la Combustion des mélanges Gazeux,* Thèse, École Polytechnique, (1996).

[P*al*96] T. Poinsot, S. Candel, and A. Trouvé, *Application of Direct Numerical Simulation to Premixed Turbulent Combustion,* Prog. Ener. Comb. Sci., **21**, (1996), pp. 531–576.

[RJ87] K. F. Roenigk and K. F. Jensen, *Low Pressure CVD of Silicon Nitride,* J. Electrochem. Soc., **134**, (1987), pp. 1777–1785.

[Ros86] D. E. Rosner, *Transport Processes in Chemically Reacting Flow Systems,* Butterworths, Boston, (1986).

[Sai95] L. Sainsaulieu, *Equilibrium Velocity Distribution Functions for a Kinetic Model of Two-Phase Fluid Flow,* Math. Mod. Meth. Appl. Sci., **5**, (1995), pp. 191–211.

[SePa98] J. H. Seinfeld and S. N. Pandis, *Atmospheric Chemistry and Physics,* John Wiley & Sons, Inc., New York, (1998).

[Smo82] M. D. Smooke, *Solution of Burner Stabilized Premixed Laminar Flames by Boundary Value Methods,* J. Comp. Phys., **48**, (1982), pp. 72–105.

[StPr71] D. R. Stull and H. Prophet, *JANAF Thermochemical Tables, Second ed.,* Washington, NBS NSRDS-NBS37 (1971).

[Van67] J. Van de Ree, *On the Definition of the Diffusion Coefficients in Reacting Gases,* Physica, **36**, (1967), pp. 118–126.

[Wal58] L. Waldmann, *Transporterscheinungen in Gasen von Mittlerem Druck,* Handbuch der Physik, S. Flügge, Ed., **12**, Springer-Verlag, Berlin, (1958), pp. 295–514.

[WT62] L. Waldmann und E. Trübenbacher, *Formale Kinetische Theorie von Gasgemischen aus Anregbaren Molekülen,* Zeitschr. Naturforschg., **17a**, (1962), pp. 363–376.

[W*al*94] J. Warnatz, M. D. Allendorf, R. J. Kee, and M. E. Coltrin, *A Model of Elementary Chemistry and Fluid Mechanics in the Combustion of Hydrogen on Platinum Surfaces,* Comb. Flame, **96** (1994), pp. 393–406.

[Wil58] F. A. Williams, *Spray Combustion and Atomization,* Phys. Fluids, **1**, (1958), pp. 541–545.

[Wil85] F. A. Williams, *Combustion Theory,* Second ed., The Benjamin/Cummings Publishing Company, Inc., Menlo Park, (1985).

[Z*al*85] Y. B. Zeldovitch, G. I. Barenblat, V. B. Librovitch, and G. M. Makhviladze, *The Mathematical Theory of Combustion and Explosions,* Consultants Bureau, New York, (1985).

3

Approximate and Simplified Models

3.1. Introduction

The system of partial differential equations governing multicomponent reactive flows has been presented in Chapter 2. These equations are derived from the kinetic theory of gases and can be used to model very different phenomena, such as laminar flames [Smo82] [Dix84] [Wil85] [EG98b], supersonic flames [DH87] [DM89] [KOB89], epitaxial growth [RJ87] [KEC87] [EGS96], or reentry problems into planetary's atmospheres [And89] [LGC92]. Some applications, such as supersonic combustion, require the full set of equations, without any simplification. In a number of different situations, however, simplifications can be introduced in the complete set of equations, following three different ideas.

The first idea is to simplify the reactive aspects of the flow under consideration. In this situation, the number of species and chemical reactions are decreased and the resulting set of partial differential equations is accordingly simplified. The second idea is to simplify the fluid dynamics aspects of the problem. This may be a geometrical simplification in the problem, a similarity assumption in the flow, or a simplification resulting from an asymptotic limit. As typical examples, we mention creeping flows, boundary layer flows, viscous shock layer flows, or mixing layer flows. Finally, the third idea is to simplify the coupling between chemistry and fluid dynamics. Such a simplification is feasible only in very particular situations, since the coupling arises through various terms in the complete equations. Of course, these ideas may also be used simultaneously, so that the whole family of resulting models is very large. In this chapter, we discuss a typical simplification associated with each of these three ideas. These simplifications are presented to fill the gap between simple models often used in classical textbooks and the full set of partial differential equations governing multicomponent flows presented in Chapter 2.

Excluding the case of unreactive flows—which can be seen as a limiting case—one may first try to simplify the chemistry taking place in the mixture. Indeed, combustion chemistry typically involves dozen of reactive species and hundreds of chemical reactions. Atmospheric chemistry and pollutant formation is even more complex and involves hundreds of reactive species. Therefore, these is a fundamental need for simplifying complex chemical mechanisms. This field is very active, and the resulting reaction schemes often involve semiglobal reactions that are no longer in the form associated with elementary reactions, as described in Chapter 2. Several simplifying procedures have been proposed and are usually associated with a distinction between fast and slow chemical time scales and the existence of low dimensional intrinsic attractive manifolds in composition space [MP92] [Pet85] [Smo91] [Pet91] [LaGo91] [LaSp97] [Spo97] [SBM98]. This area is still very active and promising, but is beyond the scope of the present book. In this chapter, we will only consider the limiting case of a single irreversible exothermic reaction in the form $\nu_{F} F + \nu_{O} O \rightarrow \nu_{P} P$. These global reactions—not to be confused with elementary reactions, which are always reversible—are qualitatively interesting, since several properties can be modeled with one reaction schemes. In flame applications, for instance, the reactant consumption in a thin internal layer or the sensitive temperature dependence of chemistry can be modeled with a single exothermic irreversible reaction. These models are also important theoretically, since they facilitate the use of analytical techniques, such as matched asymptotic expansions. When a single irreversible reaction is considered, however, further ad hoc simplifications are generally introduced in the governing equations. These simplifications are presented in Section 3.2, where we also consider the limiting case of very lean mixtures.

Numerous fluid mechanics simplifications can also be incorporated into the set of partial differential equations governing multicomponent flows. As a typical example, we present the simplifications associated with small Mach number flows in Section 3.3. In the small Mach number limit, there is a pressure decomposition into a spatially homogeneous thermodynamics pressure and a fluid dynamics perturbed pressure. In this situation, the viscous dissipation can be neglected in the energy conservation equation and the density only depends on the thermodynamic pressure. The perturbed pressure derivatives can also be neglected in the energy conservation equation. The small Mach number approximation is often used for modeling chemical vapor deposition reactors or laminar flames, for instance. As typical applications of the small Mach number limit, we then derive the governing equations for multicomponent reactive stagnation point flows and present the Shvab–Zeldovitch formulation for equidiffusive flames. Note that alternative fluid dynamic simplifications could have been one-dimensional flows, constant density flows, creeping flows, boundary layer flows, viscous shock layer flows, or mixing layer flows. These approximations could similarly be incorporated in the general equations in a straightforward way.

Finally, we discuss typical uncoupling approximations between the chemistry and fluid dynamics parts of the governing equations in Section 3.4. Couplings arise, in particular, through the density in the momentum equation, the state variable dependency of viscosities, and the convective terms in the temperature and species equations. We also observe that couplings between the species and temperature equations arise through the dependency of transport coefficients on state variables, through the density, and through cross transport effects, that is, the Soret and Dufour effects. Therefore, uncoupling can solely be advocated under extreme circumstances or by using crude approximations. Uncoupling between the compressible Navier–Stokes equations and the species equations can be used in the dilution limit, as discussed in Section 3.4.2. This situation arises in chemical reactors when a carrier gas is used with trace amount reactants and the heat production due to chemical reactions is negligible. Uncoupling between mass and momentum equations, on the one hand, and the temperature and species equations, on the other hand, can be advocated with the rather crude constant density approximation discussed in Section 3.4.3.

3.2. One-reaction chemistry

3.2.1. One-reaction kinetics

In this section, we consider the simplest reaction scheme constituted by a single global irreversible exothermic reaction in the form

$$\nu_{F} F + \nu_{O} O \to \nu_{P} P, \tag{3.2.1}$$

where F denotes the fuel, O denotes the oxidant, P denotes the products, and ν_F, ν_O, and ν_P denote the stoichiometric coefficients. We assume that this reaction takes place in the presence of a dilutant denoted by N. For such a kinetic with $n^{\mathrm{r}} = 1$ reaction and $n = 4$ species, we will use the species indexing set $\{F, O, P, N\}$, which is more convenient than $\{1, 2, 3, 4\}$.

When a single irreversible reaction is considered, however, further ad hoc simplifications of the governing equations are generally introduced, and they are presented in the following sections.

3.2.2. Approximations for one-reaction kinetics

In order to simplify the species diffusion velocities, we first assume that they are given by Fick's *empirical* law

$$\boldsymbol{\mathcal{F}}_k = \rho Y_k \boldsymbol{\mathcal{V}}_k = -\rho D_k^{\mathrm{ap}} \boldsymbol{\partial_x} Y_k, \qquad k \in \{F, O, P\}, \tag{3.2.2}$$

where D_k^{ap} denotes the empirical diffusion coefficient of the kth species. These approximations are only valid for $n-1$ species, so that only the fuel, oxidant, and product mass fractions are considered as unknowns. The dilutant mass fraction is evaluated from $Y_N = 1 - Y_F - Y_O - Y_P$, and, from mass constraints, the dilutant mass flux is $\mathcal{F}_N = -\mathcal{F}_F - \mathcal{F}_O - \mathcal{F}_P$. Note that (3.2.2) corresponds to a diagonal diffusion process for the first $n-1$ species. This diagonal diffusion problem will be rigorously investigated in Chapter 7. We also neglect Soret and Dufour effects, so that the heat flux is taken to be

$$\boldsymbol{\mathcal{Q}} = \sum_{k\in S} h_k \boldsymbol{\mathcal{F}}_k - \lambda \boldsymbol{\partial_x} T. \tag{3.2.3}$$

We further assume that the species specific heats at constant pressure are equal and independent of temperature, so that

$$c_{pk} = c_p, \qquad k \in S, \tag{3.2.4}$$

where c_p is a constant. The term $\sum_{k\in S} c_{pk}\boldsymbol{\mathcal{F}}_k \cdot \boldsymbol{\partial_x} T$, which appears in the temperature equation, then vanishes, since

$$\sum_{k\in S} c_{pk}\boldsymbol{\mathcal{F}}_k = c_p\Big(\sum_{k\in S}\boldsymbol{\mathcal{F}}_k\Big) = 0.$$

In addition, from $h_k(T) = h_k^{\mathrm{st}} + c_p(T - T^{\mathrm{st}})$, we can further simplify the heat production term $\sum_{k\in S} h_k m_k \omega_k$ by noting that

$$\sum_{k\in S} h_k m_k \omega_k = \sum_{k\in S} h_k^{\mathrm{st}} m_k \omega_k + c_p(T - T^{\mathrm{st}})\sum_{k\in S} m_k\omega_k = \sum_{k\in S} h_k^{\mathrm{st}} m_k \omega_k,$$

where h_k^{st}, $k \in S$, are the formation enthalpies at temperature T^{st}.

We further assume that the mixture transport coefficients λ, η, κ, and $\rho D_1^{\mathrm{ap}}, \ldots, \rho D_{n-1}^{\mathrm{ap}}$ are varying like a power function of temperature $T^{\mathfrak{r}}$ determined empirically, so that

$$\frac{\kappa(T)}{T^{\mathfrak{r}}} = \mathrm{Cte}, \qquad \frac{\eta(T)}{T^{\mathfrak{r}}} = \mathrm{Cte} \tag{3.2.5}$$

and

$$\frac{\lambda(T)}{T^{\mathfrak{r}}} = \mathrm{Cte}, \qquad \frac{(\rho D_k^{\mathrm{ap}})(T)}{T^{\mathfrak{r}}} = \mathrm{Cte}, \quad k \in \{F, O, P\}, \tag{3.2.6}$$

where Cte denotes a generic constant.

From the preceding assumptions, we can introduce Lewis numbers Le_F, Le_O, and Le_P for the fuel F, the oxidant O, and the reaction products P

$$Le_F = \frac{\lambda}{\rho D_F^{\mathrm{ap}} c_p}, \qquad Le_O = \frac{\lambda}{\rho D_O^{\mathrm{ap}} c_p}, \qquad Le_P = \frac{\lambda}{\rho D_P^{\mathrm{ap}} c_p}, \tag{3.2.7}$$

and we can define the Prandtl number Pr of the mixture and the viscosity ratio ι

$$Pr = \frac{\eta c_p}{\lambda}, \qquad \iota = \frac{\kappa}{\eta}, \tag{3.2.8}$$

which are constants.

Finally, we also assume that the only external force is gravity $\boldsymbol{g}$, so that the simplified equations of Section 2.2.5 can be used.

3.2.3. Simplified equations

Under the preceding approximations, the total mass conservation and the momentum conservation equation read as

$$\partial_t \rho + \boldsymbol{\partial_x} \cdot (\rho \boldsymbol{v}) = 0, \tag{3.2.9}$$

$$\begin{aligned} \partial_t(\rho \boldsymbol{v}) + \boldsymbol{\partial_x} \cdot (\rho \boldsymbol{v} \otimes \boldsymbol{v}) = & - \boldsymbol{\partial_x} p + \rho \boldsymbol{g} \\ & + Pr \boldsymbol{\partial_x} \cdot \Big\{ \frac{\lambda}{c_p} \big(\boldsymbol{\partial_x} \boldsymbol{v} + \boldsymbol{\partial_x} \boldsymbol{v}^t + (\iota - \tfrac{2}{3})(\boldsymbol{\partial_x} \cdot \boldsymbol{v}) \boldsymbol{I} \big) \Big\}, \end{aligned} \tag{3.2.10}$$

whereas the species equations read as

$$\rho \partial_t Y_F + \rho \boldsymbol{v} \cdot \boldsymbol{\partial_x} Y_F = \frac{1}{Le_F} \boldsymbol{\partial_x} \cdot \Big(\frac{\lambda}{c_p} \boldsymbol{\partial_x} Y_F \Big) - m_F \nu_F \, \omega, \tag{3.2.11}$$

$$\rho \partial_t Y_O + \rho \boldsymbol{v} \cdot \boldsymbol{\partial_x} Y_O = \frac{1}{Le_O} \boldsymbol{\partial_x} \cdot \Big(\frac{\lambda}{c_p} \boldsymbol{\partial_x} Y_O \Big) - m_O \nu_O \, \omega, \tag{3.2.12}$$

$$\rho \partial_t Y_P + \rho \boldsymbol{v} \cdot \boldsymbol{\partial_x} Y_P = \frac{1}{Le_P} \boldsymbol{\partial_x} \cdot \Big(\frac{\lambda}{c_p} \boldsymbol{\partial_x} Y_P \Big) + m_P \nu_P \, \omega, \tag{3.2.13}$$

and the last concentration is evaluated from $Y_N = 1 - Y_F - Y_O - Y_P$. Finally, the temperature equation is now in the form

$$\rho \partial_t T + \rho \boldsymbol{v} \cdot \boldsymbol{\partial_x} T = \boldsymbol{\partial_x} \cdot \Big(\frac{\lambda}{c_p} \boldsymbol{\partial_x} T \Big) + \frac{m_F \nu_F Q}{c_p} \omega + \frac{1}{c_p} \Big(\partial_t p + \boldsymbol{v} \cdot \boldsymbol{\partial_x} p - \boldsymbol{\Pi} : \boldsymbol{\partial_x} \boldsymbol{v} \Big), \tag{3.2.14}$$

where Q denotes the heat release rate per unit mass of fuel

$$Q = h_F^{\rm st} + \frac{m_O \nu_O}{m_F \nu_F} h_O^{\rm st} - \frac{m_P \nu_P}{m_F \nu_F} h_P^{\rm st}, \tag{3.2.15}$$

which is positive since the chemical reaction (3.2.1) is assumed to be exothermic. The temperature equation will be further simplified for small Mach number flows, as investigated in Section 3.3. The rate of progress ω for reaction (3.2.1) is usually approximated in the form

$$\omega = \mathfrak{A} \, T^{\mathfrak{b}} \exp\Big(-\frac{\mathfrak{E}}{RT}\Big) \Big(\frac{\rho Y_F}{m_F} \Big)^{\nu_F} \Big(\frac{\rho Y_O}{m_O} \Big)^{\nu_O}, \tag{3.2.16}$$

where $\mathfrak{A}$, $\mathfrak{b}$, and $\mathfrak{E}$ are empirical constants.

Remark 3.2.1. It may also be assumed that the species molar masses are equal, so that

$$m_k = \overline{m}, \qquad k \in S, \tag{3.2.17}$$

which introduces further simplifications in the state law. In this situation, mole fractions and mass fractions coincide. This approximation is often used implicitly in the literature. ∎

Remark 3.2.2. Introducing a Lewis number Le_N for the dilutant N would lead to a conflict between $\sum_{k\in S} Y_k = 1$ and $\sum_{k\in S} \mathcal{F}_k = 0$, except in the special situation where all Lewis numbers are equal. In particular, deducing from the constraint $\sum_{k\in S} \mathcal{F}_k = 0$ that we have $\boldsymbol{\partial_x}(\sum_{k\in S} Y_k/Le_k) = 0$, so that $\sum_{k\in S} Y_k/Le_k = \mathrm{Cte}$, would be a major mistake when Lewis numbers are distinct. ∎

3.2.4. Deficient reactants

In this section, we further assume that the fuel F and the combustion products P are present only by trace amounts in the mixture. Correspondingly, we assume that the oxidant O and the dilutant N are in excess. In this situation, we can assume that Y_O and Y_N are constant in the flow and write that

$$Y_O = Y_O^0, \tag{3.2.18}$$

$$Y_N = Y_N^0, \tag{3.2.19}$$

$$Y_F \ll Y_N^0 + Y_O^0, \tag{3.2.20}$$

$$Y_P \ll Y_N^0 + Y_O^0, \tag{3.2.21}$$

where Y_N^0 and Y_O^0 are constants. The oxidant and product balance equation then uncouple and can be discarded, so that we only have to solve the fuel equation (3.2.11). The fluid equations (3.2.9), (3.2.10), and the temperature equation (3.2.14) are unchanged. The dependent unknowns are then ρ, $\boldsymbol{v}$, p, Y_F, and T, and the production rate ω takes the simpler form

$$\omega = \mathfrak{A}' T^{\mathfrak{b}'} \exp(-\frac{\mathfrak{E}}{RT}) \left(\frac{\rho Y_F}{m_F}\right)^{\nu_F} \rho^{\nu_O}, \tag{3.2.22}$$

where $\mathfrak{A}'$ and $\mathfrak{b}'$ are constants. This model is very practical for describing various effects due to preferential diffusion of the deficient reactant by varying the Lewis number Le_F of the fuel F. Note that the final model obtained with a one-reaction chemistry and a deficient reactant is one of the simplest models describing reactive flows.

Remark 3.2.3. Lean one-reaction flow equations can also be *formally* obtained by considering the reaction $\nu_F F \to \nu_P P$ within the binary mixture $\{F, P\}$. Note that such a derivation is only *formal* and that the fictitious reaction $\nu_F F \to \nu_P P$ rather represents an *alchemy* process than a realistic chemical reaction. ■

3.3. Small Mach number flows

In this section we present the simplifications arising in the equations governing multicomponent flows when the Mach number is assumed to be small. We first investigate the order of magnitude of each term appearing in the conservation equations with respect to the Mach number. Letting the Mach number go to zero, we then obtain the isobaric equations for multicomponent flows. Mathematical aspects concerning the small Mach number limit have been discussed by Majda [Maj84] and Lions [Lio96] [Lio98]. However, these aspects are beyond the scope of this book, and we only present here a formal analysis.

We then present two typical applications of the resulting isobaric model. We first derive the equations governing strained flames in stagnation point flows as exact similar solutions of the isobaric equations. We then present the Shvab–Zeldovitch formulation for equidiffusive flows.

3.3.1. Orders of magnitude

In order to derive an asymptotic model for small Mach number flows, we have to investigate the order of magnitude of various terms appearing in the governing equations. To this purpose, for each quantity ϕ in the equations, we introduce a typical order of magnitude denoted by $<\phi>$. In particular, we introduce a characteristic velocity $<v>$, viscosity $<\eta>$, pressure $<p>$, gravity $<g>$, and density $<\rho>$. The order of magnitude of the sound velocity c is then $<c>^2 = <p>/<\rho>$. The typical hydrodynamic length $<x>$ upon consideration is also a viscous length, that is, a length such that the corresponding Reynolds number is unity. We thus have the typical viscous–diffusion length $<x> = <\eta>/<\rho><v>$ and introduce the corresponding time $<t> = <x>/<v>$. We assume that the only external force acting on the species is gravity and that the associated Froude number $Fr = <v>^2/<x><g>$ is of order unity or larger.

Upon defining the reduced quantities $\widehat{\phi} = \phi/<\phi>$ associated with each quantity ϕ of the mixture, and denoting by

$$\epsilon = \frac{<v>}{<c>} \ll 1, \tag{3.3.1}$$

a typical Mach number, we can now estimate the order of magnitude of each term in the governing partial differential equations.

3.3.2. Momentum equation and pressure splitting

Upon using the notation defined in the previous section, we obtain—after a little algebra—the reduced momentum equation

$$\epsilon^2\Big(\partial_{\widehat{t}}(\widehat{\rho}\,\widehat{\boldsymbol{v}}) + \boldsymbol{\partial}_{\widehat{\boldsymbol{x}}}\cdot(\widehat{\rho}\,\widehat{\boldsymbol{v}}\otimes\widehat{\boldsymbol{v}}) + \boldsymbol{\partial}_{\widehat{\boldsymbol{x}}}\cdot\widehat{\boldsymbol{\Pi}} - \tfrac{1}{Fr}\widehat{\rho}\,\widehat{\boldsymbol{g}}\Big) = -\boldsymbol{\partial}_{\widehat{\boldsymbol{x}}}\,\widehat{p}. \tag{3.3.2}$$

This reduced equation reveals that the spatial gradient of the pressure is of order $O(\epsilon^2)$, so that

$$\widehat{p}(t,\boldsymbol{x}) = \frac{p}{<p>} = \widehat{p}_0(t) + \epsilon^2\widehat{p}_2(t,\boldsymbol{x}). \tag{3.3.3}$$

In terms of the original variables, it yields the classical splitting

$$p(t,\boldsymbol{x}) = p_{\rm u}(t) + \widetilde{p}(t,\boldsymbol{x}), \tag{3.3.4}$$

where $p_{\rm u} = p_{\rm u}(t)$ is spatially uniform and $\widetilde{p} = p_{\rm u}O(\epsilon^2)$ is the fluid dynamic perturbation.

Upon introducing this pressure splitting in the governing equations, we obtain for $\epsilon \to 0$ the simplified state law

$$\rho = \frac{p_{\rm u}\overline{m}}{RT}, \tag{3.3.5}$$

and the simplified momentum conservation equation

$$\partial_t(\rho\boldsymbol{v}) + \boldsymbol{\partial}_{\boldsymbol{x}}\cdot(\rho\boldsymbol{v}\otimes\boldsymbol{v}) + \boldsymbol{\partial}_{\boldsymbol{x}}\widetilde{p} + \boldsymbol{\partial}_{\boldsymbol{x}}\cdot\boldsymbol{\Pi} = \rho\boldsymbol{g}, \tag{3.3.6}$$

which has been rewritten in terms of the original variables.

3.3.3. Energy equation

Proceeding in a similar way for the energy equation—taken in the enthalpy form for convenience—we obtain that

$$\partial_{\widehat{t}}(\widehat{\rho}\,\widehat{h}) + \boldsymbol{\partial}_{\widehat{\boldsymbol{x}}}\cdot(\widehat{\rho}\,\widehat{\boldsymbol{v}}\widehat{h}) + \boldsymbol{\partial}_{\widehat{\boldsymbol{x}}}\cdot\widehat{\boldsymbol{\mathcal{Q}}} = \partial_{\widehat{t}}\widehat{p}_0 + \epsilon^2\Big(\partial_{\widehat{t}}\widehat{p}_2 + \widehat{\boldsymbol{v}}\cdot\boldsymbol{\partial}_{\widehat{\boldsymbol{x}}}\widehat{p}_2 - \widehat{\boldsymbol{\Pi}}:\boldsymbol{\partial}_{\widehat{\boldsymbol{x}}}\widehat{\boldsymbol{v}}\Big), \tag{3.3.7}$$

making use of the various estimates $<h> = <c>^2$, $<\lambda> = <\eta><c_p>$ and $<\eta> = <\rho><D>$. The classical estimate $<h> = <c>^2$ reminds us that internal energy is typically associated with molecular velocity fluctuations. The reduced equation (3.3.7) now reveals that the terms $\partial_t\widetilde{p}$, $\boldsymbol{v}\cdot\boldsymbol{\partial}_{\boldsymbol{x}}p$

and $\boldsymbol{\Pi}:\boldsymbol{\partial_x v}$ can be neglected in the energy conservation equation, written in energy or enthalpy form.

As a consequence, we obtain for $\epsilon \to 0$ the simplified energy conservation equation

$$\partial_t(\rho h) + \boldsymbol{\partial_x}\cdot(\rho h \boldsymbol{v}) + \boldsymbol{\partial_x}\cdot\boldsymbol{\mathcal{Q}} = \partial_t p_{\mathrm{u}}, \tag{3.3.8}$$

presented in terms of the original variables, which appears as a thermal balance.

3.3.4. Isobaric equations

In this section, we assume that the Mach number is small and regroup the limiting equations obtained by formally letting $\epsilon \to 0$. We further assume for simplicity that the spatially uniform pressure p_{u} is independent of time. In this situation, we obtain the classical isobaric equations

$$\partial_t \rho + \boldsymbol{\partial_x}\cdot(\rho \boldsymbol{v}) = 0, \tag{3.3.9}$$

$$\partial_t(\rho Y_k) + \boldsymbol{\partial_x}\cdot(\rho Y_k \boldsymbol{v}) + \boldsymbol{\partial_x}\cdot\boldsymbol{\mathcal{F}}_k = m_k\omega_k, \qquad k \in S, \tag{3.3.10}$$

$$\partial_t(\rho \boldsymbol{v}) + \boldsymbol{\partial_x}\cdot(\rho \boldsymbol{v}\otimes\boldsymbol{v}) + \boldsymbol{\partial_x}\widetilde{p} + \boldsymbol{\partial_x}\cdot\boldsymbol{\Pi} = \rho \boldsymbol{g}, \tag{3.3.11}$$

$$\partial_t(\rho h) + \boldsymbol{\partial_x}\cdot(\rho \boldsymbol{v} h) + \boldsymbol{\partial_x}\cdot\boldsymbol{\mathcal{Q}} = 0, \tag{3.3.12}$$

which can be used in conjunction with the simplified state law

$$\rho = \frac{p_{\mathrm{u}}\overline{m}}{RT}. \tag{3.3.13}$$

Note that the energy conservation equation can also be written in temperature form

$$\begin{aligned}\rho c_p \partial_t T + \rho c_p \boldsymbol{v}\cdot\boldsymbol{\partial_x} T + \boldsymbol{\partial_x}\cdot\Big(\boldsymbol{\mathcal{Q}} - \sum_{k\in S} h_k \boldsymbol{\mathcal{F}}_k\Big) &= \\ - \sum_{k\in S} c_{pk}\boldsymbol{\mathcal{F}}_k\cdot\boldsymbol{\partial_x} T - \sum_{k\in S} h_k m_k \omega_k&.\end{aligned} \tag{3.3.14}$$

In (3.3.9)–(3.3.14) the fundamental unknowns are the density ρ, velocity $\boldsymbol{v}$, perturbed pressure $\widetilde{p}$, temperature T, and mass fractions $Y_1, \ldots, Y_n$. However, we still have the usual alternative concerning the species mass fractions. The first possibility is to consider only $n-1$ mass fractions as unknowns, say, the first $n-1$ mass fractions $Y_1, \ldots, Y_{n-1}$, and to evaluate the mass fraction of the remaining species Y_n from the relation $Y_n = 1 - \sum_{k\in\{1,\ldots,n-1\}} Y_k$. In this situation, only the first $n-1$ species conservation equations are solved. This method is generally used for simplified chemistry flows, but is not very accurate for the last species. For complex chemistry flows, a symmetric role is often given to all mass fractions by considering them as fundamental unknowns. These mass fractions are thus considered to be independent, and the relation $\sum_{k\in S} Y_k = 1$, which was an a priori consequence of the definition of the mass fractions, must be an a posteriori consequence of the governing equations.

3.3.5. Vorticity–velocity formulation

We examine in this section a reformulation of the isobaric flame equations, which eliminates the perturbed pressure $\widetilde{p}$ from the governing equations. To this end, we assume that the flow is steady and introduce the vorticity

$$\boldsymbol{w} = \boldsymbol{\partial_x} \wedge \boldsymbol{v}, \tag{3.3.15}$$

where $\wedge$ is the classical alternating product. The perturbed pressure is then eliminated by using the vorticity–velocity formulation of the Navier–Stokes equations for compressible flows [ES93].

The vorticity–velocity formulation was first introduced for incompressible flows [Bat67], but has recently been used successfully for compressible flows as natural convection problems, chemical vapor deposition reactors, and laminar flames [ES93] [EGS96] [EG98b].

This formulation consists in a transport equation for the vorticity $\boldsymbol{w}$ obtained by applying the curl operator $\boldsymbol{\partial_x}\wedge$ to the momentum equation and an elliptic second-order equation for the velocity vector $\boldsymbol{v}$ obtained by applying the curl operator $\boldsymbol{\partial_x}\wedge$ to definition (3.3.15). For three-dimensional flows, these equations can be written in the following form [ES93] [Ern96] [Ern98] :

$$\begin{aligned}\boldsymbol{\partial_x} \wedge \Big(\boldsymbol{w} \wedge (\rho \boldsymbol{v}) + \rho \boldsymbol{\partial_x}(\boldsymbol{v}\cdot\boldsymbol{v}/2)\Big) &= \boldsymbol{\partial_x} \wedge (\rho \boldsymbol{g}) \\ &+ \boldsymbol{\partial_x} \wedge \Big(\big(\boldsymbol{\partial_x} \boldsymbol{v} + (\boldsymbol{\partial_x}\boldsymbol{v})^t - 2(\boldsymbol{\partial_x}\cdot\boldsymbol{v})\boldsymbol{I}\big)\cdot\boldsymbol{\partial_x}\eta\Big) \\ &- \boldsymbol{\partial_x} \wedge (\eta \boldsymbol{\partial_x} \wedge \boldsymbol{w}) + \boldsymbol{\partial_x}(\eta \boldsymbol{\partial_x}\cdot\boldsymbol{w})\end{aligned} \tag{3.3.16}$$

and

$$\boldsymbol{\partial_x}\cdot(\boldsymbol{\partial_x}\boldsymbol{v}) + \boldsymbol{\partial_x} \wedge \boldsymbol{w} + \boldsymbol{\partial_x}\big(\boldsymbol{v}\cdot\boldsymbol{\partial_x} \log\rho\big) = 0. \tag{3.3.17}$$

It is worthwhile to note that steadiness has been used for substituting the expression $-\boldsymbol{v}\cdot\boldsymbol{\partial_x}\log\rho$ in place of $\boldsymbol{\partial_x}\cdot\boldsymbol{v}$ in the velocity equation. A simpler formulation can be obtained in two dimensions [Ern96] [Ern98].

3.3.6. Strained flows

As a first application of the isobaric flow model, we investigate the equations governing reactive mixtures in stagnation point flows, with an emphasis on strained flames. One may obtain such flames by directing two impinging jets, one on each other or against a flat plate. These models are important in order to describe the effect of strain on flame structures [LLW83] [LW82] [LW83] [LW84] [GS87] [GS89]. In order to derive the corresponding equations, a first possibility is to use a boundary layer flow approximation. In planar or cylindrical geometries, however, one may also obtain the strained flow equations as *exact* solutions of the isobaric model [Gio88] [K*al*88] by looking for self-similar solutions, as presented in the remaining part of this section. In practice, these solutions are valid in a domain when the corresponding boundary conditions are compatible with the similarity assumption.

We denote by (x, y, z) a system of Cartesian coordinates, such that y is a coordinate normal to the flame surface, Oxz being the stagnation plane. We denote by $\boldsymbol{v} = (u, v, 0)^t$ and $\boldsymbol{\mathcal{V}}_k = (\mathcal{U}_k, \mathcal{V}_k, 0)^t$, $k \in S$, the corresponding coordinates of the mass velocity $\boldsymbol{v}$ and of the diffusion velocities $\boldsymbol{\mathcal{V}}_k$, $k \in S$, respectively. We also assume that gravity is either oriented towards the normal axis, that is, $\boldsymbol{g} = (0, g, 0)^t$, or that it can be neglected. We now seek for a solution of the isobaric flow equations in the form [Gio88] [LiSm97]

$$\rho = \rho(t, y), \tag{3.3.18}$$

$$u = x\,\widehat{u}(t, y), \tag{3.3.19}$$

$$v = v(t, y), \tag{3.3.20}$$

$$\widetilde{p} = -J(t)\frac{x^2}{2} + \widehat{p}(t, y), \tag{3.3.21}$$

$$T = T(t, y), \tag{3.3.22}$$

$$Y = Y(t, y), \qquad k \in S. \tag{3.3.23}$$

Note that only the tangential velocity u and the pressure $\widetilde{p}$ depend on the tangential coordinate x in this model, with linear and quadratic dependencies. In these equations $J(t)$ represents a pressure curvature in the tangential direction.

The expression for diffusion velocities then yields

$$\boldsymbol{\mathcal{V}}_k = \left(0, \mathcal{V}_k(t, y), 0\right)^t, \tag{3.3.24}$$

so that there are no contributions along the tangential x axis. Substitution of (3.3.18)–(3.3.24) into (3.3.9)–(3.3.14) now yields the equations

$$\partial_t \rho + \rho\widehat{u} + \partial_y(\rho v) = 0, \tag{3.3.25}$$

$$\rho\partial_t\widehat{u}\,x + \rho\widehat{u}^2\,x + \rho v\partial_y\widehat{u}\,x = J\,x + \partial_y\left(\eta\partial_y\widehat{u}\right)x, \tag{3.3.26}$$

$$\rho\partial_t v + \rho v\partial_y v = -\partial_y\widehat{p} + \eta\partial_y\widehat{u} + \partial_y\left((\kappa + \tfrac{4}{3}\eta)\partial_y v\right) + \partial_y\left((\kappa - \tfrac{2}{3}\eta)\widehat{u}\right) + \rho g, \tag{3.3.27}$$

$$\rho \partial_t Y_k + \rho v \partial_y Y_k = m_k \omega_k - \partial_y(\rho Y_k \mathcal{V}_k), \qquad k \in S, \tag{3.3.28}$$

$$\begin{aligned}\rho c_p \partial_t T + \rho v c_p \partial_y T = \partial_y \Big(\lambda \partial_y T - p \sum_{k\in S} \chi_k \mathcal{V}_k\Big) \\ - \Big(\sum_{k\in S} \rho Y_k \mathcal{V}_k c_{pk}\Big) \partial_y T - \sum_{k\in S} h_k m_k \omega_k,\end{aligned} \tag{3.3.29}$$

together with the state law

$$\rho = \frac{p_{\mathrm{u}} \overline{m}}{RT}. \tag{3.3.30}$$

The normal momentum equation (3.3.27), which yields the vertical pressure correction $\widehat{p}$, uncouples from the remaining governing equations. In addition, the tangential momentum conservation equation (3.3.26) can be divided by the tangential coordinate x and yields

$$\rho \partial_t \widehat{u} + \rho v \partial_y \widehat{u} = \partial_y \Big(\eta \partial_y \widehat{u}\Big) + (J - \rho \widehat{u}^2) = 0. \tag{3.3.31}$$

The strained flow equations are then constituted by (3.3.25), (3.3.31), (3.3.28), (3.3.29), and (3.3.30).

Typical boundary conditions in the upper mixture, at $y = +\infty$, are in the form

$$T(t, +\infty) = T^{\mathrm{up}}, \qquad Y_k(t, +\infty) = Y_k^{\mathrm{up}}, \quad k \in S, \qquad \widehat{u}(t, +\infty) = \widehat{u}^{\mathrm{up}}(t). \tag{3.3.32}$$

The quantity $\alpha(t) = \widehat{u}^{\mathrm{up}}(t)$ is the strain rate imposed by the upper mixture. Note that the thermochemical state of the upper mixture at $+\infty$ cannot depend on time, since, by letting $y \to \infty$ in the species and temperature equations, we obtain $\partial_t T^{\mathrm{up}} = 0$ and $\partial_t Y_k^{\mathrm{up}} = 0$, $k \in S$. Similarly, by letting $y \to +\infty$, the momentum conservation equation (3.3.31) yields the fundamental compatibility relation

$$\partial_t \alpha(t) + \alpha^2(t) = J(t)/\rho^{\mathrm{up}}, \tag{3.3.33}$$

between the strain rate α and the pressure curvature J. This compatibility condition is required only when the domain extends up to $+\infty$. We can also choose the origin at the stagnation plane

$$v(t, 0) = 0, \tag{3.3.34}$$

and similar boundary conditions can be used for the lower mixture at $-\infty$

$$T(t, -\infty) = T^{\mathrm{lo}}, \qquad Y_k(t, -\infty) = Y_k^{\mathrm{lo}}, \quad k \in S, \qquad \widehat{u}(t, -\infty) = \widehat{u}^{\mathrm{lo}}(t). \tag{3.3.35}$$

As for the upper boundary, the thermochemical state of the mixture at $-\infty$ cannot depend on time and the momentum conservation equation yields a compatibility relation for $\widehat{u}^{\rm lo}(t) = \widehat{u}(t, -\infty)$ in the form

$$\partial_t \widehat{u}^{\rm lo}(t) + \big(\widehat{u}^{\rm lo}(t)\big)^2 = J(t)/\rho^{\rm lo}. \tag{3.3.36}$$

These equations can also be used for a stagnation point flow against a flat plate or for symmetric jets with straightforward modifications of the boundary conditions.

Remark 3.3.1. One may also use the strained flow equations (3.3.25), (3.3.31), (3.3.28), (3.3.29), and (3.3.30) for two concentric jets that are at finite distances from the stagnation plane. In the situation of finite length domains, the mixtures flowing out of the jets and their velocities can be chosen arbitrarily. More specifically, the boundary conditions at the upper end, for $y = L^{\rm up}$, then read

$$T(t, L^{\rm up}) = T^{\rm up}(t), \qquad Y_k(t, L^{\rm up}) = Y_k^{\rm up}(t), \quad k \in S, \tag{3.3.37}$$

$$\widehat{u}(t, L^{\rm up}) = \widehat{u}^{\rm up}(t), \qquad v(t, L^{\rm up}) = v^{\rm up}(t), \tag{3.3.38}$$

and at the lower end

$$T(t, -L^{\rm lo}) = T^{\rm lo}(t), \qquad Y_k(t, -L^{\rm lo}) = Y_k^{\rm lo}(t), \quad k \in S, \tag{3.3.39}$$

$$\widehat{u}(t, -L^{\rm lo}) = \widehat{u}^{\rm lo}(t), \qquad v(t, -L^{\rm lo}) = v^{\rm lo}(t), \tag{3.3.40}$$

where $L^{\rm up}$ and $-L^{\rm lo}$ are the vertical coordinates of the tube ends. In this situation, the pressure curvature J becomes an eigenvalue that solves the overdetermination in the normal velocity v equation, which is only first order but has two boundary conditions. Note that the stagnation plane defined by the condition $v = 0$ is no longer located at the origin $y = 0$ for this type of flow and that no compatibility conditions are required among the flow parameters. The reduced tangential velocity $\widehat{u}$ is typically chosen to be zero at the jet ends. ■

Remark 3.3.2. It is traditional to introduce the rescaled quantities

$$\widetilde{u} = \frac{\widehat{u}}{\alpha} = \frac{u}{\alpha x}, \tag{3.3.41}$$

$$\widetilde{v} = \rho v, \tag{3.3.42}$$

and to reformulate the previous equations in terms of $\widetilde{u}$ and $\widetilde{v}$. The corresponding conservation equations are easily obtained after a little algebra. For an infinite domain, using the compatibility condition (3.3.33), we obtain, for instance,

$$\partial_t \rho + \rho\alpha\widetilde{u} + \partial_y \widetilde{v} = 0, \tag{3.3.43}$$

$$\rho \partial_t \widetilde{u} + \widetilde{v} \partial_y \widetilde{u} = \partial_y \Big(\eta \partial_y \widetilde{u} \Big) + \alpha(\rho^{\mathrm{up}} - \rho \widetilde{u}^2) + (\partial_t \alpha / \alpha)(\rho^{\mathrm{up}} - \rho \widetilde{u}), \quad (3.3.44)$$

$$\rho \partial_t Y_k + \widetilde{v} \partial_y Y_k = -\partial_y (\rho Y_k \mathcal{V}_k) + m_k \omega_k, \qquad k \in S, \quad (3.3.45)$$

$$\rho c_p \partial_t T + \widetilde{v} c_p \partial_y T = \partial_y \Big(\lambda \partial_y T - p \sum_{k \in S} \chi_k \mathcal{V}_k \Big)$$

$$- \Big(\sum_{k \in S} \rho Y_k \mathcal{V}_k c_{pk} \Big) \partial_y T - \sum_{k \in S} h_k m_k \omega_k, \quad (3.3.46)$$

The boundary conditions for the temperature and the species are unchanged, whereas the boundary conditions for the variable $\widetilde{u}$ are

$$\widehat{u}(t, +\infty) = 1, \quad (3.3.47)$$

$$\partial_t \widetilde{u}(t, -\infty) + \alpha(\widetilde{u}^2(t, -\infty) - \rho^{\mathrm{up}}/\rho^{\mathrm{lo}}) + (\partial_t \alpha / \alpha)(\widetilde{u}(t, -\infty) - \rho^{\mathrm{up}}/\rho^{\mathrm{lo}}) = 0. \quad (3.3.48)$$

These boundary conditions are easily adapted to the case of burners located at finite length. In this situation, Dirichlet boundary conditions can be used for the variables $\widetilde{u}$ and $\widetilde{v}$. ▪

Remark 3.3.3. For axisymmetric geometries, we have to modify the equations (3.3.25) and (3.3.27). More specifically, we have to multiply by a factor of 2 the term $\rho \widehat{u}$ in (3.3.25) and the terms $\eta \partial_y \widehat{u}$ and $\partial_y \big((\kappa - \frac{2}{3}\eta) \partial_y \widehat{u} \big)$ in (3.3.27). The later equation still uncouples from the other equations. The preceding formulation can also be generalized to swirling, tubular, or spherical flows and we refer the reader to [D*al*91], [SG90], and [SG92]. ▪

3.3.7. Shvab–Zeldovitch formulation

As a second application of the isobaric approximation, we investigate the Shvab–Zeldovitch formulation for equidiffusive flows. For simplicity, we again consider, in this section, a single global irreversible exothermic reaction in the form (3.2.1). We thus combine two simplifying ideas here, one pertaining to chemistry and one pertaining to fluid mechanics.

We assume that the mixture is equidiffusive in such a way that

$$Le_{F} = 1, \qquad Le_{O} = 1, \qquad Le_{P} = 1, \quad (3.3.49)$$

which means that species diffusion and heat diffusion occur at the same rate. In particular, all Fick empirical diffusion coefficients are equal $D_F^{\mathrm{ap}} = D_O^{\mathrm{ap}} = D_P^{\mathrm{ap}} = D^{\mathrm{ap}}$, where D^{ap} denotes their common value.

A consequence of (3.3.49) is that the balance equations for Y_F, Y_O, Y_P, and T can be written in the form

$$\rho \partial_t Y_F + \rho \boldsymbol{v} \cdot \boldsymbol{\partial_x} Y_F = \boldsymbol{\partial_x} \cdot \big(\frac{\lambda}{c_p} \boldsymbol{\partial_x} Y_F \big) - m_F \nu_F \, \omega, \quad (3.3.50)$$

$$\rho\partial_t Y_O + \rho \boldsymbol{v}\cdot\boldsymbol{\partial_x} Y_O = \boldsymbol{\partial_x}\cdot(\frac{\lambda}{c_p}\boldsymbol{\partial_x} Y_O) - m_O \nu_O\, \omega, \tag{3.3.51}$$

$$\rho\partial_t Y_P + \rho \boldsymbol{v}\cdot\boldsymbol{\partial_x} Y_P = \boldsymbol{\partial_x}\cdot(\frac{\lambda}{c_p}\boldsymbol{\partial_x} Y_P) + m_P \nu_P\, \omega, \tag{3.3.52}$$

and

$$\rho\partial_t T + \rho \boldsymbol{v}\cdot\boldsymbol{\partial_x} T = \boldsymbol{\partial_x}\cdot(\frac{\lambda}{c_p}\boldsymbol{\partial_x} T) + \frac{m_F \nu_F Q}{c_p}\, \omega, \tag{3.3.53}$$

where Q denotes the heat release rate per unit mass of fuel, so that all of these equations are proportional. It is thus interesting to introduce the composite variables ξ_F, ξ_O, and ξ_P, given by

$$\xi_F = Y_F + \frac{c_p}{Q}\, T, \qquad \xi_O = Y_O + \frac{c_p}{Q}\frac{m_O\nu_O}{m_F\nu_F}\, T, \qquad \xi_P = Y_P - \frac{c_p}{Q}\frac{m_P\nu_P}{m_F\nu_F}\, T, \tag{3.3.54}$$

and termed the "Shvab–Zeldovitch variables," since they are governed by the balance equation

$$\rho\partial_t \xi + \rho \boldsymbol{v}\cdot\boldsymbol{\partial_x} \xi = \boldsymbol{\partial_x}\cdot(\frac{\lambda}{c_p}\boldsymbol{\partial_x} \xi), \tag{3.3.55}$$

which does not contain any production term. Note that Y_N is also governed by this equation, since $\boldsymbol{\mathcal{F}}_N = -\boldsymbol{\mathcal{F}}_F - \boldsymbol{\mathcal{F}}_O - \boldsymbol{\mathcal{F}}_P$ yields for equidiffusive flows that $\boldsymbol{\mathcal{F}}_N = D^{\mathrm{ap}}\boldsymbol{\partial_x}(Y_F + Y_O + Y_P)$, so that $\boldsymbol{\mathcal{F}}_N = -D^{\mathrm{ap}}\boldsymbol{\partial_x} Y_N$.

This generic equation (3.3.55) obviously has constant solutions. Nevertheless, the quantities ξ_F, ξ_O, ξ_P, and Y_N will effectively be constants provided that the boundary conditions are compatible with such an assumption. In this section, we assume that this is indeed the case and we thus obtain that

$$Y_F = Y_F^0 - \frac{c_p}{Q}\,(T - T^0), \tag{3.3.56}$$

$$Y_O = Y_O^0 - \frac{c_p}{Q}\frac{m_O\nu_O}{m_F\nu_F}\,(T - T^0), \tag{3.3.57}$$

$$Y_P = Y_P^0 + \frac{c_p}{Q}\frac{m_P\nu_P}{m_F\nu_F}\,(T - T^0), \tag{3.3.58}$$

$$Y_N = Y_N^0, \tag{3.3.59}$$

where Y_F^0, Y_O^0, Y_P^0, Y_N^0, and T^0 are constants that can be determined from the boundary conditions.

The governing equations can thus be reduced to the balance equations

$$\partial_t\rho + \boldsymbol{\partial_x}\cdot(\rho\boldsymbol{v}) = 0, \tag{3.3.60}$$

$$\partial_t(\rho\boldsymbol{v}) + \boldsymbol{\partial_x}\cdot(\rho\boldsymbol{v}\otimes\boldsymbol{v}) + \boldsymbol{\partial_x}\widetilde{p} + \boldsymbol{\partial_x}\cdot\boldsymbol{\Pi} = \rho\boldsymbol{g}, \tag{3.3.61}$$

$$\rho\partial_t T + \rho \boldsymbol{v}\cdot\boldsymbol{\partial_x} T - \boldsymbol{\partial_x}\cdot(\frac{\lambda}{c_p}\boldsymbol{\partial_x} T) = \frac{m_F \nu_F Q}{c_p}\, \omega, \tag{3.3.62}$$

with unknowns ρ, $\boldsymbol{v}$, $\widetilde{p}$, and T. In these equations, the density ρ and the production term ω are simply functions of temperature. Note that, for equidiffusive mixtures, it is generally assumed that

$$Pr = 1, \qquad \iota = 0. \tag{3.3.63}$$

Remark 3.3.4. The equidiffusive model is even simpler than the deficient reactant model. Such simple models are usually quite crude quantitatively, but are important theoretically, since numerous mathematical and asymptotic investigations have been conducted in this framework. ■

3.4. Coupling

In this section we discuss the coupling between the fluid and thermochemistry parts of the governing equations presented in Chapter 2 or in the previous sections of Chapter 3. We first review all of the coupling terms between the momentum, energy, and species equations. We then discuss the simplifications associated with the dilution limit and the constant density approximation.

3.4.1. Coupling of partial differential equations

There are a number of couplings between the fluid part and the thermochemistry part of the partial differential equations governing multicomponent reactive mixtures.

The first direct influence of the thermochemistry part over the fluid part is through the density ρ and the state law. For instance, any heat release due to chemical reactions will increase the temperature and correspondingly decrease the density, and this will directly influence the velocity fields. Another coupling arises through the temperature and mass fraction dependencies of the viscosities.

Conversely, the velocity fields directly influence the convective terms in the temperature and species equations.

We also observe that the coupling between the species and temperature equations arises through the dependency of transport coefficients on state variables, through the density ρ, and through the cross transport effects, that is, the Soret and Dufour effects.

As a consequence, any uncoupling between all of these partial differential equations is generally not feasible in multicomponent reactive flows. Only very special flows allow this type of simplifications.

In the following sections, we examine two special cases, namely, the dilution approximation and the constant density approximation. In the dilution approximation, one must first solve the compressible fluid equations and then the species equations. In the constant density approximation, one must first solve the incompressible fluid equations and then the temperature and species equations.

3.4.2. Dilution approximation

The dilution approximation is an asymptotic limit that arises in several models of practical importance as, for instance, chemical vapor deposition models. In this limit, the mixture is composed of an excess species, while all other species are in trace amounts.

In the dilution limit, the last species is in large excess, so that in a first approximation we have $Y_k = 0$, $k \in \{1, \ldots, n-1\}$, and $Y_n = 1$ uniformly. As a consequence, the species variables do not appear in the total mass, momentum, and energy equations, say, in enthalpy form. We can thus solve the total mass, momentum, and energy equations under this approximation and then solve the $n-1$ species equations, assuming that the density, temperature and velocity fields are given by the previous calculation.

If we try now to estimate the domain of validity of this approximation, we note that a small variation of trace amount species will not seriously influence the density, provided that the mass ratios of trace species with respect to the dilutant are of order unity. Similarly, replacing the mixture enthalpy by the dilutant enthalpy is reasonably accurate, provided that the specific heats are of the same order of magnitude—which is often the case—and that the mixture heat formation can be replaced by that of the dilutant, that is, we have to assume that the heat release or heat consumption due to chemical reactions is negligible.

A straightforward generalization is then to consider a mixture that differs by trace amounts from a spatially uniform inert state. The corresponding modifications are straightforward to derive, and the details are omitted. This situation applies, for instance, to a multicomponent carrier gas or to a carrier gas and one heavier reactant, provided that all remaining species are in trace amounts with respect to both of these gases.

3.4.3. Constant density approximation

The constant density approximation is an unrealistic limit used in reactive flow theory to suppress the analytical complexities due to gas expansion. These models often give a correct qualitative description of diffusional–thermal stability effects, as thermodiffusive instabilities in flames [Bal62] [Siv77]. These models can be obtained by various asymptotic methods [MS79] and are usually termed thermo–diffusive. In this section, we discuss the corresponding uncouplings in the governing equations.

The constant density approximation simply consists in letting the density be a constant in the governing equations. In this situation, one must solve the incompressible Navier–Stokes equations and then the coupled temperature and species equations, once the velocity field is known.

If we now try to estimate the domain of validity of the constant density approximation, we first see that small density variations must involve small temperature variations, so that the heat release must be small, which is unrealistic for most flames. Finally, compressibility effects have to be small, so that the Mach number has to be small.

3.5. Notes

3.1. Another chemistry approximation is that of chemical equilibrium flows. In this situation, the reactions taking place in the mixture are assumed to be so fast that chemical equilibrium holds locally. The governing equations then express the conservation of atomic elements, momentum, and energy. The resulting set of governing equations will be presented and discussed in Chapter 10.

3.6. References

[And89] J. D. Anderson, Jr., *Hypersonics and High Temperature Gas Dynamics,* McGraw-Hill Book Company, New York, (1989).

[Bat67] G. K. Batchelor, *An Introduction to Fluid Dynamics,* Cambridge University Press, Cambridge, (1967).

[B*al*62] G. I. Barenblat, Y. B. Zeldovitch, and A. G. Istratov, *On Diffusional–Thermal Stability of a Laminar Flame,* (in Russian), Prikl. Mekh. Tekh. Fiz., **4**, (1962), pp. 21–31.

[Dix84] G. Dixon-Lewis, *Computer Modeling of Combustion Reactions in Flowing Systems with Transport,* in W. C. Gardiner, Ed., *Combustion Chemistry.* Springer-Verlag, New York, (1984), pp. 21–125.

[D*al*91] G. Dixon-Lewis , V. Giovangigli, R. J. Kee, J. A. Miller, B. Rogg, M. D. Smooke, G. Stahl, and J. Warnatz, *Numerical Modeling of the Structure and Properties of Tubular Strained Laminar Premixed Flames,* Dynamics of Deflagration and Reactive Systems: Flames, A. L. Kuhl, J. C. Leyer, A. A. Borisov, and W. A. Sirignano, Eds., Progress in Astronautics and Aeronautics, **131**, AIAA, Washington, DC, (1991), pp. 125–144.

[DH87] J. P. Drumond and M. Y. Hussaini, *Numerical Simulation of a Supersonic Reacting Mixing Layer,* AIAA 19th Fluid Dynamics, Plasma Physics and Laser Conference, Honolulu, Hawaii, Paper AIAA-87-1325, (1987).

[DM89] J. P. Drumond and H. S. Mukunda, *A Numerical Study of Mixing Enhancement in Supersonic Reacting Flow Fields,* 3rd International Conference on Numerical Combustion, Sophia-Antipolis (Antibes), A. Dervieux and B. Larouturrou, Eds., Lecture Notes in Physics, Springer-Verlag **351**, (1989), pp. 36–64.

[Ern96] A. Ern, *Sur la Formulation Tourbillon-Vitesse des Équations de Navier-Stokes avec Densité et Viscosité Variables,* C. R. Acad. Sci. Paris, Série II, **323**, (1996), pp. 813–818.

[Ern98] A. Ern, *Vorticity-Velocity Formulation of the Stokes Problem with Variable Density and Viscosity,* Math. Models Meth. Appl. Sciences, **8**, (1998), pp. 203–218.

[EG98b] A. Ern and V. Giovangigli, *Thermal Diffusion Effects in Hydrogen/Air and Methane/Air Flames,* Comb. Theor. Mod., **2**, (1998), pp. 349–372.

[EGS96] A. Ern, V. Giovangigli, and M. Smooke, *Numerical Study of a Three-Dimensional Chemical Vapor Deposition Reactor with Detailed Chemistry,* J. Comp. Phys., **126**, (1996), pp. 21–39.

[ES93] A. Ern and M. Smooke, *The Vorticity-Velocity Formulation,* J. Comp. Phys., **105**, (1993), pp. 58–68.

[Gio88] V. Giovangigli, *Structure et Extinction de Flammes Laminaires Prémélangées,* Doctorat ès Sciences, Université Paris 6, (1988).

[GS87] V. Giovangigli and M. Smooke, *Extinction Limits of Strained Premixed Laminar Flames with Complex Chemistry*, Comb. Sci. Tech., **53**, (1987), pp. 23–49.

[GS89] V. Giovangigli and M. Smooke, *Adaptive Continuation Algorithms with Application to Combustion Problems*, Appl. Numer. Math., **5**, (1989), pp. 305–331.

[KOB89] K. Kailasanath, E. S. Oran, and J. P. Boris, *Numerical Simulation of Flames and Detonations,* 3rd International Conference on Numerical Combustion, Sophia-Antipolis (Antibes), A. Dervieux and B. Larouturrou, Eds., Lecture Notes in Physics, Springer-Verlag **351**, (1989), pp. 82–97.

[KEC87] R. J. Kee, G. H. Evans, and M. E. Coltrin, *Application of Supercomputers to Model Fluid Transport and Chemical Kinetics in Chemical Vapor Deposition Reactors,* Supercomputer Research in Chemistry and Chemical Engineering, K. F. Jensen and D. G. Truhlar, Eds., ACS Symposium Series, **353**, (1987), pp. 334–352.

[K*al*88] R. J. Kee, J. A. Miller, G. H. Evans, and G. Dixon-Lewis, *A Computational Model of the Structure and Extinction of Strained, Opposed Flow, Premixed Methane-Air Flames,* 22nd Symposium International on Combustion, The Combustion Institute, Pittsburgh, (1988), pp. 1479–1494.

[LGC92] B. Laboudigue, V. Giovangigli, and S. Candel, *Numerical Solution of a Free-Boundary Problem in Hypersonic Flow Theory: Nonequilibrium Viscous Shock Layers,* J. Comp. Phys., **102**, (1992), pp. 297–309.

[LaGo91] S. H. Lam and D. A. Goussis, *Conventional and Computational Singular Perturbation for Simplified Kinetics Modelling,* Mech. Aero. Eng. Dept., Princeton University, Technical Report 1864(b)-MAE, (1991).

[LaSp97] B. Larrouturou and B. Sportisse, *Some Mathematical and Numerical Aspects of reduction in Chemical Kinetics,* Computational Science for the 21st Century, Conference in honor of Professor R. Glowinsky, M. O. Bristeau, G. Etgen, W. Fitzgibbon, J. L. Lions, J. Périaux, M. F. Wheeler, Eds., J. Wiley & Sons, Inc., Chichester, (1997), pp. 422–432.

[LLW83] P. A. Libby, A. Liñan, and F. A. Williams, *Strained Premixed Laminar Flames with Nonunity Lewis Numbers,* Comb. Sci. Tech., **34**, (1983), pp. 257–293.

[LW82] P. A. Libby and F. A. Williams, *Structure of Laminar Flamelets in Premixed Turbulent Flames,* Comb. Flame, **44**, (1982), pp. 287–303.

[LW83] P. A. Libby and F. A. Williams, *Strained Premixed Laminar Flames Under Nonadiabatic Conditions,* Comb. Sci. Tech., **31**, (1983), pp. 1–42.

[LW84] P. A. Libby and F. A. Williams, *Strained Premixed Laminar Flames with Two Reaction Zones,* Comb. Sci. Tech., **37**, (1984), pp. 221–252.

[LiSm97] P. A. Libby and M. D. Smooke, *The Computation of Flames in Stagnation Flows,* Comb. Sci. Tech., **127**, (1997), pp. 197–211.

[Lio96] P. L. Lions, *Mathematical Topics in Fluid Mechanics,* Volume 1: Imcompressible models, Oxford Lecture Series in Mathematics and its Applications, **3**, Clarendon Press, Oxford, (1996).

[Lio98] P. L. Lions, *Mathematical Topics in Fluid Mechanics,* Volume 2: Compressible models, Oxford Lecture Series in Mathematics and its Applications, **3**, Clarendon Press, Oxford, (1998).

[MP92] U. Maas and S. B. Pope, *Simplifying Chemical Kinetics : Intrinsic Low-Dimensional Manifolds in Composition Space,* Comb. Flame **88**, (1992), pp. 239–264.

[Maj84] A. Majda, *Compressible Fluid Flow and Systems of Conservation laws in Several Space Variables,* Applied Mathematical Science, **53**, Springer-Verlag, New–York, (1984).

[MS79] B. J. Matkowsky and G. I. Sivashinsky, *An Asymptotic Derivation of Two Models in Flame Theory Associated with the Constant Density Approximation,* SIAM J. Appl. Math., **37**, (1979), pp. 686–699.

[Pet85] N. Peters, *Numerical and Asymptotic Analysis of a Systematically Reduced Reaction Schemes for Hydrocarbon Flames,* in: R. Glowinsky, B. Larrouturou, and R. Temam, Eds., Numertical Simulation of Combustion Phenomena, Springer-Verlag, Berlin, (1985).

[Pet91] N. Peters, *Reducing Mechanisms,* in: M. D. Smooke, Ed., Reduced Kinetic Mechanisms and Asymptotic Approximation for Methane-Air Flames, Lectures Notes in Physics, **384**, Springer-Verlag, Berlin (1991).

[RJ87] K. F. Roenigk and K. F. Jensen, *Low Pressure CVD of Silicon Nitride,* J. Electrochem. Soc., **134**, (1987), pp. 1777–1785.

[SBM98] D. Schmidt, T. Blasenbrey, and U. Mass, *Intrinsic Low-dimensional Manifolds of Strained and Unstrained Flames,* Comb. Theor. Model., **2**, (1998), pp. 135–152.

[Siv77] G. I. Sivashinsky, *Diffusional–Thermal Theory of Cellular Flames,* Comb. Sci. Tech., **15**, (1977), pp. 137–146.

[Smo82] M. D. Smooke, *Solution of Burner-Stabilized Premixed Laminar Flames by Boundary Value Methods,* J. Comp. Phys., **48**, (1982), pp. 72–105.

[Smo91] M. D. Smooke, Ed. *Reduced Kinetics Mechanism and Asymptotic Approximations for Methane-Air Flames,* Lecture Notes in Physics, **384**, Springer-Verlag, Berlin, (1991).

[SG90] M. D. Smooke and V. Giovangigli, *Structure and Extinction of Premixed Tubular Flames,* 23rd Symposium International on Combustion, The Combustion Institute, Pittsburgh, (1990), pp. 447–454.

[SG92] M. D. Smooke and V. Giovangigli, *A Comparison between Experimental Measurements and Numerical Calculations of the Structure of Premixed Rotating Counterflow Methane-Air Flames,* 24rd Symposium International on Combustion, The Combustion Institute, Pittsburgh, (1992), pp. 161–168.

[Spo97] B. Sportisse, Ed. *Numerical Aspects of Rediction in Chemical Kinetics,* Proceedings of the September 1997 Workshop held at Marne-la-Vallée, (1997).

[Wil85] F. A. Williams, *Combustion Theory,* Second ed., The Benjamin/Cummings Publishing Company, Inc., Menlo Park, (1985).

4

Derivation from the Kinetic Theory

4.1. Introduction

In this chapter, we summarize the derivation of multicomponent reactive flow equations from the kinetic theory of gases. This derivation can be found in numerous textbooks, usually for monatomic species mixtures and/or inert species. In this chapter, we consider the general case of dilute polyatomic reactive gas mixtures. We refer the reader, however, to classical textbooks on kinetic theory [CC70] [FK72] [Mal90] [Mal91] for detailed comments or intermediate calculations since we only give here a brief description. In addition, mathematical aspects of Boltzmann-type equations are out of the scope of this book, and we refer the reader to [Cer88], [DL89], [Gru93], [CIP94], [GG95], and [GM97].

We first present the concept of distribution function and write generalized Boltzmann equations in a semiclassical framework. The reactive source term is written for arbitrary chemical mechanisms, but specific reactions are considered for illustration. We introduce the kinetic entropy and obtain the kinetic entropy equation. We show, in particular, that entropy production is nonnegative, which yields the Boltzmann H–theorem. In particular, entropy production due to either inert or reactive collisions is nonnegative [LH60] [Gru93] [EG98a].

We then consider the Enskog expansion and focus on the Maxwellian reaction regime, in which chemical characteristic times are larger than the mean free times of the molecules. We establish the corresponding macroscopic equations, that is, the conservation equations, thermodynamic properties, transport fluxes, and chemical production rates. We further obtain the macroscopic entropy conservation equation and discuss the link between the kinetic entropy and the macroscopic entropy.

We also express the transport coefficients in terms of bracket products involving solutions of integral equations arising from linearized Boltzmann equations. Upon using a Galerkin variational procedure, evaluating transport coefficients is then reduced to solving linear systems. Various linear systems of practical interest will be explicited in Chapter 5 for each transport coefficient.

4.2. Kinetic framework

In this section, we present a reactive kinetic theory framework. We first introduce the species distribution functions and discuss the link between this molecular description and the macroscopic point of view. We then present the reactive generalized Boltzmann equations first introduced by Ludwig and Heil [LH60] [Kus91] [Gru93] [EG94] [ACG94].

4.2.1. Distribution functions

We consider a dilute reactive mixture consisting of n chemical species having internal degrees of freedom. We assume a semiclassical framework in which the translational motion of the particles is treated classically and the internal degrees of freedom are treated quantum mechanically. We also restrict our analysis to isotropic gas mixtures, thus excluding the case of strong external magnetic fields.

The state of the mixture is described by the species distribution functions denoted by $f_i(t, \boldsymbol{x}, \boldsymbol{c}_i, \mathrm{I})$, $i \in S$, where t is the time, $\boldsymbol{x}$ is the three-dimensional spatial coordinate, $\boldsymbol{c}_i$ is the velocity of the ith species, I is the index for the quantum internal energy states of the ith species, and $S = \{1, \dots, n\}$ is the set of species indices. We denote by $\mathfrak{Q}_i$ the indexing set of the quantum internal energy states I of the ith species. The quantity $f_i(t, \boldsymbol{x}, \boldsymbol{c}_i, \mathrm{I})\delta\boldsymbol{x}\delta\boldsymbol{c}_i$ represents the expected number of molecules of type i in quantum state I in the volume element $\delta\boldsymbol{x}$ located at $\boldsymbol{x}$, whose velocities lie in $\delta\boldsymbol{c}_i$ about velocity $\boldsymbol{c}_i$ at time t. The species distribution functions are governed by Boltzmann equations written in the next sections.

The family of distribution functions will be written for convenience in the form $(f_i)_{i\in S}$. More generally, for a family of functions ξ_i, $i \in S$, where ξ_i depends on $\boldsymbol{c}_i$ and I, we will often use the compact notation $\xi = (\xi_i)_{i\in S}$, the dependence on $(t, \boldsymbol{x})$ being left implicit.

4.2.2. Macroscopic properties

The macroscopic properties of a gas are easily obtained from the molecular description by species distribution functions. More specifically, the number density of the ith species—the number of molecules per unit volume—is naturally given by

$$n_i = \sum_{I \in \Omega_i} \int f_i d\boldsymbol{c}_i, \tag{4.2.1}$$

so that the mass density of the ith species is

$$\rho_i = \mathfrak{m}_i n_i = \sum_{I \in \Omega_i} \int \mathfrak{m}_i f_i d\boldsymbol{c}_i, \tag{4.2.2}$$

where $\mathfrak{m}_i$ denotes the mass of the molecules of the ith species. Similarly, the hydrodynamic velocity $\boldsymbol{v}$ is given by

$$\rho \boldsymbol{v} = \sum_{\substack{i \in S \\ I \in \Omega_i}} \int \mathfrak{m}_i \boldsymbol{c}_i f_i d\boldsymbol{c}_i, \tag{4.2.3}$$

where $\rho = \sum_{i \in S} \rho_i$ is the total density, and the internal energy $\mathcal{E}$ per unit volume is given by

$$\tfrac{1}{2}\rho \boldsymbol{v} \cdot \boldsymbol{v} + \mathcal{E} = \sum_{\substack{i \in S \\ I \in \Omega_i}} \int (\tfrac{1}{2}\mathfrak{m}_i \boldsymbol{c}_i \cdot \boldsymbol{c}_i + E_{iI}) f_i d\boldsymbol{c}_i, \tag{4.2.4}$$

where E_{iI} is the internal energy of the molecules of the ith species in the Ith quantum state.

In the next section, we write the Boltzmann equations governing the species distribution functions f_i, $i \in S$. The macroscopic conservation equations for the densities ρ_i, $i \in S$, velocity $\boldsymbol{v}$, and internal energy $\mathcal{E}$ will then be obtained by taking moments of these Boltzmann equations.

4.2.3. Boltzmann equations

In this section, we present the equations governing the species distribution functions $f = (f_i)_{i \in S}$. These equations are generalized Boltzmann equations that take into account the reactive aspects of the mixture. For detailed presentations and discussions about Boltzmann-type equations, we refer to the textbooks [CC70] [FK72].

The generalized Boltzmann equations are in the form

$$\mathcal{D}_i(f_i) = \mathcal{S}_i(f) + \mathcal{C}_i(f), \qquad i \in S, \tag{4.2.5}$$

where $\mathcal{D}_i$ denotes the usual streaming differential operator

$$\mathcal{D}_i(f_i) = \partial_t f_i + \boldsymbol{c}_i \cdot \boldsymbol{\partial_x} f_i + \boldsymbol{b}_i \cdot \boldsymbol{\partial}_{\boldsymbol{c}_i} f_i. \tag{4.2.6}$$

In these equations, $\partial_t f_i$ is the accumulation term, $\boldsymbol{c}_i \cdot \boldsymbol{\partial_x} f_i$ is associated with the change in location due to the velocity $\boldsymbol{c}_i$ of the molecules, $\boldsymbol{b}_i \cdot \boldsymbol{\partial}_{\boldsymbol{c}_i} f_i$ is associated with the change in velocity due to the external force $\boldsymbol{b}_i$ acting on the ith species, $\mathcal{S}_i(f)$ is the nonreactive or scattering source term, and $\mathcal{C}_i(f)$ is the reactive or chemistry source term.

The species Boltzmann equations thus express the conservation of particles in the phase space and can also be derived from the BBGKY-hierarchy [Wal58] [CC70] [FK72]. In the next section, we discuss the structure of the source terms $\mathcal{S}_i(f)$ and $\mathcal{C}_i(f)$.

4.2.4. Scattering source terms

The term $\mathcal{S}_i(f)$ represents the rate at which the velocity distribution of the ith species is altered by inert molecular collisions. This nonreactive collision term is derived in numerous textbooks and can be written as

$$\mathcal{S}_i(f) = \sum_{j \in S} \sum_{\substack{I' \in \mathfrak{Q}_i \\ J \in \mathfrak{Q}_j \\ J' \in \mathfrak{Q}_j}} \int \Big(f_i' f_j' \frac{a_{iI} a_{jJ}}{a_{iI'} a_{jJ'}} - f_i f_j \Big) W_{ij}^{IJI'J'} d\boldsymbol{c}_j d\boldsymbol{c}_i' d\boldsymbol{c}_j', \qquad (4.2.7)$$

where I and J are the indices for the quantum energy states of the ith and j^{th} species before collision, respectively, I' and J' are the corresponding numbers after collision, a_{iI} is the degeneracy of the Ith quantum energy shell of the ith species, $W_{ij}^{IJI'J'}$ is the transition probability for nonreactive collisions, and f_i' and f_j' denote, respectively, $f_i' = f_i(t, \boldsymbol{x}, \boldsymbol{c}_i', I')$ and $f_j' = f_j(t, \boldsymbol{x}, \boldsymbol{c}_j', J')$ [Wal58] [WT62] [EG94]. Only binary collisions are considered since the system is dilute. The transition probabilities satisfy the reciprocity relations

$$W_{ij}^{IJI'J'} a_{iI} a_{jJ} = W_{ij}^{I'J'IJ} a_{iI'} a_{jJ'}. \qquad (4.2.8)$$

Note that we have chosen to work with transition probabilities rather than with classical collision cross sections. The reason for using transition probabilities is that the reactive collision term $\mathcal{C}_i(f)$ is then much easier to write. For binary collisions, denoting by $\sigma_{ij}^{IJI'J'}$ the collision cross section, we have the identity

$$g_{ij} \sigma_{ij}^{IJI'J'} d\boldsymbol{e}_{ij}' = W_{ij}^{IJI'J'} d\boldsymbol{c}_i' d\boldsymbol{c}_j', \qquad (4.2.9)$$

where g_{ij} is the absolute value of the relative velocity $\boldsymbol{c}_i - \boldsymbol{c}_j$ of the incoming particles and $\boldsymbol{e}_{ij}'$ is the unit vector in the direction of the relative velocity $\boldsymbol{c}_i' - \boldsymbol{c}_j'$ after collision [LH60] [Gru93] [ACG94] [GG95]. Conservation of mass, momentum, and energy holds during collisions and is taken into account in the model by using dirac delta functions in transition probabilities [LH60] [Gru93] [ACG94] [GG95].

4.2.5. Reactive source terms

The reactive source term $\mathfrak{C}_i(f)$ represents the rate at which the distribution function of the ith species is altered by reactive molecular collisions. We consider, in particular, a chemical reaction mechanism composed by an arbitrary number of elementary reactions. Note that both bimolecular and trimolecular chemical reactions are taken into account although triple nonreactive collisions have been neglected in the nonreactive source term (4.2.7). It is indeed important in the applications to consider triple reactive collisions, since recombination reactions cannot often proceed otherwise [Kus91] [ACG94] [GG95]. Some of these triple collisions may also correspond to successive binary collisions [ACG94].

The chemical reactions are indexed by r and written for convenience in the form

$$\sum_{i\in\mathfrak{F}^r} \mathfrak{M}_i \rightleftharpoons \sum_{k\in\mathfrak{B}^r} \mathfrak{M}_k, \qquad r \in R, \tag{4.2.10}$$

where $\mathfrak{F}^r$ and $\mathfrak{B}^r$ are, respectively, the indices for the reactant and product species in the rth elementary reaction, counted with their order of multiplicity, and R is the set of reaction indices. The letters $\mathfrak{F}^r$ and $\mathfrak{B}^r$ are mnemonics for the forward and backward directions, respectively. We denote by ν_{ir}^{f} and ν_{ir}^{b} the stoichiometric coefficients of the ith species among reactants and products, respectively, and we also denote by F^r and B^r the indices of internal energy states for reactants and products, respectively. In other words, the forward and backward coefficients ν_{ir}^{f} and ν_{ir}^{b} are the order of multiplicity of species i in $\mathfrak{F}^r$ and $\mathfrak{B}^r$, respectively. For a given $i \in S$, $\mathfrak{F}_i^r$ denotes the set of reactant indices in which the index for the ith species has been removed only once, and we introduce a similar notation for $\mathfrak{B}_i^r$, F_I^r and B_I^r. The reactive source term in (4.2.5) now reads as

$$\mathfrak{C}_i(f) = \sum_{r\in R} \mathfrak{C}_i^r(f), \tag{4.2.11}$$

where $\mathfrak{C}_i^r(f)$ is the source term for the ith species due to the rth elementary reaction (4.2.10), which is given by [LH60] [Kus91] [Gru93] [ACG94] [EG98a]

$$\begin{aligned}\mathfrak{C}_i^r(f) = \nu_{ir}^{\text{f}} \sum_{F_I^r, B^r} \int \Big(\prod_{k\in\mathfrak{B}^r} f_k \frac{\prod_{k\in\mathfrak{B}^r} \beta_{kK}}{\prod_{j\in\mathfrak{F}^r} \beta_{jJ}} - \prod_{j\in\mathfrak{F}^r} f_j \Big) \mathcal{W}_{\mathfrak{F}^r\mathfrak{B}^r}^{F^r B^r} \prod_{j\in\mathfrak{F}_i^r} d\boldsymbol{c}_j \prod_{k\in\mathfrak{B}^r} d\boldsymbol{c}_k \\ + \nu_{ir}^{\text{b}} \sum_{F^r, B_I^r} \int \Big(\prod_{j\in\mathfrak{F}^r} f_j - \frac{\prod_{k\in\mathfrak{B}^r} \beta_{kK}}{\prod_{j\in\mathfrak{F}^r} \beta_{jJ}} \prod_{k\in\mathfrak{B}^r} f_k \Big) \mathcal{W}_{\mathfrak{F}^r\mathfrak{B}^r}^{F^r B^r} \prod_{j\in\mathfrak{F}^r} d\boldsymbol{c}_j \prod_{k\in\mathfrak{B}_i^r} d\boldsymbol{c}_k .\end{aligned} \tag{4.2.12}$$

In these expressions, the statistical weight β_{iI} is given by

$$\beta_{iI} = \frac{h_{\mathrm{P}}^3}{a_{iI}\mathfrak{m}_i^3}, \tag{4.2.13}$$

where h_{P} is the Planck constant, $\mathfrak{m}_i$ is the mass of the particles of the ith species, and $\mathcal{W}_{\mathfrak{F}^r\mathfrak{B}^r}^{F^rB^r}$ is the transition probability for a reactive collision in which the reactants $\mathfrak{F}^r$ with internal energy states F^r are transformed into products $\mathfrak{B}^r$ with internal energy states B^r. The sums over F^r (or F_I^r) represent the sums over $J \in \mathfrak{Q}_j$ for all $j \in \mathfrak{F}^r$ ($j \in \mathfrak{F}_i^r$, respectively). Similarly, the sums over B^r (or B_I^r) represent the sums over $K \in \mathfrak{Q}_k$ for all $k \in \mathfrak{B}^r$ ($k \in \mathfrak{B}_i^r$, respectively). The following reciprocity relation holds for the reactive transition probabilities

$$\mathcal{W}_{\mathfrak{F}^r\mathfrak{B}^r}^{F^rB^r} \prod_{k\in\mathfrak{B}^r} \beta_{kK} = \mathcal{W}_{\mathfrak{B}^r\mathfrak{F}^r}^{B^rF^r} \prod_{j\in\mathfrak{F}^r} \beta_{jJ}, \tag{4.2.14}$$

which coincides with (4.2.8) for nonreactive collisions. For bimolecular reactions, these transition probabilities can also be expressed in terms of reactive collision cross sections as (4.2.9). For three body reactions, however, the relation between transition probabilities and collision cross sections is complex, so that expressing reactive collision sources in terms of cross sections is unduly complex [ACG94]. In order to illustrate the molecular reactive term $\mathcal{C}_i(f)$, several practical examples are given in the next section. As for inert collisions, conservation of atomic elements, mass, momentum, and energy is taken into account by dirac delta functions in transition probabilities [LH60] [Gru93] [ACG94] [GG95].

4.2.6. Examples

In order to illustrate the general expression of the reactive source term $\mathcal{C}_i^r(f)$, consider the bimolecular reaction

$$\mathfrak{M}_i + \mathfrak{M}_j \rightleftharpoons \mathfrak{M}_k + \mathfrak{M}_l, \tag{4.2.15}$$

with species indices i, j, k, l assumed to be distinct, so that $\nu_{ir}^{\mathrm{f}} = \nu_{jr}^{\mathrm{f}} = 1$ and $\nu_{kr}^{\mathrm{b}} = \nu_{lr}^{\mathrm{b}} = 1$, all other stoichiometric coefficients being zero. We then have $\mathfrak{F}^r = \{i, j\}$, $\mathfrak{B}^r = \{k, l\}$, and denoting by I, J, K, L the species internal energy states indices, the reactive source term reads as

$$\mathcal{C}_i^r(f) = \sum_{\substack{J\in\mathfrak{Q}_j\\ K\in\mathfrak{Q}_k\\ L\in\mathfrak{Q}_l}} \int \Big(f_k f_l \frac{\beta_{kK}\beta_{lL}}{\beta_{iI}\beta_{jJ}} - f_i f_j\Big) \mathcal{W}_{ijkl}^{IJKL} d\boldsymbol{c}_j d\boldsymbol{c}_k d\boldsymbol{c}_l, \tag{4.2.16}$$

where $\mathcal{W}_{ijkl}^{IJKL}$ denotes the transition probability associated with (4.2.15). Similarly, now consider a bimolecular reaction in which the colliding molecules are identical

$$\mathfrak{M}_i + \mathfrak{M}_i \rightleftharpoons \mathfrak{M}_k + \mathfrak{M}_l. \tag{4.2.17}$$

We then have $\mathfrak{F}^r = \{ i, i \}$, $\mathfrak{B}^r = \{ k, l \}$, so that $\nu_{ir}^{\mathrm{f}} = 2$ and $\nu_{kr}^{\mathrm{b}} = \nu_{lr}^{\mathrm{b}} = 1$, all other stoichiometric coefficients being zero. The reactive source term then becomes

$$\mathcal{C}_i^r(f) = 2 \sum_{\substack{\tilde{I}\in\mathfrak{Q}_i\\ K\in\mathfrak{Q}_k\\ L\in\mathfrak{Q}_l}} \int \Big(f_k f_l \frac{\beta_{kK}\beta_{lL}}{\beta_{iI}\beta_{i\tilde{I}}} - f_i \tilde{f}_i\Big) \mathcal{W}_{iikl}^{I\tilde{I}KL} d\tilde{\boldsymbol{c}}_i d\boldsymbol{c}_k d\boldsymbol{c}_l, \tag{4.2.18}$$

where $\tilde{f}_i$ denotes $f_i(t, \boldsymbol{x}, \tilde{\boldsymbol{c}}_i, \tilde{I})$ and the symbol $\tilde{}$ is used to distinguish one of the collision partners in (4.2.17).

In the case of a chemical reaction involving three products

$$\mathfrak{M}_i + \mathfrak{M}_j \rightleftharpoons \mathfrak{M}_k + \mathfrak{M}_l + \mathfrak{M}_m, \tag{4.2.19}$$

with all the indices i, j, k, l, m assumed to be distinct, the reactive source term reads as

$$\mathcal{C}_i(f) = \sum_{\substack{J\in\mathfrak{Q}_j\\ K\in\mathfrak{Q}_k\\ L\in\mathfrak{Q}_l\\ M\in\mathfrak{Q}_m}} \int \Big(f_k f_l f_m \frac{\beta_{kK}\beta_{lL}\beta_{mM}}{\beta_{iI}\beta_{jJ}} - f_i f_j\Big) \mathcal{W}_{ijklm}^{IJKLM} d\boldsymbol{c}_j d\boldsymbol{c}_k d\boldsymbol{c}_l d\boldsymbol{c}_m. \tag{4.2.20}$$

Finally, in the case where the ith species is present as reactant and product

$$\mathfrak{M}_i + \mathfrak{M}_j \rightleftharpoons \mathfrak{M}_i + \mathfrak{M}_l, \tag{4.2.21}$$

the source term reads as

$$\begin{aligned} \mathcal{C}_i(f) = & \sum_{\substack{J\in\mathfrak{Q}_j\\ \tilde{I}\in\mathfrak{Q}_i\\ L\in\mathfrak{Q}_l}} \int \Big(\tilde{f}_i f_l \frac{\beta_{i\tilde{I}}\beta_{lL}}{\beta_{iI}\beta_{jJ}} - f_i f_j\Big) \mathcal{W}_{ijil}^{IJ\tilde{I}L} d\boldsymbol{c}_j d\tilde{\boldsymbol{c}}_i d\boldsymbol{c}_l \\ & + \sum_{\substack{\tilde{I}\in\mathfrak{Q}_i\\ J\in\mathfrak{Q}_j\\ L\in\mathfrak{Q}_l}} \int \Big(\tilde{f}_i f_j - \frac{\beta_{iI}\beta_{lL}}{\beta_{i\tilde{I}}\beta_{jJ}} f_i f_l\Big) \mathcal{W}_{ijil}^{\tilde{I}JIL} d\tilde{\boldsymbol{c}}_i d\boldsymbol{c}_j d\boldsymbol{c}_l, \end{aligned} \tag{4.2.22}$$

keeping in mind that the forward and backward events are distinct in (4.2.21) at variance with inert collisions.

4.3. Kinetic entropy

A fundamental aspect of the kinetic theory of gases is the Boltzmann H-theorem, which is equivalent to the second law of thermodynamics. In this section, we introduce the kinetic entropy and establish the kinetic entropy balance equation and the H-theorem. We establish, in particular, that both the nonreactive source term (4.2.7) and the reactive source term (4.2.12) yield positive entropy productions [LH60] [CC70] [FK72] [dGM84] [GG95] [EG98a].

4.3.1. Definition of the kinetic entropy

The kinetic entropy $\mathcal{S}^{\mathrm{kin}}$ per unit volume is defined by

$$\mathcal{S}^{\mathrm{kin}} = -k_{\mathrm{B}} \sum_{\substack{i\in S \\ I\in\mathfrak{Q}_i}} \int f_i \big(\log(\beta_{iI} f_i) - 1\big) d\boldsymbol{c}_i, \tag{4.3.1}$$

where the weights β_{iI} are associated with the reciprocity relations of transition probabilities (4.2.8) (4.2.14) and k_{B} is the Boltzmann constant.

4.3.2. Kinetic entropy equation

Multiplying the Boltzmann equation (4.2.5) by $\log(\beta_{iI} f_i)$, integrating with respect to $\boldsymbol{c}_i$, and summing over i and I yields, after some algebra, the kinetic entropy conservation equation

$$\partial_t \mathcal{S}^{\mathrm{kin}} + \boldsymbol{\partial_x} \cdot (\mathcal{S}^{\mathrm{kin}} \boldsymbol{v}) + \boldsymbol{\partial_x} \cdot \boldsymbol{J}^{\mathrm{kin}} = \mathfrak{v}^{\mathrm{kin}}, \tag{4.3.2}$$

where $\boldsymbol{v}$ is the mean average velocity (4.2.3), $\boldsymbol{J}^{\mathrm{kin}}$ is the kinetic entropy diffusive flux given by

$$\boldsymbol{J}^{\mathrm{kin}} = -k_{\mathrm{B}} \sum_{\substack{i\in S \\ I\in\mathfrak{Q}_i}} \int (\boldsymbol{c}_i - \boldsymbol{v}) f_i \big(\log(\beta_{iI} f_i) - 1\big) d\boldsymbol{c}_i, \tag{4.3.3}$$

and $\mathfrak{v}^{\mathrm{kin}}$ is the entropy source term that can be split into

$$\mathfrak{v}^{\mathrm{kin}} = \mathfrak{v}^{\mathcal{S}} + \mathfrak{v}^{\mathcal{C}}, \tag{4.3.4}$$

where $\mathfrak{v}^{\mathcal{S}}$ is the nonreactive contribution

$$\mathfrak{v}^{\mathcal{S}} = -k_{\mathrm{B}} \sum_{\substack{i\in S \\ I\in\mathfrak{Q}_i}} \int \mathcal{S}_i(f) \log(\beta_{iI} f_i) d\boldsymbol{c}_i \tag{4.3.5}$$

and $\mathfrak{v}^{\mathfrak{C}}$ is the reactive contribution

$$\mathfrak{v}^{\mathfrak{C}} = -k_{\mathrm{B}} \sum_{\substack{i\in S \\ I\in \mathfrak{Q}_i}} \int \mathfrak{C}_i(f) \log(\beta_{iI} f_i) d\boldsymbol{c}_i. \tag{4.3.6}$$

In the next section, we investigate the sign of $\mathfrak{v}^{\mathrm{kin}}$ and show that both contributions $\mathfrak{v}^{\mathfrak{S}}$ and $\mathfrak{v}^{\mathfrak{C}}$ are nonnegative.

4.3.3. Positivity of entropy production

After a little algebra, the nonreactive source term $\mathfrak{v}^{\mathfrak{S}}$ can be rewritten in the form

$$\frac{4\mathfrak{v}^{\mathfrak{S}}}{k_{\mathrm{B}}} = \sum_{\substack{i,j\in S \\ I,I'\in\mathfrak{Q}_i \\ J,J'\in\mathfrak{Q}_j}} \int \log\left(\frac{a_{iI}a_{jJ}}{a_{iI'}a_{jJ'}}\frac{f_i'f_j'}{f_if_j}\right)\left(f_i'f_j'\frac{a_{iI}a_{jJ}}{a_{iI'}a_{jJ'}} - f_if_j\right) W_{ij}^{IJI'J'} d\boldsymbol{c}_i d\boldsymbol{c}_j d\boldsymbol{c}_i' d\boldsymbol{c}_j', \tag{4.3.7}$$

and the reactive source term $\mathfrak{v}^{\mathfrak{C}}$ can be rewritten in the form

$$\begin{aligned}\frac{\mathfrak{v}^{\mathfrak{C}}}{k_{\mathrm{B}}} =& \sum_{\substack{r\in R \\ F^r, B^r}} \int \log\left(\frac{\prod_{k\in\mathfrak{B}^r}\beta_{kK}f_k}{\prod_{j\in\mathfrak{F}^r}\beta_{jJ}f_j}\right) \\ &\times \left(\prod_{k\in\mathfrak{B}^r} f_k \frac{\prod_{k\in\mathfrak{B}^r}\beta_{kK}}{\prod_{j\in\mathfrak{F}^r}\beta_{jJ}} - \prod_{j\in\mathfrak{F}^r} f_j\right) \mathcal{W}_{\mathfrak{F}^r\mathfrak{B}^r}^{F^rB^r} \prod_{j\in\mathfrak{F}^r} d\boldsymbol{c}_j \prod_{k\in\mathfrak{B}^r} d\boldsymbol{c}_k.\end{aligned} \tag{4.3.8}$$

The sums over F^r in $\mathfrak{v}^{\mathfrak{C}}$ represent the sums over $J \in \mathfrak{Q}_j$ for all $j \in \mathfrak{F}^r$, and the sums over B^r represent the sums over $K \in \mathfrak{Q}_k$ for all $k \in \mathfrak{B}^r$. From these expressions, it is readily seen that both $\mathfrak{v}^{\mathfrak{S}}$ and $\mathfrak{v}^{\mathfrak{C}}$ are sums of nonnegative terms, since, for any positive x and y, we have $(x-y)\log(x/y) \geq 0$, so that, finally, $\mathfrak{v}^{\mathrm{kin}} \geq 0$.

The generalized Boltzmann equations (4.2.5) are thus compatible with the H-theorem and yield a dissipative structure. Upon defining the functional $H^{\mathrm{kin}} = -\int \mathcal{S}^{\mathrm{kin}}\, d\boldsymbol{x}$, it is readily seen by integrating equation (4.3.2) that $d_tH^{\mathrm{kin}} \leq 0$, which is equivalent to the second law of thermodynamics. As is classical, irreversibility has been introduced, not in the trajectories of particles, which obey deterministic laws, but in the statistics over trajectories [FK72]. The famous *Stosszahlansatz*, that is, the assumption that the molecules involved in a collision are uncorrelated, is the irreversible assumption.

It is fundamental to observe that both terms $\mathfrak{v}^{\mathfrak{S}}$ and $\mathfrak{v}^{\mathfrak{C}}$ are nonnegative. In other words, collisions always increase entropy regardless of the reactive aspect. As a consequence, any modeling of multicomponent flows has to reproduce this behavior. More specifically, transport fluxes and source terms must guarantee that both contributions to entropy production are nonnegative.

4.4. Enskog expansion

In this section, we present the Enskog expansion of the species distribution functions. The zero-order terms correspond to Maxwellian distributions and lead to the Euler equations. On the other hand, the first-order perturbed distribution functions are governed by linearized Boltzmann equations and lead to the Navier–Stokes equations.

4.4.1. Asymptotic orders

In this work we are concerned with chemical nonequilibrium regimes where the chemistry characteristic times are larger than the times of free flight of the particles and the characteristic times of internal energy relaxation. An approximate solution of the Boltzmann equations (4.2.5) is obtained using the Enskog expansion. To this end, we rewrite the generalized Boltzmann equations in the form

$$\mathcal{D}_i(f_i) = \frac{1}{\varepsilon}\mathcal{S}_i(f) + \varepsilon^{\mathfrak{a}}\mathcal{C}_i(f), \qquad i \in S, \tag{4.4.1}$$

where ε is the formal parameter associated with the Enskog expansion and $\mathfrak{a}$ depends on the regime under consideration. The term $1/\varepsilon$ emphasizes that the dominant effect is that of nonreactive collisions which drive the gas towards Maxwellian equilibrium. In other words, fast collisions are nonreactive collisions, so that the integral operator $\mathcal{S}$ will be termed the fast collision operator.

In addition, the species distribution functions are expanded in the form

$$f_i = f_i^0\big(1 + \varepsilon\phi_i + \mathcal{O}(\varepsilon^2)\big), \qquad i \in S, \tag{4.4.2}$$

and an approximate solution of Boltzmann equations is constructed by substituting (4.4.2) into (4.4.1) and equating powers of ε [CC70] [FK72] [M*al*90]. The resulting models depend on the parameter $\mathfrak{a}$, that is, on the regime under consideration. In the kinetic equilibrium regime we have $\mathfrak{a} = -1$, in the strong reaction regime we have $\mathfrak{a} = 0$, and, finally, in the Maxwellian reaction regime, we have $\mathfrak{a} = 1$. The kinetic equilibrium regime—in which chemical characteristic times are as small as collision times—is beyond the scope of this book, and we refer the reader to [LH60] and [EG98a] for more details. On the other hand, the strong reaction regime $\mathfrak{a} = 0$ and the Maxwellian reaction regime $\mathfrak{a} = 1$ will be considered simultaneously in this chapter.

4.4.2. Collisional invariants of the fast operator

Collisional invariants—also termed summational invariants—are functionals $\psi = (\psi_i)_{i\in S}$ whose value summed over the molecules involved in a collision do not change during the collision. Collisional invariants are associated with macroscopic conservation equations and therefore of fundamental importance.

We are interested in the collisional invariants of the fast collision operator $\mathcal{S}$. There are only $n+4$ linearly independent scalar collisional invariants ψ^l, $l \in [1, n+4]$, which can be taken in the form [WT62]

$$\begin{cases} \psi^k = (\delta_{ki})_{i\in S}, & k \in S, \\ \psi^{n+\nu} = (m_i c_{\nu i})_{i\in S}, & \nu = 1,2,3, \\ \psi^{n+4} = (\tfrac{1}{2} m_i \boldsymbol{c}_i \cdot \boldsymbol{c}_i + E_{i\text{I}})_{i\in S}, & \end{cases} \tag{4.4.3}$$

where $c_{\nu i}$ is the component of $\boldsymbol{c}_i$ in the νth spatial coordinate and $E_{i\text{I}}$ is the internal energy of the ith species in the Ith quantum energy state, including the energy of formation. These invariants are associated with the conservation of molecule type, momentum, and energy in nonreactive collisions. We will denote by $\mathcal{I}$ the linear space spanned by these $n+4$ scalar collisional invariants

$$\mathcal{I} = \text{Span}\,\{\ \psi^l,\ l = 1, \ldots, n+4\ \}. \tag{4.4.4}$$

The momentum invariants $\psi^{n+\nu}$, $\nu = 1,2,3$, and the energy invariant ψ^{n+4} are also collisional invariants for the reactive collision operator $\mathcal{C}$. This is not the case for the species type invariants ψ^k, $k \in S$, but still is for the atomic invariants $\widetilde{\psi}^l = (\mathfrak{a}_{il})_{i\in S}$, $l \in \mathfrak{A}$, where $\mathfrak{a}_{kl}$ is the number of lth atom in the kth species, $\mathfrak{A} = \{1, \ldots, n^{\mathfrak{a}}\}$ is the set of atom indices, and $n^{\mathfrak{a}}$ is the number of atoms—or elements—in the mixture [EG98a].

4.4.3. Macroscopic equations

We introduce for convenience the scalar product

$$\langle\!\langle \xi, \zeta \rangle\!\rangle = \sum_{\substack{i\in S \\ \text{I}\in\mathfrak{Q}_i}} \int \xi_i \zeta_i \, d\boldsymbol{c}_i, \tag{4.4.5}$$

which naturally arises when expressing macroscopic expressions. In the following, we will also have to use this scalar product for tensorial quantities. In the situation where ξ and ζ are tensors, the integrated quantity simply becomes $\xi_i \odot \zeta_i$, where $\odot$ denotes the maximum contracted product between ξ_i and ζ_i [EG94].

The fast collision operator $\mathcal{S} = (\mathcal{S}_i(f))_{i\in S}$ and the fast collisional invariants (4.4.3) then satisfy the relations

$$\langle\langle \psi^l, \mathcal{S}(f)\rangle\rangle = 0, \qquad l \in \{1, \ldots, n+4\}. \tag{4.4.6}$$

More specifically, using inverse collisions and symmetrizing between i and j, one can show [CC70] [FK72] [M*al*90] that, for any family $\xi = (\xi_i)_{i\in S}$, we have

$$\begin{aligned} 4\langle\langle \xi, \mathcal{S}(f)\rangle\rangle = \sum_{\substack{i,j\in S\\ \mathrm{I},\mathrm{I}'\in\mathcal{Q}_i\\ \mathrm{J},\mathrm{J}'\in\mathcal{Q}_j}} \int (\xi_i + \xi_j - \xi_i' - \xi_j') \\ \times \Big(f_i' f_j' \frac{a_{i\mathrm{I}} a_{j\mathrm{J}}}{a_{i\mathrm{I}'} a_{j\mathrm{J}'}} - f_i f_j\Big) W_{ij}^{\mathrm{IJI'J'}} d\boldsymbol{c}_i d\boldsymbol{c}_j d\boldsymbol{c}_i' d\boldsymbol{c}_j', \end{aligned} \tag{4.4.7}$$

where $\xi_i' = \xi_i(\boldsymbol{c}_i', \mathrm{I}')$ and $\xi_j' = \xi_j(\boldsymbol{c}_j', \mathrm{J}')$. By definition of collisional invariants, we have $\xi_i + \xi_j - \xi_i' - \xi_j' = 0$ during collisions and the relations (4.4.6) follow.

Using (4.4.6), exact macroscopic conservation equations are now obtained from (4.4.1) by taking the scalar product $\langle\langle\, , \rangle\rangle$ with ψ^l

$$\langle\langle \psi^l, \mathcal{D}(f)\rangle\rangle = \varepsilon^{\mathfrak{a}} \langle\langle \psi^l, \mathcal{C}(f)\rangle\rangle, \qquad l \in \{1, \ldots, n+4\}, \tag{4.4.8}$$

where we have introduced the families $\mathcal{D}(f) = \big(\mathcal{D}_i(f_i)\big)_{i\in S}$ and $\mathcal{C}(f) = \big(\mathcal{C}_i(f)\big)_{i\in S}$. Upon substituting the expansion (4.4.2) into (4.4.8), equating the powers of ε and integrating by parts the first term, we can then construct approximate macroscopic conservation equations.

4.5. Zero-order approximation

In this section we discuss the Euler regime resulting from the zero-order Enskog expansion.

4.5.1. Maxwellian distributions

Upon equating the powers of ε^{-1} in (4.4.1), the family of zero-order distribution functions $f^0 = (f_i^0)_{i\in S}$ is found to be the solution of the system

$$\mathcal{S}_i(f^0) = 0, \qquad i \in S. \tag{4.5.1}$$

Multiplying (4.5.1) by $\log(\beta_{iI} f_i^0)$, integrating with respect to the velocity $\boldsymbol{c}_i$, and summing over i and I yields that the nonreactive entropy production of order minus one $\mathfrak{v}^{(-1)\mathcal{S}}$ is zero

$$\mathfrak{v}^{(-1)\mathcal{S}} = -k_{\mathrm{B}} \sum_{\substack{i\in S\\ I\in\mathfrak{Q}_i}} \int \mathcal{S}_i(f^0) \log(\beta_{iI} f_i^0) d\boldsymbol{c}_i = 0. \tag{4.5.2}$$

Proceeding now as in Section 4.3.3, one can easily rewrite $\mathfrak{v}^{(-1)\mathcal{S}}$ as a sum of nonnegative terms (4.3.7) with f^0 in place of f. As a consequence, $\mathfrak{v}^{(-1)\mathcal{S}}$ vanishes if and only if we have $f_i^{0\prime} f_j^{0\prime}/a_{iI'}a_{jJ'} = f_i^0 f_j^0/a_{iI}a_{jJ}$ during nonreactive collisions, that is, if and only if $\big(\log(\beta_{iI} f_i^0)\big)_{i\in S}$ is a collisional invariant of the fast collision operator. We may therefore write

$$\big(\log(\beta_{iI} f_i^0)\big)_{i\in S} \in \mathcal{I}, \tag{4.5.3}$$

where $\mathcal{I}$ is the linear space spanned by ψ^l, $l \in \{1,\dots,n+4\}$, introduced in (4.4.4). As a consequence, there exist constants α_i, $i \in S$, $\boldsymbol{w} \in \mathbb{R}^3$, and γ, such that

$$\log(\beta_{iI} f_i^0) = \alpha_i - \boldsymbol{w}\cdot \mathfrak{m}_i \boldsymbol{c}_i - \gamma(\tfrac{1}{2}\mathfrak{m}_i \boldsymbol{c}_i\cdot\boldsymbol{c}_i + E_{iI}), \qquad i \in S. \tag{4.5.4}$$

The constants α_i, $i \in S$, $\boldsymbol{w}$, and γ are determined from macroscopic constraints. More specifically, in order to determine f^0 uniquely, it is classical in the Enskog expansion to impose that f^0 yields the same local macroscopic properties as the exact solution f, so that

$$n_i = \sum_{I\in\mathfrak{Q}_i} \int f_i d\boldsymbol{c}_i = \sum_{I\in\mathfrak{Q}_i} \int f_i^0 d\boldsymbol{c}_i, \qquad i \in S, \tag{4.5.5}$$

$$\rho\boldsymbol{v} = \sum_{\substack{i\in S\\ I\in\mathfrak{Q}_i}} \int \mathfrak{m}_i \boldsymbol{c}_i f_i d\boldsymbol{c}_i = \sum_{\substack{i\in S\\ I\in\mathfrak{Q}_i}} \int \mathfrak{m}_i \boldsymbol{c}_i f_i^0 d\boldsymbol{c}_i, \tag{4.5.6}$$

and

$$\tfrac{1}{2}\rho\boldsymbol{v}\cdot\boldsymbol{v} + \mathcal{E} = \sum_{\substack{i\in S\\ I\in\mathfrak{Q}_i}} \int (\tfrac{1}{2}\mathfrak{m}_i \boldsymbol{c}_i\cdot\boldsymbol{c}_i + E_{iI}) f_i d\boldsymbol{c}_i = \sum_{\substack{i\in S\\ I\in\mathfrak{Q}_i}} \int (\tfrac{1}{2}\mathfrak{m}_i \boldsymbol{c}_i\cdot\boldsymbol{c}_i + E_{iI}) f_i^0 d\boldsymbol{c}_i. \tag{4.5.7}$$

These constraints on f^0 may be written more elegantly in the compact form $\langle\!\langle f^0, \psi^l \rangle\!\rangle = \langle\!\langle f, \psi^l \rangle\!\rangle$, for $l \in \{1,\dots,n+4\}$.

Substituting (4.5.4) into the constraints (4.5.5)–(4.5.7) yields, after some algebra, the following expressions:

$$f_i^0 = \frac{1}{\beta_{iI}} \frac{n_i}{Q_i} \exp\left\{ -\frac{\mathfrak{m}_i}{2k_{\mathrm{B}}T}(\boldsymbol{c}_i - \boldsymbol{v})^2 - \frac{E_{iI}}{k_{\mathrm{B}}T} \right\}, \tag{4.5.8}$$

or, equivalently,

$$f_i^0 = \left(\frac{\mathfrak{m}_i}{2\pi k_{\mathrm{B}} T}\right)^{3/2} \frac{a_{iI} n_i}{Q_i^{\mathrm{int}}} \exp\left\{-\frac{\mathfrak{m}_i}{2k_{\mathrm{B}}T}(\boldsymbol{c}_i - \boldsymbol{v})^2 - \frac{E_{iI}}{k_{\mathrm{B}}T}\right\}, \tag{4.5.9}$$

where we have introduced the partition function for internal energy Q_i^{int} of the ith species

$$Q_i^{\mathrm{int}} = \sum_{I\in\mathfrak{Q}_i} a_{iI} \exp\Big(-\frac{E_{iI}}{k_{\mathrm{B}}T}\Big), \tag{4.5.10}$$

as well as the translational and full partition functions per unit volume

$$Q_i^{\mathrm{tr}} = \left(\frac{2\pi \mathfrak{m}_i k_{\mathrm{B}} T}{h_{\mathrm{P}}^2}\right)^{3/2}, \qquad Q_i = Q_i^{\mathrm{tr}}\, Q_i^{\mathrm{int}}. \tag{4.5.11}$$

The zero-order distribution functions $f^0 = (f_i^0)_{i\in S}$ finally appear as generalized Maxwellian distribution functions.

4.5.2. Zero-order macroscopic equations

Substituting now the expansion (4.4.2) into (4.4.8) and keeping the terms in ε^0, we obtain the zero-order macroscopic equations

$$\langle\!\langle \psi^l, \mathcal{D}(f^0)\rangle\!\rangle = \delta_{\mathfrak{a}0}\langle\!\langle \psi^l, \mathcal{C}(f^0)\rangle\!\rangle, \qquad l\in\{1,\dots,n+4\}, \tag{4.5.12}$$

where $\mathcal{D}(f^0) = \big(\mathcal{D}_i(f_i^0)\big)_{i\in S}$ and $\mathcal{C}(f^0) = \big(\mathcal{C}_i(f^0)\big)_{i\in S}$. After some algebra, the relations (4.5.12) are rewritten in the form

$$\partial_t \rho_i + \boldsymbol{\partial_x}\cdot(\rho_i \boldsymbol{v}) = \delta_{\mathfrak{a}0}\mathfrak{m}_i\overline{\omega}_i^0, \qquad i\in S, \tag{4.5.13}$$

$$\partial_t(\rho\boldsymbol{v}) + \boldsymbol{\partial_x}\cdot(\rho\boldsymbol{v}\otimes\boldsymbol{v} + p\boldsymbol{I}) = \sum_{i\in S}\rho_i \boldsymbol{b}_i, \tag{4.5.14}$$

$$\partial_t(\tfrac{1}{2}\rho\boldsymbol{v}\cdot\boldsymbol{v} + \mathcal{E}) + \boldsymbol{\partial_x}\cdot\big((\tfrac{1}{2}\rho\boldsymbol{v}\cdot\boldsymbol{v} + \mathcal{E} + p)\boldsymbol{v}\big) = \sum_{i\in S}\rho_i \boldsymbol{v}\cdot\boldsymbol{b}_i, \tag{4.5.15}$$

where $\rho_i = \mathfrak{m}_i n_i$ is the mass density of the ith species, $\overline{\omega}_i^0$ is the zero-order molecular production rate of the ith species

$$\overline{\omega}_i^0 = \langle\!\langle \psi^i, \mathcal{C}(f^0)\rangle\!\rangle = \sum_{I\in\mathfrak{Q}_i}\int \mathcal{C}_i(f^0)\,d\boldsymbol{c}_i, \tag{4.5.16}$$

$p = (\sum_{i\in S} n_i)k_{\mathrm{B}}T$ is the thermodynamic pressure, and $\mathcal{E}$ is the internal energy per unit volume

$$\mathcal{E} = \sum_{\substack{i\in S\\ I\in\mathfrak{Q}_i}}\int \big(\tfrac{1}{2}\mathfrak{m}_i(\boldsymbol{c}_i - \boldsymbol{v})\cdot(\boldsymbol{c}_i - \boldsymbol{v}) + E_{iI}\big) f_i^0\, d\boldsymbol{c}_i. \tag{4.5.17}$$

The zero-order macroscopic equations thus appear as the Euler reactive compressible equations. The source term will be investigated more closely in Section 4.6.

4.5.3. Zero-order time derivatives

After some algebra, it is easily obtained from (4.5.17) and (4.5.9) that

$$\mathcal{E} = \sum_{i \in S} n_i \big(\tfrac{3}{2} k_{\mathrm{B}} T + \overline{E}_i\big), \tag{4.5.18}$$

where

$$\overline{E}_i = \sum_{I \in \mathfrak{Q}_i} \frac{a_{iI} E_{iI}}{Q_i^{\mathrm{int}}} \exp\left\{ -\frac{E_{iI}}{k_{\mathrm{B}} T} \right\}. \tag{4.5.19}$$

An equation for the temperature is then easily recovered from the reactive Euler equations by eliminating the terms corresponding to the kinetic energy. A straightforward calculation yields

$$\Big(\sum_{i \in S} n_i\Big) \mathfrak{c}_v \big(\partial_t T + \boldsymbol{v} \cdot \boldsymbol{\partial_x} T\big) = -p\, \boldsymbol{\partial_x} \cdot \boldsymbol{v} - \delta_{\mathfrak{a}0} \sum_{i \in S} \big(\tfrac{3}{2} k_{\mathrm{B}} T + \overline{E}_i\big) \overline{\omega}_i^0, \tag{4.5.20}$$

where the constant volume heat capacity per molecule $\mathfrak{c}_v$ reads as

$$\mathfrak{c}_v = \tfrac{3}{2} k_{\mathrm{B}} + \sum_{i \in S} X_i \mathfrak{c}_i^{\mathrm{int}}. \tag{4.5.21}$$

The first term accounts for the translational heat capacity, while $\mathfrak{c}_i^{\mathrm{int}}$ is the internal heat capacity of the molecules of the ith species $\mathfrak{c}_i^{\mathrm{int}} = d\overline{E}_i / dT$ and X_i is the mole fraction of the ith species.

From (4.5.20), (4.5.13), and (4.5.14), we can thus estimate the zero-order time derivatives of n_i, $i \in S$, $\boldsymbol{v}$, and T in terms of source terms and macroscopic variables gradients, as will be needed in the next section.

4.6. First-order approximation

In this section we discuss the first-order Enskog expansion leading to the multicomponent reactive flow macroscopic equations.

4.6.1. Linearized Boltzmann operator

In order to investigate the first-order macroscopic equations, we have to discuss the properties of the linearized fast collision operator around Maxwellian distributions. This linearized operator is defined from

$$\mathfrak{I}^{\mathfrak{S}} = (\mathfrak{I}_i^{\mathfrak{S}})_{i \in S}, \tag{4.6.1}$$

where $\mathfrak{I}_i^{\mathrm{S}}$ is the linearized fast collision operator of the ith species

$$\mathfrak{I}_i^{\mathrm{S}}(\phi) = \sum_{j\in S} \sum_{\substack{\mathrm{J}\in\mathfrak{Q}_j \\ \mathrm{I}'\in\mathfrak{Q}_i \\ \mathrm{J}'\in\mathfrak{Q}_j}} \int f_j^0(\phi_i + \phi_j - \phi_i' - \phi_j') W_{ij}^{\mathrm{IJI'J'}} d\boldsymbol{c}_j d\boldsymbol{c}_i' d\boldsymbol{c}_j'. \tag{4.6.2}$$

We then introduce the associated integral bracket operator

$$[\![\xi,\zeta]\!] = \langle\!\langle f^0\xi, \mathfrak{I}^{\mathrm{S}}(\zeta)\rangle\!\rangle, \tag{4.6.3}$$

defined for families $\xi = (\xi_i)_{i\in S}$ and $\zeta = (\zeta_i)_{i\in S}$, where ξ_i and ζ_i only depend on t, $\boldsymbol{x}$, $\boldsymbol{c}_i$, and I. It is classically shown that the bracket operator is symmetric $[\![\xi,\zeta]\!] = [\![\zeta,\xi]\!]$, positive semidefinite $[\![\xi,\xi]\!] \geq 0$, and its kernel is spanned by the collisional invariants, that is, $[\![\xi,\xi]\!] = 0$ if and only if $\xi \in \mathcal{I}$ [Wal58] [CC70] [FK72] [M*al*90]. These definitions of the bracket operator are easily extended to tensor families by replacing scalar multiplication by full tensor contraction.

4.6.2. Linearized Boltzmann equations

Upon equating the powers of ε^0 in (4.4.1), the first-order perturbations $\phi = (\phi_i)_{i\in S}$ are solutions of nonhomogeneous integral equations in the form

$$\mathfrak{I}_i^{\mathrm{S}}(\phi) = \Psi_i, \qquad i \in S, \tag{4.6.4}$$

where $\mathfrak{I}_i^{\mathrm{S}}$ is the linearized fast collision Boltzmann operator of the ith species and the right member Ψ_i uniquely depends on f^0 and reads as

$$\Psi_i = -\mathfrak{D}_i(\log f_i^0) + \delta_{\mathrm{a}0}\mathfrak{C}_i(f^0)/f_i^0, \qquad i \in S. \tag{4.6.5}$$

In order to determine uniquely the perturbation $\phi = (\phi_i)_{i\in S}$, the integral equations (4.6.4) are completed with the $n+4$ scalar constraints obtained from (4.5.5)–(4.5.7)

$$\langle\!\langle f^0\phi, \psi^l\rangle\!\rangle = 0, \qquad l \in \{1,\dots,n+4\}. \tag{4.6.6}$$

The right member Ψ_i in (4.6.5) may now be evaluated, using the zero-order macroscopic conservation equations to replace the zero-order time derivatives by spatial gradients and source terms. After lengthy calculations, it is found that Ψ_i can be decomposed in the form [LH60] [CC70] [FK72] [EG94] [ACG94]

$$\Psi_i = \Psi_i^{\mathrm{S}} + \delta_{\mathrm{a}0}\Psi_i^{\mathfrak{C}}, \qquad i \in S, \tag{4.6.7}$$

where Ψ_i^{S} is the frozen—or nonreactive—right member and $\Psi_i^{\mathfrak{C}}$ is solely due to chemical reactions.

The ith component of $\varPsi^{\mathrm{S}} = (\varPsi_i^{\mathrm{S}})_{i\in S}$ is found to be in the form

$$\varPsi_i^{\mathrm{S}} = -\boldsymbol{\varPsi}_i^{\eta}:\boldsymbol{\partial}_{\boldsymbol{x}}\boldsymbol{v} - \tfrac{1}{3}\varPsi_i^{\kappa}\,\boldsymbol{\partial}_{\boldsymbol{x}}\boldsymbol{\cdot}\boldsymbol{v} - \sum_{j\in S}\boldsymbol{\varPsi}_i^{D_j}\boldsymbol{\cdot}\Big(\boldsymbol{\partial}_{\boldsymbol{x}}p_j - \rho_j\boldsymbol{b}_j\Big) - \boldsymbol{\varPsi}_i^{\widehat{\lambda}}\boldsymbol{\cdot}\boldsymbol{\partial}_{\boldsymbol{x}}\Big(\frac{1}{k_{\mathrm{B}}T}\Big), \tag{4.6.8}$$

where $p_i = n_i k_{\mathrm{B}} T$ is the partial pressure of the ith species, in such a way that the macroscopic variable gradients act as driving forces. In this expression, we also have

$$\boldsymbol{\varPsi}_i^{\eta} = \frac{\mathfrak{m}_i}{k_{\mathrm{B}}T}\Big((\boldsymbol{c}_i-\boldsymbol{v})\otimes(\boldsymbol{c}_i-\boldsymbol{v}) - \tfrac{1}{3}(\boldsymbol{c}_i-\boldsymbol{v})\boldsymbol{\cdot}(\boldsymbol{c}_i-\boldsymbol{v})\boldsymbol{I}\Big), \tag{4.6.9}$$

$$\varPsi_i^{\kappa} = \frac{2\mathfrak{c}^{\text{int}}}{\mathfrak{c}_v k_{\mathrm{B}}T}\big(\tfrac{1}{2}\mathfrak{m}_i(\boldsymbol{c}_i-\boldsymbol{v})\boldsymbol{\cdot}(\boldsymbol{c}_i-\boldsymbol{v}) - \tfrac{3}{2}k_{\mathrm{B}}T\big) + \frac{2\mathfrak{c}_v^{\text{tr}}}{\mathfrak{c}_v k_{\mathrm{B}}T}(\overline{E}_i - E_{iI}), \tag{4.6.10}$$

$$\boldsymbol{\varPsi}_i^{D_j} = \frac{1}{p_i}(\delta_{ij} - Y_i)(\boldsymbol{c}_i-\boldsymbol{v}), \qquad j\in S, \tag{4.6.11}$$

$$\boldsymbol{\varPsi}_i^{\widehat{\lambda}} = \Big(\frac{5k_{\mathrm{B}}T}{2} - \tfrac{1}{2}\mathfrak{m}_i(\boldsymbol{c}_i-\boldsymbol{v})\boldsymbol{\cdot}(\boldsymbol{c}_i-\boldsymbol{v}) + \overline{E}_i - E_{iI}\Big)(\boldsymbol{c}_i-\boldsymbol{v}), \tag{4.6.12}$$

where $\boldsymbol{I}$ is the unit operator, $\mathfrak{c}_v^{\text{tr}} = \frac{3}{2}k_{\mathrm{B}}$ is the translational constant-volume specific heat per molecule, $\mathfrak{c}^{\text{int}} = \sum_{i\in S} X_i \mathfrak{c}^{\text{int}}$ is the mixture internal specific heat per molecule, and $\mathfrak{c}_v = \mathfrak{c}_v^{\text{tr}} + \mathfrak{c}^{\text{int}}$ is the mixture constant-volume specific heat per molecule.

On the other hand, the term $\varPsi_i^{\mathcal{C}}$ is found to be

$$\begin{aligned}\varPsi_i^{\mathcal{C}} = \frac{\mathcal{C}_i(f^0)}{f_i^0} - \frac{\overline{\omega}_i^0}{n_i} - &\frac{\sum_{j\in S}(\frac{3}{2}k_{\mathrm{B}}T + \overline{E}_j)\overline{\omega}_j^0}{p\mathfrak{c}_v T} \\ &\times\big(\tfrac{3}{2}k_{\mathrm{B}}T - \tfrac{1}{2}\mathfrak{m}_i(\boldsymbol{c}_i-\boldsymbol{v})\boldsymbol{\cdot}(\boldsymbol{c}_i-\boldsymbol{v}) + \overline{E}_i - E_{iI}\big),\end{aligned} \tag{4.6.13}$$

where $\overline{\omega}_i^0$ is the molecular Maxwellian production rate of the ith species (4.5.16). In order to expand $\varPsi_i^{\mathcal{C}}$, we now write that

$$\varPsi_i^{\mathcal{C}} = \sum_{r\in R}\varPsi_i^{r}\,\overline{\tau}_r, \tag{4.6.14}$$

where $\overline{\tau}_r$ is the zero-order macroscopic rate of progress of the rth reaction

$$\overline{\tau}_r = \mathcal{K}_r\Big\{\prod_{k\in S}\Big(\frac{n_k}{Q_k}\Big)^{\nu_{kr}^{\mathrm{f}}} - \prod_{k\in S}\Big(\frac{n_k}{Q_k}\Big)^{\nu_{kr}^{\mathrm{b}}}\Big\}, \tag{4.6.15}$$

with

$$\mathcal{K}_r = \sum_{F^r,B^r}\int \mathfrak{D}_r \prod_{\mathfrak{F}^r} d\boldsymbol{c}_j \prod_{\mathfrak{B}^r} d\boldsymbol{c}_k, \tag{4.6.16}$$

$$\mathfrak{D}_r = \prod_{j\in\mathfrak{F}^r} \exp\Big(-\frac{\mathfrak{m}_j}{2k_{\mathrm{B}}T}(\boldsymbol{c}_j - \boldsymbol{v})\cdot(\boldsymbol{c}_j - \boldsymbol{v}) - \frac{E_{jJ}}{k_{\mathrm{B}}T}\Big) \frac{\mathcal{W}^{F^r B^r}_{\mathfrak{F}^r \mathfrak{B}^r}}{\prod_{j\in\mathfrak{F}^r} \beta_{jJ}}, \tag{4.6.17}$$

or, equivalently,

$$\mathfrak{D}_r = \prod_{k\in\mathfrak{B}^r} \exp\Big(-\frac{\mathfrak{m}_k}{2k_{\mathrm{B}}T}(\boldsymbol{c}_k - \boldsymbol{v})\cdot(\boldsymbol{c}_k - \boldsymbol{v}) - \frac{E_{kK}}{k_{\mathrm{B}}T}\Big) \frac{\mathcal{W}^{B^r F^r}_{\mathfrak{B}^r \mathfrak{F}^r}}{\prod_{k\in\mathfrak{B}^r} \beta_{kK}}, \tag{4.6.18}$$

and where

$$\Psi_i^r = \frac{\widetilde{\Psi}_i^r}{f_i^0} - \frac{\nu_{ir}^{\mathrm{b}} - \nu_{ir}^{\mathrm{f}}}{n_i} - \frac{\sum_{j\in S}(\frac{3}{2}k_{\mathrm{B}}T + \overline{E}_j)(\nu_{jr}^{\mathrm{b}} - \nu_{jr}^{\mathrm{f}})}{pc_{\mathrm{v}}T}$$
$$\times \left(\tfrac{3}{2}k_{\mathrm{B}}T - \tfrac{1}{2}\mathfrak{m}_i(\boldsymbol{c}_i - \boldsymbol{v})\cdot(\boldsymbol{c}_i - \boldsymbol{v}) + \overline{E}_i - E_{iI}\right), \tag{4.6.19}$$

with

$$\widetilde{\Psi}_i^r = \frac{1}{\mathcal{K}_r}\Big(-\nu_{ir}^{\mathrm{f}} \sum_{F_I^r, B^r} \int \mathfrak{D}_r \prod_{\mathfrak{F}_i^r} d\boldsymbol{c}_j \prod_{\mathfrak{B}^r} d\boldsymbol{c}_k + \nu_{ir}^{\mathrm{b}} \sum_{F^r, B_I^r} \int \mathfrak{D}_r \prod_{\mathfrak{F}^r} d\boldsymbol{c}_j \prod_{\mathfrak{B}_i^r} d\boldsymbol{c}_k\Big). \tag{4.6.20}$$

This expansion (4.6.14) of $\Psi^{\mathfrak{C}}$ plays a similar role as the expansion (4.6.8) for $\Psi^{\mathfrak{S}}$, with the zero-order macroscopic rates of progress $\overline{\tau}_r$, $r \in R$, acting as driving forces, in place of macroscopic variable gradients.

4.6.3. Expansion of perturbed distributions

From the decomposition $\Psi = \Psi^{\mathfrak{S}} + \delta_{\mathfrak{a}0}\Psi^{\mathfrak{C}}$ of $\Psi = (\Psi_i)_{i\in S}$, we obtain a decomposition of $\phi = (\phi_i)_{i\in S}$ in the form

$$\phi_i = \phi_i^{\mathfrak{S}} + \delta_{\mathfrak{a}0}\phi_i^{\mathfrak{C}}, \qquad i \in S. \tag{4.6.21}$$

From (4.6.8), $\phi_i^{\mathfrak{S}}$ is then expanded in the form

$$\phi_i^{\mathfrak{S}} = -\boldsymbol{\phi}_i^{\eta} : \boldsymbol{\partial_x v} - \tfrac{1}{3}\phi_i^{\kappa}\, \boldsymbol{\partial_x}\cdot\boldsymbol{v} - \sum_{j\in S} \boldsymbol{\phi}_i^{D_j}\cdot(\boldsymbol{\partial_x} p_j - \rho_j \boldsymbol{b}_j) - \boldsymbol{\phi}_i^{\widehat{\lambda}}\cdot\boldsymbol{\partial_x}\Big(\frac{1}{k_{\mathrm{B}}T}\Big), \tag{4.6.22}$$

and, similarly, from (4.6.14), $\phi_i^{\mathfrak{C}}$ is expanded in the form

$$\phi_i^{\mathfrak{C}} = \sum_{r\in R} \phi^r\, \overline{\tau}_r. \tag{4.6.23}$$

For each of these expansion coefficients $\phi^\mu = (\phi_i^\mu)_{i\in S}$, where μ stands for η, κ, D_j, $j \in S$, $\widehat{\lambda}$, or $r \in R$, we denote by $\Psi^\mu = (\Psi_i^\mu)_{i\in S}$ the corresponding right member and ϕ^μ is then the solution of the constrained integral equations

$$\begin{cases} \mathfrak{I}^{\mathfrak{S}}(\phi^\mu) = \Psi^\mu, \\ \langle\!\langle f^0\phi^\mu, \psi^l \rangle\!\rangle = 0, \qquad l \in \{1,\dots n+4\}. \end{cases} \tag{4.6.24}$$

From the conservation of mass, momentum, and energy during collisions, it is shown that $\langle\!\langle \Psi^\mu, \psi^l \rangle\!\rangle = 0$, for $\mu = \eta$, κ, D_j, $j \in S$, $\widehat{\lambda}$, or $\mu = r$, $r \in R$. As a consequence, these systems (4.6.24) are well posed for all μ, that is, the right member $\Psi^\mu = (\Psi_i^\mu)_{i\in S}$ is in the range of the integral operator $\mathfrak{I}^S$ and the solution $\phi^\mu = (\phi_i^\mu)_{i\in S}$ is unique [FK72] [EG94]. Note that these linear integral equations (4.6.24) are now tensorial equations, since ϕ^μ and Ψ^μ may have scalar components for $\mu = \kappa$ or $\mu = r$, $r \in R$, vector components for $\mu = D_j$, $j \in S$, or $\mu = \widehat{\lambda}$, and, finally, traceless matrix components for $\mu = \eta$.

4.6.4. Macroscopic equations and transport fluxes

Substituting the expansion (4.4.2) into (4.4.8) and keeping terms in ε^0 and ε^1, we obtain the macroscopic equations in the Navier–Stokes regime

$$\langle\!\langle \psi^l, \mathcal{D}(f^0) + \mathcal{D}(f^0\phi) \rangle\!\rangle = \langle\!\langle \psi^l, \mathcal{C}(f^0) + \delta_{\mathfrak{a}0}\partial_f\mathcal{C}(f^0)f^0\phi \rangle\!\rangle, \quad l \in \{1, \ldots, n+4\}, \tag{4.6.25}$$

where we have defined $f^0\phi = (f_i^0\phi_i)_{i\in S}$, $\mathcal{D}(f^0) = (\mathcal{D}_i(f_i^0))_{i\in S}$, $\mathcal{D}(f^0\phi) = (\mathcal{D}_i(f_i^0\phi_i))_{i\in S}$, and $\partial_f\mathcal{C}(f^0)f^0\phi = (\partial_f\mathcal{C}_i(f^0)f^0\phi)_{i\in S}$.

After some algebra, the macroscopic conservation equations are found in the form

$$\partial_t\rho_i + \boldsymbol{\partial_x}\cdot(\rho_i\boldsymbol{v}) + \boldsymbol{\partial_x}\cdot(\rho_i\boldsymbol{\mathcal{V}}_i) = \mathfrak{m}_i\overline{\omega}_i, \qquad i \in S, \tag{4.6.26}$$

$$\partial_t(\rho\boldsymbol{v}) + \boldsymbol{\partial_x}\cdot(\rho\boldsymbol{v}\otimes\boldsymbol{v} + p\boldsymbol{I}) + \boldsymbol{\partial_x}\cdot\boldsymbol{\Pi} = \sum_i \rho_i\boldsymbol{b}_i, \tag{4.6.27}$$

$$\partial_t(\tfrac{1}{2}\rho\boldsymbol{v}\cdot\boldsymbol{v} + \mathcal{E}) + \boldsymbol{\partial_x}\cdot\big((\tfrac{1}{2}\rho\boldsymbol{v}\cdot\boldsymbol{v} + \mathcal{E} + p)\boldsymbol{v}\big) + \boldsymbol{\partial_x}\cdot(\boldsymbol{\mathcal{Q}} + \boldsymbol{\Pi}\cdot\boldsymbol{v}) = \sum_i \rho_i\boldsymbol{b}_i\cdot(\boldsymbol{v} + \boldsymbol{\mathcal{V}}_i), \tag{4.6.28}$$

where we have introduced the species diffusion velocities $\boldsymbol{\mathcal{V}}_i$, $i \in S$,

$$n_i\boldsymbol{\mathcal{V}}_i = \sum_{I\in\mathfrak{Q}_i} \int (\boldsymbol{c}_i - \boldsymbol{v}) f_i^0\phi_i d\boldsymbol{c}_i, \qquad i \in S, \tag{4.6.29}$$

viscous tensor

$$\boldsymbol{\Pi} = \sum_{\substack{i\in S\\ I\in\mathfrak{Q}_i}} \int \mathfrak{m}_i(\boldsymbol{c}_i - \boldsymbol{v}) \otimes (\boldsymbol{c}_i - \boldsymbol{v}) f_i^0\phi_i d\boldsymbol{c}_i, \tag{4.6.30}$$

heat flux

$$\boldsymbol{\mathcal{Q}} = \sum_{\substack{i\in S\\ I\in\mathfrak{Q}_i}} \int (\tfrac{1}{2}\mathfrak{m}_i(\boldsymbol{c}_i - \boldsymbol{v})\cdot(\boldsymbol{c}_i - \boldsymbol{v}) + E_{iI})(\boldsymbol{c}_i - \boldsymbol{v}) f_i^0\phi_i d\boldsymbol{c}_i, \tag{4.6.31}$$

and macroscopic molecule production rate

$$\overline{\omega}_i = \sum_{I\in\mathfrak{Q}_i} \int \Big(\mathfrak{C}_i(f^0) + \delta_{\mathrm{a}0}\partial_f \mathfrak{C}_i(f^0) f^0\phi\Big)\, d\boldsymbol{c}_i, \qquad i \in S. \tag{4.6.32}$$

The species diffusion fluxes appearing in the conservation equations are thus given by $\boldsymbol{\mathcal{F}}_i = \rho_i \boldsymbol{\mathcal{V}}_i$, $i \in S$. In the next sections, we express the species diffusion velocities $\boldsymbol{\mathcal{V}}_i$, $i \in S$, heat flux vector $\boldsymbol{\mathcal{Q}}$, and viscous tensor $\boldsymbol{\Pi}$ in terms of macroscopic variable gradients and transport coefficients, and we also investigate the chemistry source terms.

4.6.5. Transport coefficients

Using the definitions (4.6.29)–(4.6.31) of transport fluxes, the definitions (4.6.9)–(4.6.12) of the expansion coefficients Ψ^μ, with $\mu = \eta$, κ, D_j, $j \in S$, or $\widehat{\lambda}$, and after lengthy algebra, the transport fluxes can be expressed in terms of the species perturbed distribution function ϕ as [WT62] [FK72]

$$\boldsymbol{\mathcal{V}}_i = k_{\mathrm{B}} T \langle\!\langle \boldsymbol{\Psi}^{D_i}, f^0\phi \rangle\!\rangle, \qquad i \in S, \tag{4.6.33}$$

$$\boldsymbol{\Pi} = k_{\mathrm{B}} T \langle\!\langle \boldsymbol{\Psi}^{\eta}, f^0\phi \rangle\!\rangle + \tfrac{1}{3} k_{\mathrm{B}} T \langle\!\langle \Psi^{\kappa}, f^0\phi \rangle\!\rangle \boldsymbol{I}, \tag{4.6.34}$$

$$\boldsymbol{\mathcal{Q}} = -\langle\!\langle \boldsymbol{\Psi}^{\widehat{\lambda}}, f^0\phi \rangle\!\rangle + \sum_{i\in S} \big(\tfrac{5}{2} k_{\mathrm{B}} T + \overline{E}_i\big) n_i \boldsymbol{\mathcal{V}}_i. \tag{4.6.35}$$

The transport fluxes can now be expressed in terms of macroscopic variable gradients, source terms, and transport coefficients by simply substituting the expansions (4.6.21), (4.6.22), and (4.6.23) of the species perturbed distribution functions ϕ into (4.6.33)–(4.6.35), making use of the symmetry of the integral bracket operator $[\![\xi, \zeta]\!] = \langle\!\langle f^0\xi, \mathfrak{I}^{\mathrm{S}}\zeta \rangle\!\rangle$ and space isotropy [WT62] [CC70] [FK72] [EG94].

Defining the species multicomponent diffusion coefficients D_{ij}, $i, j \in S$, and the thermal diffusion coefficients θ_i, $i \in S$, by

$$\begin{cases} D_{ij} = \dfrac{p\, k_{\mathrm{B}} T}{3}\, [\![\boldsymbol{\phi}^{D_i}, \boldsymbol{\phi}^{D_j}]\!], & i, j \in S, \\[2ex] \theta_i = -\dfrac{1}{3}\, [\![\boldsymbol{\phi}^{\widehat{\lambda}}, \boldsymbol{\phi}^{D_i}]\!], & i \in S, \end{cases} \tag{4.6.36}$$

the species diffusion velocities $\boldsymbol{\mathcal{V}}_i$, $i \in S$, are then expressed in the form

$$\boldsymbol{\mathcal{V}}_i = -\sum_{j\in S} D_{ij} \widehat{\boldsymbol{d}}_j - \theta_i \boldsymbol{\partial}_{\boldsymbol{x}} \log T, \qquad i \in S, \tag{4.6.37}$$

where we have introduced the unconstrained species diffusion driving forces

$$\widehat{\boldsymbol{d}}_j = \frac{1}{p} \left(\boldsymbol{\partial}_{\boldsymbol{x}} p_j - \rho_j \boldsymbol{b}_j\right), \qquad j \in S. \tag{4.6.38}$$

Similar expressions are obtained for the diffusion fluxes $\mathcal{F}_i = \rho_i \mathcal{V}_i$, $i \in S$, which naturally appear in the conservation equations.

Upon defining the partial thermal conductivity $\widehat{\lambda}$ by

$$\widehat{\lambda} = \frac{1}{3k_B T^2} [\![\phi^{\widehat{\lambda}}, \phi^{\widehat{\lambda}}]\!], \tag{4.6.39}$$

the heat flux vector $\mathcal{Q}$ is then found in the form

$$\mathcal{Q} = -\widehat{\lambda}\boldsymbol{\partial_x} T - p\sum_{i\in S} \theta_i \widehat{\boldsymbol{d}}_i + \sum_{i\in S} (\tfrac{5}{2}k_B T + \overline{E}_i) n_i \mathcal{V}_i. \tag{4.6.40}$$

Finally, defining the shear viscosity η and the volume viscosity κ from

$$\eta = \frac{k_B T}{10} [\![\phi^{\eta}, \phi^{\eta}]\!], \tag{4.6.41}$$

$$\kappa = \frac{k_B T}{9} [\![\phi^{\kappa}, \phi^{\kappa}]\!], \tag{4.6.42}$$

and the chemical pressure coefficients $\mathfrak{y}_r$, $r \in R$, with

$$\mathfrak{y}_r = \frac{k_B T}{3} [\![\phi^{\kappa}, \phi^{r}]\!], \qquad r \in R,$$

the viscous stress tensor $\boldsymbol{\Pi}$ then reads as

$$\boldsymbol{\Pi} = \delta_{\mathfrak{a}0} p^{\mathrm{reac}} \boldsymbol{I} - \kappa(\boldsymbol{\partial_x}\cdot\boldsymbol{v})\boldsymbol{I} - \eta\big(\boldsymbol{\partial_x}\boldsymbol{v} + \boldsymbol{\partial_x}\boldsymbol{v}^t - \tfrac{2}{3}(\boldsymbol{\partial_x}\cdot\boldsymbol{v})\boldsymbol{I}\big), \tag{4.6.43}$$

where the chemical pressure is given by

$$p^{\mathrm{reac}} = \sum_{r\in R} \mathfrak{y}_r \overline{\tau}_r. \tag{4.6.44}$$

It is also possible to rewrite the species diffusion velocities $\mathcal{V}_i$, $i \in S$, and the heat flux $\mathcal{Q}$ in terms of the thermal diffusion ratios χ_i, $i \in S$, and the thermal conductivity λ. More specifically, the thermal diffusion ratios χ_i, $i \in S$, are defined as the unique solution of the constrained singular system

$$\begin{cases} \displaystyle\sum_{j\in S} D_{ij}\chi_j = \theta_i, \quad i \in S, \\ \displaystyle\sum_{j\in S} \chi_j = 0, \end{cases} \tag{4.6.45}$$

while the thermal conductivity λ is defined as

$$\lambda = \widehat{\lambda} - \frac{p}{T}\sum_{j\in S} \theta_j \chi_j. \tag{4.6.46}$$

Using these new transport coefficients, the species diffusion velocities and the heat flux now are easily rewritten in the form

$$\boldsymbol{\mathcal{V}}_i = -\sum_{j\in S} D_{ij}(\widehat{\boldsymbol{d}}_j + \chi_j \boldsymbol{\partial_x} \log T), \qquad i \in S, \tag{4.6.47}$$

$$\boldsymbol{\mathcal{Q}} = -\lambda \boldsymbol{\partial_x} T + p\sum_{i\in S} \chi_i \boldsymbol{\mathcal{V}}_i + \sum_{i\in S} (\tfrac{5}{2} k_{\mathrm{B}} T + \overline{E}_i) n_i \boldsymbol{\mathcal{V}}_i. \tag{4.6.48}$$

However, one may want to calculate directly the coefficients λ and χ_i, $i \in S$, without the intermediate evaluations of $\widehat{\lambda}$ and θ_i, $i \in S$. To this end, we define

$$\boldsymbol{\phi}^{\lambda} = \boldsymbol{\phi}^{\widehat{\lambda}} + p k_{\mathrm{B}} T \sum_{j\in S} \chi_j \boldsymbol{\phi}^{D_j}, \tag{4.6.49}$$

and one can establish that

$$\lambda = \frac{1}{3 k_{\mathrm{B}} T^2} [\![\boldsymbol{\phi}^{\lambda}, \boldsymbol{\phi}^{\lambda}]\!] \tag{4.6.50}$$

and

$$\chi_k = \frac{\mathfrak{m}_k}{3 p k_{\mathrm{B}} T^2} [\![\mathfrak{V}^k, \boldsymbol{\phi}^{\lambda}]\!], \qquad k \in S, \tag{4.6.51}$$

where the family $\mathfrak{V}^k = (\mathfrak{V}_i^k)_{i\in S}$ is defined by $\mathfrak{V}_i^k = (\boldsymbol{c}_k - \boldsymbol{v})\delta_{ki}$. The first of these expressions is classical, and the second was established in [EG94]. From these expressions, one can then derive practical explicit expressions for these coefficients. The thermal diffusion ratios may also be interpreted as Lagrange multipliers [EG94].

The transport coefficients introduced above satisfy several important properties. These properties can be directly obtained from the properties of the bracket integral operator or be deduced from the study of transport linear systems [EG94]. In particular, from the symmetry of the bracket product, we first obtain that the matrix $D = (D_{ij})_{i,j\in S}$ is symmetric. Similarly, from $\sum_{i\in S} Y_i \boldsymbol{\Psi}^{D_i} = 0$, we deduce that

$$\sum_{i\in S} Y_i \boldsymbol{\phi}^{D_i} = 0, \tag{4.6.52}$$

so that from (4.6.36) we have

$$\sum_{i\in S} Y_i D_{ij} = 0, \qquad j \in S, \tag{4.6.53}$$

and

$$\sum_{i\in S} Y_i \theta_i = 0. \tag{4.6.54}$$

In addition, the diffusion matrix $D = (D_{ij})_{i,j\in S}$ is positive semidefinite, since the bracket integral operator is nonnegative. Moreover, its kernel is exactly spanned by the mass fraction vector $Y = (Y_i)_{i\in S}$, since there are no collisional invariants spanned by the $\boldsymbol{\phi}^{D_i}$, $i \in S$, apart from the zero vector and the rank of ϕ^{D_i}, $i \in S$, is $n-1$, as for $\boldsymbol{\Psi}^{D_i}$, $i \in S$. Similarly, the partial thermal conductivity $\widehat{\lambda}$ and the thermal conductivity λ are shown to be positive. The shear viscosity η is positive, and the volume viscosity κ is nonnegative. From the mass constraints, it is also easily seen that the constrained diffusion driving forces $\boldsymbol{d}_i$, $i \in S$, defined in (2.5.4), can be used in place of the $\widehat{\boldsymbol{d}}_i$, $i \in S$, from the relations (2.5.11).

Finally, note that there is a chemical pressure in the viscous tensor (4.6.43) in the regime $\mathfrak{a} = 0$, as was first pointed out by Ludwig and Heil [LH60]. This term is generally neglected in the governing equations without complete justification [EG94].

4.6.6. Chemistry source terms

The macroscopic molecule production rate (4.6.32) appearing in the ith species governing equation (4.6.26) is easily rewritten in the form

$$\overline{\omega}_i = \overline{\omega}_i^0 + \delta_{\mathfrak{a}0}\mathfrak{w}_i, \tag{4.6.55}$$

where $\overline{\omega}_i^0$ is estimated by using Maxwellian distributions, whereas the perturbed term $\mathfrak{w}_i$ arises from the linearization $\partial_f\mathcal{C}(f^0)$ of the chemistry source term. The zero-order term $\overline{\omega}_i^0$ is easily rewritten in the form

$$\overline{\omega}_i^0 = \sum_{r\in R}(\nu_{ir}^{\mathrm{b}}-\nu_{ir}^{\mathrm{f}})\overline{\tau}_r = \sum_{r\in R}(\nu_{ir}^{\mathrm{b}}-\nu_{ir}^{\mathrm{f}})\Big(\mathcal{K}_r^{\mathrm{f}}\prod_{k\in S} n_k^{\nu_{kr}^{\mathrm{f}}} - \mathcal{K}_r^{\mathrm{b}}\prod_{k\in S} n_k^{\nu_{kr}^{\mathrm{b}}}\Big), \tag{4.6.56}$$

where $\mathcal{K}_r^{\mathrm{f}} = \mathcal{K}_r/\prod_{k\in S} Q_k^{\nu_{kr}^{\mathrm{f}}}$ and $\mathcal{K}_r^{\mathrm{b}} = \mathcal{K}_r/\prod_{k\in S} Q_k^{\nu_{kr}^{\mathrm{b}}}$. In particular, these rates are compatible with the law of mass action, all elementary reactions are reversible and $\mathcal{K}_r^{\mathrm{f}} = \mathcal{K}_r^{\mathrm{b}}\mathcal{K}_r^{\mathrm{e}}$, $r \in R$, where the equilibrium constants—with respect to number densities—are given by

$$\mathcal{K}_r^{\mathrm{e}} = \prod_{k\in S} Q_k^{\nu_{kr}^{\mathrm{b}}-\nu_{kr}^{\mathrm{f}}}, \qquad r \in R. \tag{4.6.57}$$

These expressions are also compatible with statistical mechanics at equilibrium [Wil85]. Note that the reciprocity relations between the forward and backward reaction rates constants are direct consequences of reciprocity relations of transition probabilities. These expressions for the reaction rates of progress are easily adapted to other units than number densities.

In order to estimate the perturbed source terms, we define the forward and backward partial rates $\overline{\tau}_r^{\mathrm{f}}$ and $\overline{\tau}_r^{\mathrm{b}}$ from

$$\overline{\tau}_r^{\mathrm{f}} = \mathcal{K}_r\prod_{k\in S}\Big(\frac{n_k}{Q_k}\Big)^{\nu_{kr}^{\mathrm{f}}}, \qquad \overline{\tau}_r^{\mathrm{b}} = \mathcal{K}_r\prod_{k\in S}\Big(\frac{n_k}{Q_k}\Big)^{\nu_{kr}^{\mathrm{b}}}, \tag{4.6.58}$$

in such a way that $\overline{\tau}_r = \overline{\tau}_r^{\rm f} - \overline{\tau}_r^{\rm b}$, $r \in R$, and we introduce

$$\Psi_i^{r\rm f} = \nu_{ir}^{\rm f} \sum_{F_I^r, B^r} \int \mathfrak{D}_r \prod_{\mathfrak{F}_i^r} dc_j \prod_{\mathfrak{B}^r} dc_k \Big/ \mathcal{K}_r f_i^0, \tag{4.6.59}$$

$$\Psi_i^{r\rm b} = \nu_{ir}^{\rm b} \sum_{F^r, B_I^r} \int \mathfrak{D}_r \prod_{\mathfrak{F}^r} dc_j \prod_{\mathfrak{B}_i^r} dc_k \Big/ \mathcal{K}_r f_i^0, \tag{4.6.60}$$

where $\mathfrak{D}_r$ is defined in (4.6.17), in such a way that $\Psi^r - (\Psi^{r\rm b} - \Psi^{r\rm f}) \in \mathcal{I}$. The perturbed terms $\mathfrak{w}_i$, $i \in S$, can then be rewritten in the form

$$\mathfrak{w}_i = \sum_{r \in R} (\nu_{ir}^{\rm b} - \nu_{ir}^{\rm f}) \Big(\overline{\tau}_r^{\rm f} \langle\!\langle \psi^{r\rm f}, f^0 \phi \rangle\!\rangle - \overline{\tau}_r^{\rm b} \langle\!\langle \psi^{r\rm b}, f^0 \phi \rangle\!\rangle \Big), \qquad i \in S. \tag{4.6.61}$$

Upon substituting expansions (4.6.22) and (4.6.23) into (4.6.61) and making use of space isotropy, we obtain

$$\langle\!\langle \psi^{r\rm f}, f^0 \phi \rangle\!\rangle = -\tfrac{1}{3} \langle\!\langle \psi^{r\rm f}, f^0 \phi^\kappa \rangle\!\rangle \boldsymbol{\partial}_{\boldsymbol{x}} \cdot \boldsymbol{v} + \sum_{s \in R} \langle\!\langle \psi^{r\rm f}, f^0 \phi^s \rangle\!\rangle \overline{\tau}_s \tag{4.6.62}$$

and

$$\langle\!\langle \psi^{r\rm b}, f^0 \phi \rangle\!\rangle = -\tfrac{1}{3} \langle\!\langle \psi^{r\rm b}, f^0 \phi^\kappa \rangle\!\rangle \boldsymbol{\partial}_{\boldsymbol{x}} \cdot \boldsymbol{v} + \sum_{s \in R} \langle\!\langle \psi^{r\rm b}, f^0 \phi^s \rangle\!\rangle \overline{\tau}_s. \tag{4.6.63}$$

As a consequence, the perturbed source term $\mathfrak{w}_i$ is the sum of a quadratics in $\overline{\tau}_r^{\rm f}$ and $\overline{\tau}_r^{\rm b}$, $r \in R$, and of a linear form with respect to $\overline{\tau}_r^{\rm f}$ and $\overline{\tau}_r^{\rm b}$, $r \in R$, multiplied by $\boldsymbol{\partial}_{\boldsymbol{x}} \cdot \boldsymbol{v}$. In particular, when there are at most three reactants or products in chemical reactions, $\mathfrak{w}_i$ is the sum of a polynomial in n_i, $i \in S$, of degree lower than three multiplied by $\boldsymbol{\partial}_{\boldsymbol{x}} \cdot \boldsymbol{v}$, and of a polynomial in n_i, $i \in S$, of degree lower than six.

Note that the source term is much simpler in the Maxwellian reaction regime $\mathfrak{a} = 1$ than in the regime $\mathfrak{a} = 0$. The perturbed terms were first discussed by Prigogine and Xhrouet for monatomic species [PX49] and by Ludwig and Heil for polyatomic species [LH60]. These terms have been estimated in a number of simplified situations by Prigogine and Xhrouet [PX49], Prigogine and Mathieu [PM50], Takayanagi [Tak51], Present [Pre59], and Shizgal and Karplus [SK70] [SK71a] [SK71b], and they are generally believed to be small [HCB54]. Although complete estimates of $\mathfrak{w}_i$, $i \in \mathcal{S}$, are still missing for realistic reactive mixtures, these terms are usually neglected. With these simplifications, the mass production rates $\overline{\omega}_i$, $i \in \mathcal{S}$, appearing in the species conservation equations (4.6.26), then reduce to the Maxwellian production rates $\overline{\omega}_i = \overline{\omega}_i^0$, $i \in S$. In particular, upon neglecting both the perturbed chemistry production terms and the chemical pressure, the governing equations for the two regimes $\mathfrak{a} = 0$ and $\mathfrak{a} = 1$ coincide.

4.6.7. Thermodynamics

At the first-order in the Enskog expansion, the entropy may be evaluated by using the zero-order Maxwellian distribution functions [dGM84]. More specifically, at the first-order, the entropy per unit volume is given by

$$\mathcal{S}^{\mathrm{kin}} = -k_{\mathrm{B}} \sum_{\substack{i\in S \\ I\in\mathfrak{Q}_i}} \int f_i^0 \big(\log(\beta_{iI} f_i^0) - 1\big) d\boldsymbol{c}_i + \mathcal{O}(\varepsilon^2), \tag{4.6.64}$$

in such a way that $\mathcal{S}^{\mathrm{kin}} = \sum_{i\in S} n_i S_i^{\mathrm{molec}} + \mathcal{O}(\varepsilon^2)$, where the species zero-order entropies per molecule are given by

$$S_i^{\mathrm{molec}} = \tfrac{5}{2} k_{\mathrm{B}} + \frac{\overline{E}_i}{T} - k_{\mathrm{B}} \log \frac{n_i}{Q_i}. \tag{4.6.65}$$

As a consequence, the macroscopic entropy $\mathcal{S}$ per unit volume is taken to be

$$\mathcal{S} = \sum_{i\in S} n_i S_i^{\mathrm{molec}} = \sum_{i\in S} \rho_i s_i, \tag{4.6.66}$$

where $s_i = S_i^{\mathrm{molec}}/\mathfrak{m}_i$ is the entropy per unit mass of the ith species and $\mathcal{S}$ coincides up to second order with the kinetic entropy $\mathcal{S}^{\mathrm{kin}}$. Similarly, we introduce the volumetric energy $\mathcal{E}$, enthalpy $\mathcal{H}$, and Gibbs free energy $\mathcal{G}$, which are found in the form

$$\mathcal{E} = \sum_{i\in S} \rho_i e_i, \qquad e_i = \tfrac{3}{2}\frac{k_{\mathrm{B}} T}{\mathfrak{m}_i} + \frac{\overline{E}_i}{\mathfrak{m}_i}, \tag{4.6.67}$$

$$\mathcal{H} = \sum_{i\in S} \rho_i h_i, \qquad h_i = \tfrac{5}{2}\frac{k_{\mathrm{B}} T}{\mathfrak{m}_i} + \frac{\overline{E}_i}{\mathfrak{m}_i}, \tag{4.6.68}$$

$$\mathcal{G} = \sum_{i\in S} \rho_i g_i, \qquad g_i = \frac{k_{\mathrm{B}} T}{\mathfrak{m}_i} \log \frac{n_i}{Q_i}. \tag{4.6.69}$$

The above thermodynamic functions depend on the temperature T and the mass densities ρ_i, $i \in S$, or, equivalently, on T, p, and the mass fractions Y_i, $i \in S$, where $Y_i = \rho_i/\rho$.

We can further define the reduced chemical potentials

$$\mu_i = \frac{g_i}{RT} = \frac{1}{\mathcal{N}\mathfrak{m}_i} \log \frac{n_i}{Q_i}, \qquad i \in S, \tag{4.6.70}$$

where $R = \mathcal{N} k_{\mathrm{B}}$ is the gas constant and $\mathcal{N}$ is the Avogadro number, in such a way that, from (4.6.15), we obtain

$$\overline{\tau}_r = \mathcal{K}_r \Big(\exp\big(\sum_{k\in S} \mu_k m_k \nu_{kr}^{\mathrm{f}}\big) - \exp\big(\sum_{k\in S} \mu_k m_k \nu_{kr}^{\mathrm{b}}\big) \Big), \qquad r \in R, \tag{4.6.71}$$

where $m_k = \mathcal{N}\mathfrak{m}_k$ is the molar mass of the kth species. This is the natural symmetric expression of the reaction rates of progress $\overline{\tau}_r$, $r \in R$.

A conservation equation for $\mathcal{S}$ is then obtained from the conservation equations (4.6.26)–(4.6.28) and the relation $T\mathbb{D}\mathcal{S} = \mathbb{D}\mathcal{E} - \sum_{i\in S} g_i \mathbb{D}\rho_i$, where $\mathbb{D}$ denotes the total derivative, which is easily derived. After some algebra, it is found that

$$\partial_t \mathcal{S} + \boldsymbol{\partial_x}\cdot(\boldsymbol{v}\mathcal{S}) + \boldsymbol{\partial_x}\cdot\boldsymbol{J} = \mathfrak{v}, \tag{4.6.72}$$

where $\boldsymbol{J}$ is the entropy flux vector given by

$$\boldsymbol{J} = \frac{1}{T}\Big(\boldsymbol{\mathcal{Q}} - \sum_{i\in S} g_i\rho_i\boldsymbol{\mathcal{V}}_i\Big). \tag{4.6.73}$$

When $\mathfrak{a} = 1$, a straightforward calculation yields the entropy source term $\mathfrak{v}$ in the form

$$\begin{aligned}\mathfrak{v} = {} & \lambda\frac{\boldsymbol{\partial_x}T\cdot\boldsymbol{\partial_x}T}{T^2} + \frac{p}{T}\sum_{i,j\in S} D_{ij}(\boldsymbol{d}_i + \chi_i\boldsymbol{\partial_x}\log T)\cdot(\boldsymbol{d}_j + \chi_j\boldsymbol{\partial_x}\log T)\\ & + \frac{\eta}{2T}\big(\boldsymbol{\partial_x v} + (\boldsymbol{\partial_x v})^t - \tfrac{2}{3}(\boldsymbol{\partial_x}\cdot\boldsymbol{v})\boldsymbol{I}\big):\big(\boldsymbol{\partial_x v} + (\boldsymbol{\partial_x v})^t - \tfrac{2}{3}(\boldsymbol{\partial_x}\cdot\boldsymbol{v})\boldsymbol{I}\big)\\ & + \frac{\kappa}{T}(\boldsymbol{\partial_x}\cdot\boldsymbol{v})^2 - \sum_{k\in S}\frac{\mathfrak{m}_k g_k \overline{\omega}_k^0}{T}.\end{aligned} \tag{4.6.74}$$

From the properties of the transport coefficients stated in Section 4.3, we readily obtain that the entropy production term $\mathfrak{v}$ is indeed a sum of nonnegative terms. The entropy production due to chemical reactions reads as $-\sum_{k\in S}\mathfrak{m}_k g_k\overline{\omega}_k^0/T$ and is nonnegative from thermochemistry properties, as will be seen in Chapter 6.

This entropy equation can also be obtained directly from the kinetic entropy equation (4.3.2). Upon expanding the scattering and chemical kinetic production terms $\mathfrak{v}^{\mathfrak{S}}$ and $\mathfrak{v}^{\mathfrak{C}}$ in powers of ε and then letting $\varepsilon = 1$, we obtain that

$$\mathfrak{v}^{\mathfrak{S}} = \mathfrak{v}^{(-1)\mathfrak{S}} + \mathfrak{v}^{0\mathfrak{S}} + \mathfrak{v}^{1\mathfrak{S}} + O(\varepsilon^2) \tag{4.6.75}$$

and

$$\mathfrak{v}^{\mathfrak{C}} = \mathfrak{v}^{0\mathfrak{C}} + \mathfrak{v}^{1\mathfrak{C}} + O(\varepsilon^2). \tag{4.6.76}$$

Then, in the regime $\mathfrak{a} = 1$, one easily finds that $\mathfrak{v}^{(-1)\mathfrak{S}} = 0$ and $\mathfrak{v}^{0\mathfrak{S}} = 0$ and that $\mathfrak{v}^{1\mathfrak{S}} = k_{\mathrm{B}}[\![\phi,\phi]\!] = k_{\mathrm{B}}[\![\phi^{\mathfrak{S}},\phi^{\mathfrak{S}}]\!]$. Similarly, in this regime, we find that $\mathfrak{v}^{0\mathfrak{C}} = 0$ and $\mathfrak{v}^{1\mathfrak{C}} = -\sum_{k\in S}\mathfrak{m}_k g_k\overline{\omega}_k^0/T$. In particular, the gradient terms of $\mathfrak{v}$ in (4.6.74) correspond to $\mathfrak{v}^{1\mathfrak{S}}$, whereas the chemistry term corresponds to $\mathfrak{v}^{1\mathfrak{C}}$. In this situation, both inert and reactive contributions to $\mathfrak{v}$ are of the same order of magnitude and nonnegative.

In the regime $\mathfrak{a} = 0$, complexities arise with the chemical pressure and the perturbed chemical sources. The expansions (4.6.75) and (4.6.76) are still valid, but the coefficients are changed. We still have $\mathfrak{v}^{(-1)\mathrm{S}} = 0$ and $\mathfrak{v}^{0\mathrm{S}} = 0$, but the first nonzero term $\mathfrak{v}^{1\mathrm{S}}$ is now given by $\mathfrak{v}^{1\mathrm{S}} = k_{\mathrm{B}}[\![\phi, \phi]\!] = k_{\mathrm{B}}[\![\phi^{\mathrm{S}} + \phi^{\mathrm{C}}, \phi^{\mathrm{S}} + \phi^{\mathrm{C}}]\!]$. There is thus a first difference $\mathrm{d}\mathfrak{v}^{\mathrm{S}}$ in the entropy production (4.6.74) in the form

$$\mathrm{d}\mathfrak{v}^{\mathrm{S}} = k_{\mathrm{B}}[\![\phi^{\mathrm{S}} + \phi^{\mathrm{C}}, \phi^{\mathrm{S}} + \phi^{\mathrm{C}}]\!] - k_{\mathrm{B}}[\![\phi^{\mathrm{S}}, \phi^{\mathrm{S}}]\!]. \tag{4.6.77}$$

On the other hand, since $\mathfrak{a} = 0$, we now have $\mathfrak{v}^{0\mathrm{C}} = -\sum_{k\in S} \mathfrak{m}_k g_k \overline{\omega}_k^0 / T$, whereas the next term $\mathfrak{v}^{1\mathrm{C}}$ is obtained after lengthy calculations and yields a second difference $\mathfrak{v}^{1\mathrm{C}} = \mathrm{d}\mathfrak{v}^{\mathrm{C}}$ in the entropy production (4.6.74) in the form

$$\mathrm{d}\mathfrak{v}^{\mathrm{C}} = \frac{1}{T} p^{\mathrm{reac}} \boldsymbol{\partial}_{\boldsymbol{x}} \cdot \boldsymbol{v} - k_{\mathrm{B}}[\![\phi^{\mathrm{C}}, \phi^{\mathrm{C}}]\!] - \sum_{k\in S} \mathfrak{m}_k g_k \mathfrak{w}_k / T, \tag{4.6.78}$$

where $\mathfrak{w}_k$, $k \in S$, are the perturbed source terms. Summing both terms $\mathrm{d}\mathfrak{v}^{\mathrm{C}}$ and $\mathrm{d}\mathfrak{v}^{\mathrm{C}}$, we finally obtain—after some algebra—the entropy production difference $\mathrm{d}\mathfrak{v}$ between the regimes $\mathfrak{a} = 0$ and $\mathfrak{a} = 1$

$$\mathrm{d}\mathfrak{v} = -\frac{1}{T} p^{\mathrm{reac}} \boldsymbol{\partial}_{\boldsymbol{x}} \cdot \boldsymbol{v} - \sum_{k\in S} \mathfrak{m}_k g_k \mathfrak{w}_k / T. \tag{4.6.79}$$

It is readily seen then that, in this regime, the reactive contribution $\mathfrak{v}^{0\mathrm{C}} + \mathfrak{v}^{1\mathrm{C}}$ to entropy production has no clear sign, in contradiction with the underlying kinetic framework, so that the regime $\mathfrak{a} = 0$ is in some sense ill posed. Note also that, in this regime, the leading inert and reactive contributions to entropy production are not of the same order. As a consequence, in the following, we only consider either the regime $\mathfrak{a} = 1$ or the regime $\mathfrak{a} = 0$ without the perturbed source terms and the chemical pressure, so that both resulting models coincide.

4.7. Transport linear systems

In this section we discuss the evaluation of transport coefficients defined in Section 4.6.5. We investigate, in particular, the general structure of transport linear systems arising from a Galerkin approximate solution of the linearized Boltzmann equations (4.6.24). These approximate solutions are associated with the Chapman–Enskog series expansion of transport coefficients [CC70].

4.7.1. Galerkin method

The linear integral equations (4.6.24) are solved approximately by using a variational procedure [WT62] [CC70] [FK72]. More specifically, a finite dimensional functional space Ξ^μ is first selected

$$\Xi^\mu = \mathrm{span}\{\ \xi^{\mathrm{p}k},\ (\mathrm{p},k) \in \mathcal{B}^\mu\ \}, \tag{4.7.1}$$

where $\xi^{\mathrm{p}k}$, $(\mathrm{p},k) \in \mathcal{B}^\mu$, are basis functions and $\mathcal{B}^\mu$ is a set of basis function indices. We denote by $\mathfrak{n}$ the dimension of the Galerkin variational approximation space Ξ^μ. The distribution function ϕ^μ is then expanded in the form

$$\phi^\mu = \sum_{(\mathrm{p},k)\in\mathcal{B}^\mu} \alpha_k^{\mathrm{p}\mu} \xi^{\mathrm{p}k}, \tag{4.7.2}$$

where the $\alpha_k^{\mathrm{p}\mu}$ are scalars. In the notation $\alpha_k^{\mathrm{p}\mu}$, the superscript μ refers to the coefficient μ, the superscript p refers to the type of function that is considered, and the subscript k refers to the species.

The variational procedure applied to the integral equation (4.6.24a) yields the system

$$[\xi^{\mathrm{p}k}, \phi^\mu] = \langle\!\langle \xi^{\mathrm{p}k}, f^0 \Psi^\mu \rangle\!\rangle, \qquad (\mathrm{p},k) \in \mathcal{B}^\mu, \tag{4.7.3}$$

which must be solved under the constraints (4.6.24b). The relations (4.7.3) yield a linear system of size $\mathfrak{n}$ in the form

$$\sum_{(\mathrm{q},l)\in\mathcal{B}^\mu} G_{kl}^{\mathrm{pq}} \alpha_l^{\mathrm{q}\mu} = \beta_k^{\mathrm{p}\mu}, \qquad (\mathrm{p},k) \in \mathcal{B}^\mu, \tag{4.7.4}$$

where $G_{kl}^{\mathrm{pq}} = [\xi^{\mathrm{p}k}, \xi^{\mathrm{q}l}]$, $\beta_k^{\mathrm{p}\mu} = \langle\!\langle \xi^{\mathrm{p}k}, f^0 \Psi^\mu \rangle\!\rangle$, and the unknowns are the $\mathfrak{n}$ coefficients $\alpha_k^{\mathrm{p}\mu}$, $(\mathrm{p},k) \in \mathcal{B}^\mu$. Similarly, the $n{+}4$ tensorial constraints (4.6.24b) also yield the $(n{+}4)\tau_\mu$ scalar constraints

$$\sum_{(\mathrm{p},k)\in\mathcal{B}^\mu} \mathcal{G}_k^{\mathrm{p}l\nu} \alpha_k^{\mathrm{p}\mu} = 0, \qquad (l,\nu) \in [1, n{+}4]\times[1, \tau_\mu], \tag{4.7.5}$$

where τ_μ is the tensorial dimension of ϕ^μ and Ψ^μ, $\mathcal{G}_k^{\mathrm{p}l\nu} = \langle\!\langle \xi^{\mathrm{p}k}, f^0 \mathcal{T}_\nu \widehat{\psi}^l \rangle\!\rangle$, for $l \in [1, n{+}4]$, $\nu \in [1, \tau_\mu]$, and $(\mathrm{p},k) \in \mathcal{B}^\mu$, where $\mathcal{T}_\nu$, $\nu \in [1, \tau_\mu]$ is the corresponding canonical basis of tensors.

In most practical applications, owing to the orthogonality properties of the basis functions $\xi^{\mathrm{p}k}$, $(\mathrm{p},k) \in \mathcal{B}^\mu$, and the tensorial collisional invariants $\mathcal{T}_\nu \widehat{\psi}^l$, $l \in [1, n{+}4]$, $\nu \in \{1, \ldots, \tau_\mu\}$, either all of these relations are found to be trivial, i.e., yield zero constraint coefficients $\mathcal{G}_k^{\mathrm{p}l\nu} = 0$, $(\mathrm{p},k) \in \mathcal{B}^\mu$, or yield at most one nontrivial constraint.

As a consequence, we may write the transport linear systems in the compact form

$$\begin{cases} G\alpha^{\mu} = \beta^{\mu}, \\ \langle \mathcal{G}, \alpha^{\mu} \rangle = 0, \end{cases} \tag{4.7.6}$$

where $G \in \mathbb{R}^{\mathrm{n},\mathrm{n}}$ is given by $G_{kl}^{\mathrm{pq}} = [\xi^{\mathrm{p}k}, \xi^{\mathrm{q}l}]$, $(\mathrm{p},k),(\mathrm{q},l) \in \mathcal{B}^{\mu}$, α^{μ} and β^{μ} are vectors of $\mathbb{R}^{\mathrm{n}}$ and compact notation for the coefficients $\alpha_k^{\mathrm{p}\mu}$ and $\beta_k^{\mathrm{p}\mu}$, $(\mathrm{p},k) \in \mathcal{B}^{\mu}$, and $\mathcal{G} \in \mathbb{R}^{\mathrm{n}}$ is the constraint vector, which may be zero, depending on the transport coefficient μ under consideration.

Finally, solving the system (4.7.6) yields the products $[\phi^{\mu}, \phi^{\mu}]$ from the relation

$$[\phi^{\mu}, \phi^{\mu}] = \langle\!\langle \Psi^{\mu}, f^0 \phi^{\mu} \rangle\!\rangle = \sum_{(\mathrm{p},k)\in\mathcal{B}^{\mu}} \alpha_k^{\mathrm{p}\mu} \beta_k^{\mathrm{p}\mu}, \tag{4.7.7}$$

from which the transport coefficients are easily evaluated.

4.7.2. Basis functions

The basis functions $\xi^{\mathrm{p}k}$, $(\mathrm{p},k) \in \mathcal{B}^{\mu}$, are generally chosen as simple linear combinations of the functions ϕ^{a0cdk} defined by

$$\phi^{a0cdk}(c_k, K) = \Big(S_{a+\frac{1}{2}}^{c}(\boldsymbol{w}_k\cdot\boldsymbol{w}_k)\, W_k^d(\epsilon_{kK})\, \overline{\otimes^a \boldsymbol{w}_k}\, \delta_{ki} \Big)_{i\in S}, \tag{4.7.8}$$

where $\boldsymbol{w}_k = \sqrt{\mathrm{m}_k/2k_{\mathrm{B}}T}\,(\boldsymbol{c}_k - \boldsymbol{v})$ is the reduced relative velocity of the molecules of the kth species and $\epsilon_{kK} = E_{kK}/k_{\mathrm{B}}T$ is the reduced internal energy of species k in state K. In addition, a, c, and d are integers, $S_{a+1/2}^{c}$ is the Laguerre and Sonine polynomial of order c with parameter $a+1/2$, W_k^d is the Wang Chang and Uhlenbeck polynomial of order d for the kth species, and $\overline{\otimes^a \boldsymbol{w}_k}$ is a tensor of rank a with respect to the three-dimensional space, given by $\overline{\otimes^0 \boldsymbol{w}_k} = 1$, $\overline{\otimes^1 \boldsymbol{w}_k} = \boldsymbol{w}_k$, and $\overline{\otimes^2 \boldsymbol{w}_k} = \boldsymbol{w}_k \otimes \boldsymbol{w}_k - \frac{1}{3}\boldsymbol{w}_k\cdot\boldsymbol{w}_k \boldsymbol{I}$ [WT62]. In the notation ϕ^{abcdk}, the first index a thus refers to the tensorial rank with respect to $\mathbb{R}^3$, the second index $b = 0$ refers to the absence of polarization effects [M*al*90], the third index c refers to the Laguerre and Sonine polynomial, the fourth index d refers to the Wang Chang and Uhlenbeck polynomial, and the last index k refers to the species. These functions have important orthogonality properties since we have the relations [WT62] [M*al*90] [M*al*91]

$$\langle\!\langle f^0 \phi^{a0cdk}, \phi^{a'0c'd'l} \rangle\!\rangle = \langle\!\langle f^0 \phi^{a0cdk}, \phi^{a0cdk} \rangle\!\rangle \delta_{aa'}\delta_{cc'}\delta_{dd'}\delta_{kl}, \tag{4.7.9}$$

for $a, a', c, c', d, d' \geq 0$ and $k, l \in S$. Various properties of the Laguerre and Sonine polynomials, Wang Chang and Uhlenbeck polynomials, and functions ϕ^{a0cdk} are summarized in [FK72] [M*al*90] [EG94].

The transport linear systems in their natural and symmetric form associated with classical orthogonal polynomial expansions have been evaluated in [EG94]. The following scalar basis functions have been used for the scalar integral equation in $\phi^\kappa = (\phi_i^\kappa)_{i\in S}$:

$$\begin{cases} \phi^{0010k} = \big((\tfrac{3}{2} - \boldsymbol{w}_k\cdot\boldsymbol{w}_k)\delta_{ki}\big)_{i\in S}, & k\in S, \\ \phi^{0001k} = \big((\bar\epsilon_k - \epsilon_{kK})\delta_{ki}\big)_{i\in S}, & k\in S_{\mathrm{pol}}, \end{cases} \tag{4.7.10}$$

where S_{pol} denotes the set of polyatomic molecules, n^{p} is the number of polyatomic molecules, and $\bar\epsilon_k = \overline{E}_k/k_{\mathrm{B}}T$. The following vector basis functions have also been used for the vector integral equations in $\boldsymbol{\phi}^{D_l} = (\boldsymbol{\phi}_i^{D_l})_{i\in S}$, $l\in S$, $\boldsymbol{\phi}^{\widehat\lambda} = (\boldsymbol{\phi}_i^{\widehat\lambda})_{i\in S}$, and $\boldsymbol{\phi}^{\lambda} = (\boldsymbol{\phi}_i^{\lambda})_{i\in S}$:

$$\begin{cases} \phi^{1000k} = (\boldsymbol{w}_k\delta_{ki})_{i\in S}, & k\in S, \\ \phi^{1010k} = \big((\tfrac{5}{2} - \boldsymbol{w}_k\cdot\boldsymbol{w}_k)\boldsymbol{w}_k\delta_{ki}\big)_{i\in S}, & k\in S, \\ \phi^{1001k} = \big((\bar\epsilon_k - \epsilon_{kK})\boldsymbol{w}_k\delta_{ki}\big)_{i\in S}, & k\in S_{\mathrm{pol}}, \end{cases} \tag{4.7.11}$$

and the following traceless symmetric matrix basis functions have been used for the traceless symmetric matrix integral equation in $\boldsymbol{\phi}^{\eta} = (\boldsymbol{\phi}_i^{\eta})_{i\in S}$:

$$\phi^{2000k} = \big((\boldsymbol{w}_k\otimes\boldsymbol{w}_k - \tfrac{1}{3}\boldsymbol{w}_k\cdot\boldsymbol{w}_k\boldsymbol{I})\delta_{ki}\big)_{i\in S}, \qquad k\in S. \tag{4.7.12}$$

Explicit calculations of transport linear systems in terms of molecular parameters and state variables are presented in Chapter 5.

However, one may also consider simplified formulations of the transport linear systems associated with the use of smaller variational approximation spaces Ξ^μ for the perturbed distribution functions $\phi^\mu = (phi_i^\mu)_{i\in S}$. Using functional spaces of lower dimension will indeed reduce the size of the transport linear systems $\mathfrak{n}$ and hence simplify the transport algorithms [T*al*79] [VK88] [M*al*90] [EG94]. In particular, for the evaluation of the volume viscosity, the following basis functions were considered in [EG94] :

$$\widehat\phi^{0001k} = \phi^{0001k} - X_k\frac{\mathfrak{c}_k^{\mathrm{int}}}{\mathfrak{c}_v}\Big(\sum_{l\in S}\phi^{0010l} + \sum_{l\in S_{\mathrm{pol}}}\phi^{0001l}\Big), \qquad k\in S_{\mathrm{pol}}, \tag{4.7.13}$$

whereas, for the thermal conductivities, thermal diffusion coefficients, and diffusion coefficients, the following basis functions have been used [EG94] :

$$\phi^{10ek} = \phi^{1010k} + \delta_{kS_{\mathrm{pol}}}\phi^{1001k}, \qquad k\in S, \tag{4.7.14}$$

where $\delta_{kS_{\mathrm{pol}}} = 1$ if the kth species is polyatomic, whereas $\delta_{kS_{\mathrm{pol}}} = 0$ if the kth species is monatomic. Note that the basis functions ϕ^{10ek}, $k\in S$, are associated with the sum of the kinetic and internal energy of the molecules [T*al*79] [VK88] [EG94].

4.7.3. Structure of transport linear systems

By using the kinetic theory framework, one can establish the following properties for the transport linear systems [EG94] :

$$\begin{cases} G \ \text{ is symmetric positive semidefinite,} \\ N(G) \oplus \mathcal{G}^{\perp} = \mathbb{R}^{\mathrm{n}}, \\ \beta^{\mu} \in R(G), \end{cases} \tag{4.7.15}$$

provided that the following perpendicularity property holds between the Galerkin approximation space Ξ^{μ} and the tensorial collisional invariant linear space $\mathcal{I}^{\mu}$ of the same tensorial rank than ϕ^{μ} and Ψ^{μ} :

$$\mathcal{I}^{\mu} = \mathcal{I}^{\mu} \cap \Xi^{\mu} \quad \oplus \quad \mathcal{I}^{\mu} \cap \Xi^{\mu\perp}, \tag{4.7.16}$$

where $\mathcal{I}^{\mu} \cap \Xi^{\mu\perp}$ denotes the elements of $\mathcal{I}^{\mu}$ that are orthogonal to Ξ^{μ} with respect to the bilinear form $\langle\!\langle f^0\,,\ \rangle\!\rangle$. Note that the nullspace of the matrix G is easily shown to be

$$N(G) = \{\ x = (x_k^{\mathrm{p}})_{(\mathrm{p},k)\in\mathcal{B}^{\mu}} \in \mathbb{R}^{\mathrm{n}}, \quad \sum_{(\mathrm{p},k)\in\mathcal{B}^{\mu}} x_k^{\mathrm{p}} \xi^{\mathrm{p}k} \in \mathcal{I}^{\mu}\ \},$$

since the nullspace of the bracket integral operator is constituted by collisional invariants.

When there are no constraints, that is, when the vector $\mathcal{G}$ is zero, the nullspace $N(G)$ reduces to zero and G is symmetric positive definite. On the other hand, when $\mathcal{G}$ is nonzero, the nullspace $N(G)$ is one dimensional. These properties will be used in Chapter 5 when investigating multicomponent transport algorithms.

4.7.4. Sparse transport matrix

The sparse transport matrix $db(G)$ is defined [EG94] as the diagonal of all rectangular blocks G^{pq}, $\mathrm{p},\mathrm{q} \in \mathcal{F}$, of G, that is,

$$db(G)_{kl}^{\mathrm{pq}} = G_{kl}^{\mathrm{pq}} \delta_{kl}, \qquad (\mathrm{p},k),(\mathrm{q},l) \in \mathcal{B}^{\mu}.$$

In order to establish the convergence of various iterative algorithms, we will need to know whether the matrices $db(G)$ and $2db(G) - G$ are positive definite on $\mathbb{R}^{\mathrm{n}}$. Within the kinetic framework, one can show that [EG94]

$$\begin{cases} 2db(G) - G \ \text{ is symmetric positive semidefinite for } n \geq 1, \\ 2db(G) - G \ \text{ is positive definite for } n \geq 3, \\ N\big(2db(G) - G\big) = \{\ x^*,\ x \in N(G)\ \} \text{ for } n = 2, \\ N\big(2db(G) - G\big) = N(G) \text{ for } n = 1, \end{cases} \tag{4.7.17}$$

and the matrix $db(G)$ satisfies

$$\begin{cases} db(G) \text{ is symmetric positive semidefinite for } n \geq 1, \\ db(G) \text{ is positive definite for } n \geq 2, \\ N\big(db(G)\big) = N(G) \text{ for } n = 1, \end{cases} \tag{4.7.18}$$

where for $n = 2$ we have defined the components of x^* by $(x^{\mathrm{p}k})^* = (-1)^k x^{\mathrm{p}k}$, $(\mathrm{p}, k) \in \mathcal{B}^\mu$.

These properties can be established by assuming that the basis functions are "localized with respect to the species," that is,

$$\xi_i^{\mathrm{p}k} = 0 \qquad \text{for} \qquad i \neq k, \tag{4.7.19}$$

and that the basis functions are orthogonal to constants [EG94]. All of these properties will be used in Chapter 5 for iterative solution of the transport linear systems.

Remark 4.7.1. The structure of the matrix $2db(G) - G$ reveals that the general case for mixtures is $n \geq 3$ and that binary mixtures are a degenerate case inadequate for a general theory. ∎

4.7.5. Vanishing mass fractions

For mathematical investigations and computational applications, it is important to understand the mathematical and numerical behavior of the transport coefficients when some mass fractions become arbitrarily small. Zero mass fractions indeed lead to artificial singularities in the transport linear systems. These artificial singularities can be eliminated by considering rescaled versions of the original systems [EG94]. In particular, provided that the diffusion matrix is replaced by the flux diffusion matrix $C_{kl} = \rho Y_k D_{kl}$, $k, l \in S$, it is proven in [EG94] that all transport coefficients are smooth rational functions of the mass fractions and admit finite limits when some mass fractions become arbitrarily small.

4.8. Notes

4.1. The kinetic theory of dilute gases with molecules having internal degrees of freedom was first developed by Wang Chang and Uhlenbeck and De Boer [WU51] [WUD64] in a semiclassical framework, in which the translational motion is treated classically and the internal motion quantum mechanically. In this treatment, one assumes a symmetry condition on the quantum cross sections, which is only valid if the molecular states are nondegenerate and is analogous to the classical assumption of the existence of inverse collisions. Waldmann [Wal58] and then Mason and Monchick [MM62] have further shown that, although the treatment required the assumption of nondegeneracy and detailed balance, its results were still valid for all molecules, provided that the quantum mechanical cross sections were replaced by degeneracy averaged quantum mechanical cross sections. The required symmetry property of the cross sections is then obtained from the invariance of the Hamiltonian under the combined operation of space inversion and time reversal, rather than by requiring the assumption of detailed balance [Wal58] [MS64].

A fully quantum mechanical treatment was given by McCourt and Snider [MS64] [MS65] using the quantum mechanical Boltzmann equation derived by Waldmann [Wal57] and Snider [Sni60]. This fully quantum mechanical transport theory was able to describe the Kagan–Affassanaev polarizations associated with the Senftleben–Beenaker effects [M*al*90] [M*al*91], that is, the effect of magnetic fields on transport properties, but the corresponding macroscopic equations are more complicated than the Navier–Stokes equations since they involve now a macroscopic angular momentum conservation equation [MS64] [FK72]. These polarization effects are only important in the presence of applied magnetic or electric fields and not addressed in this book. Under the so-called isotropic or Pidduck approximation [MVW88] valid in the absence of polarization effects, the quantum mechanical theory yields the same formal results as the semiclassical approach [MS64] [MVW88], which has been be used in this chapter. It should also be noted that a fully classical theory was later developed by Kagan and Maksimov [M*al*90]. The corresponding Boltzmann equation is the classical limit of the quantum mechanical Waldmann–Snider equation [M*al*90].

4.2. The extension of the semiclassical theory to dilute polyatomic gas mixtures was given by Waldmann and Trübenbacher [WT62] and by Monchick, Yun, and Mason [MYM63]. Although the transport coefficients obtained from both treatments are identical, there is an important difference, however, in the final structure of the linear systems that need to be solved to obtain the transport properties. Indeed, Monchick, Yun, and Mason systematically eliminated the singularities arising in the naturally singular and symmetric linear systems obtained from the variational approximation procedure. More specifically, they used explicitly the linear constraints to zero the diagonal coefficients of the system matrices, following a procedure introduced by Curtiss and Hirschfelder [CH49] [HCB54]. Although such a formulation of the linear systems may be used for direct numerical inversions, the symmetric positive definite forms of the linear systems obtained here can be inverted at half the computational cost [EG94]. Moreover, the original constrained singular symmetric forms obtained in [WT62] are preferable for iterative techniques, and the symmetric systems have simpler analytic expressions, so that they are better suited for analytic approximations of the transport coefficients. Furthermore, Waldmann and Trübenbacher [WT62] have used symmetric diffusion coefficients, which are formally compatible with Onsager reciprocal relations. Note that symmetric diffusion coefficients have also been considered by Waldmann [Wal58], Chapman and Cowling [CC70], Ferziger and Kapper [FK72], and Curtiss [Cur68], at variance with Monchick, Yun, and Mason [MYM63], Curtiss and Hirschfelder [CH49], and Hirschfelder, Curtiss, and Bird [HCB54], who have artificially destroyed this symmetry [Van67]. In this book, we have used the elegant formalism of [WT62], which fully respects the natural symmetries appearing in the model.

4.3. The kinetic equilibrium regime $\mathfrak{a} = -1$ leads to equilibrium flows which can also be obtained from nonequilibrium flows, as will be discussed in Chapter 10. The kinetic equilibrium regime has been investigated in [EG98a], where the macroscopic equations as well as the transport fluxes have been obtained. In particular, it can be shown that the same equations and transport fluxes are obtained as if chemical equilibrium is imposed into the macroscopic nonequilibrium equations. The corresponding transport coefficients, however, do not coincide [EG98a].

4.4. The kinetic theory and the Enskog expansion can also be used to derive various boundary conditions near solid surfaces [Loh86].

4.9. References

[ACG94] B. V. Alexeev, A. Chikhaoui, and I. T. Grushin, *Application of the Generalized Chapman-Enskog Method to the Transport-Coefficient Calculation in a Reacting Gas Mixture,* Phys. Review E, **49**, (1994), pp. 2809–2825.

[Cer88] C. Cercignani, *The Boltzmann Equation and Its Applications,* Applied Mathematical Sciences, Volume **67**, Springer-Verlag, Berlin, (1988).

[CIP94] C. Cercignani, R. Illner, and M. Pulvirenti, *The Mathematical Theory of Dilute Gases*, Applied Mathematical Sciences, Volume **106**, Springer-Verlag, Berlin, (1994).

[CC70] S. Chapman and T. G. Cowling, *The Mathematical Theory of Non-Uniform Gases,* Cambridge University Press, Cambridge, (1970).

[Cur68] C. F. Curtiss, *Symmetric Gaseous Diffusion Coefficients,* J. Chem. Phys., **49**, (1968), pp. 2917–2919.

[CH49] C. F. Curtiss and J. O. Hirschfelder, *Transport Properties of Multicomponent Gas Mixtures,* J. Chem. Phys., **17**, (1949), pp. 550–555.

[dGM84] S. R. de Groot and P. Mazur, *Non-Equilibrium Thermodynamics,* Dover, New York, (1984).

[DL89] R. J. DiPerna and P. L. Lions, *On the Global Existence for Boltzmann equations: Global Existence and Weak Stability,* Ann. Math., **130**, (1989), pp. 321–366.

[EG94] A. Ern and V. Giovangigli, *Multicomponent Transport Algorithms,* Lecture Notes in Physics, New Series "Monographs," **m 24**, Springer-Verlag, Berlin, 1994.

[EG98a] A. Ern and V. Giovangigli, *The Kinetic Equilibrium Regime,* Physica-A, **260**, (1998), pp. 49–72.

[FK72] J. H. Ferziger and H. G. Kaper, *Mathematical Theory of Transport Processes in Gases,* North Holland Publishing Company, Amsterdam, (1972).

[Gru93] C. Grunfeld, *On a Class of Kinetic Equations for Reacting Gas Mixtures with Multiple Collisions,* C. R. Acad. Sci. Paris, **316**, Série I, (1993), pp. 953–958.

[GG95] C. Grunfeld and E. Georgescu, *On a Class of Kinetic Equations for Reacting Gas Mixtures,* Mat. Fiz., Analiz, Geom., **2**, (1995), pp. 408–435.

[GM97] C. Grunfeld and D. Marinescu, *On the Numerical Simulation of Reactive Boltzmann Type Equations,* Transp. Theory Stat. Phys., **26**, (1997), pp. 287–318.

[HCB54] J. O. Hirschfelder, C. F. Curtiss, and R. B. Bird, *Molecular Theory of Gases and Liquids,* John Wiley & Sons, Inc., New York, (1954).

[Kus91] I. Kuščer, *Dissociation and Recombination in an Inhomogeneous Gas,* Physica A, **176**, (1991), pp. 542–556.

[Loh86] G. Lohöfer, *Navier-Stokes Boundary Conditions of a Gas Mixture,* Phys. Fluids, **29**, (1986), pp. 4025–4031.

[LH60] G. Ludwig and M. Heil, *Boundary Layer Theory with Dissociation and Ionization,* In Advances in Applied Mechanics, Volume **VI**, Academic Press, New York, (1960), pp. 39–118.

[MM62] E. A. Mason and L. Monchick, *Heat Conductivity of Polyatomic and Polar Gases.* J. Chem. Phys., **36**, (1962), pp. 1622–1639.

[M*al*90] F. R. McCourt, J. J. Beenakker, W. E. Köhler, and I. Kuščer, *Non Equilibrium Phenomena in Polyatomic Gases,* Volume I: Dilute Gases, Clarendon Press, Oxford, (1990).

[M*al*91] F. R. McCourt, J. J. Beenakker, W. E. Köhler, and I. Kuščer, *Non Equilibrium Phenomena in Polyatomic Gases,* Volume II: Cross Sections, Scattering and Rarefied Gases, Clarendon Press, Oxford, (1991).

[MS64] F. R. McCourt and R. F. Snider, *Transport Properties of Gases with Rotational States,* J. Chem. Phys., **41**, (1964), pp. 3185–3194.

[MS65] F. R. McCourt and R. F. Snider, *Transport Properties of Gases with Rotational States. II.* J. Chem. Phys., **43**, (1965), pp. 2276–2283.

[MVW88] J. Millat, V. Vesovic, and W. A. Wakeham, *On the Validity of the Simplified Expression for the Thermal Conductivity of Thijsse et al.,* Physica A, **166**, (1988), pp. 153–164.

[MYM63] L. Monchick, K. S. Yun, and E. A. Mason, *Formal Kinetic Theory of Transport Phenomena in Polyatomic Gas Mixtures,* J. Chem. Phys., **39**, (1963), pp. 654–669.

[Pre59] R. D. Present, *On the Velocity Distribution in a Chemically Reacting Gas,* J. Chem. Phys., **31**, (1959), pp. 747–750.

[PM50] I. Prigogine and M. Mathieu, *Sur la Perturbation de la Distribution de Maxwell par des Réactions Chimiques en Phase Gazeuse,* Physica., **16**, (1950), pp. 51–64.

[PX49] I. Prigogine and E. Xhrouet, *On the Perturbation of Maxwell Distribution Function by Chemical Reactions in Gases,* Physica, **15**, (1949), pp. 913–932.

[SK70] B. Shizgal and M. Karplus, *Nonequilibrium Contributions to the Rate of Reaction. I. Perturbation of the Velocity Distribution Function.* J. Chem. Phys., **52**, (1970), pp. 4262–4278.

[SK71a] B. Shizgal and M. Karplus, *Nonequilibrium Contributions to the Rate of Reaction. II. Isolated Multicomponent Systems,* J. Chem. Phys., **54**, (1971), pp. 4345–4356.

[SK71b] B. Shizgal and M. Karplus, *Nonequilibrium Contributions to the Rate of Reaction. III. Isothermal Multicomponent Systems,* J. Chem. Phys., **54**, (1971), pp. 4357–4362.

[Sni60] R. F. Snider, *Quantum-Mechanical Modified Boltzmann Equation for Degenerate Internal States,* J. Chem. Phys., **32**, (1960), pp. 1051–1060.

[Tak51] K. Takayanagi, *On the Theory of Chemically Reacting Gas,* Prog. Theor. Phys., **VI**, (1951), pp. 486–497.

[Tal79] B. J. Thijsse, G. W. 't Hooft, D. A. Coombe, H. F. P. Knaap, and J. J. M. Beenakker, *Some Simplified Expressions for the Thermal Conductivity in an External Field.* Physica A, **98**, (1979), pp. 307–312.

[VK88] R. J. Van den Oord and J. Korving, *The Thermal Conductivity of Polyatomic Molecules,* J. Chem. Phys., **89**, (1988), pp. 4333–4338.

[Van67] J. Van de Ree, *On the Definition of the Diffusion Coefficients in Reacting Gases,* Physica, **36**, (1967), 118–126.

[Wal57] L. Waldmann, *Die Boltzmann-Gleichung für Gase mit Rotierenden Molekülen,* Zeitschr. Naturforschg., **12a**, (1957), pp. 660–662.

[Wal58] L. Waldmann, *Transporterscheinungen in Gasen von Mittlerem Druck,* Handbuch der Physik, S. Flügge, Ed., **12**, Springer-Verlag, Berlin, (1958), pp. 295–514.

[WT62] L. Waldmann und E. Trübenbacher, *Formale Kinetische Theorie von Gasgemischen aus Anregbaren Molekülen.* Zeitschr. Naturforschg., **17a**, (1962), pp. 363–376.

[WU51] C. S. Wang Chang and G. E. Uhlenbeck, *Transport Phenomena in Polyatomic Gases,* University of Michigan Engineering Research Report CM-681, (1951).

[WUD64] C. S. Wang Chang, G. E. Uhlenbeck, and J. De Boer, *The Heat Conductivity and Viscosity of Polyatomic Gases,* J. De Boer and G. E. Uhlenbeck, Eds., Studies in Statistical Mechanics **2**, North Holland Publishing Company, Amsterdam, (1964), pp. 242–268.

[Wil85] F. A. Williams, *Combustion Theory,* Second ed., The Benjamin/Cummings Publishing Company, Inc., Menlo Park, (1985).

5

Transport Coefficients

5.1. Introduction

The transport fluxes appearing in the multicomponent flows equations have been expressed in terms of macroscopic variable gradients and transport coefficients in Chapters 2 and 4. These transport coefficients, however, are not explicitly given by the kinetic theory of gases. Evaluation of these coefficients requires solving linear systems termed "transport linear systems" and arising from Galerkin approximate solutions of linearized Boltzmann equations, as detailed in Chapter 4. Evaluation of transport coefficients thus appears to be a difficult task in the modeling and in the numerical simulation of multicomponent reactive flows and is the scope of this chapter.

We first examine various mathematical properties of the transport linear systems. Subsequently, we deduce various strategies for evaluating the transport coefficients. In particular, we obtain symmetric positive definite formulations which can be inverted by the Cholesky algorithm. We further obtain convergent iterative techniques that yield by truncation accurate approximate expressions of transport coefficients [EG94]. Empirical relations are also discussed, but the resulting expressions are generally less accurate. We also examine which data are needed for evaluating transport coefficients.

We then specifically consider the coefficients η, κ, C, D, θ, $\widehat{\lambda}$, χ, and λ. For each coefficient, we present a transport linear system of reduced dimension in a rescaled form. For larger transport linear systems, leading to more accurate coefficients, the reader is refered to [EG94]. We also indicate various formulae obtained from iterative techniques and various empirical relations [EG94]. Numerical implementation of these algorithms is finally discussed in [EG95b], [EG96b], and [EG96d].

5.2. Transport algorithms

In this section we examine a general theory that applies to each of the transport coefficients [EG94] [EG96a]. We obtain a general framework that allows the evaluation of transport coefficients either by a direct method or by using iterative techniques.

5.2.1. Transport linear systems

Denote by μ any of the transport coefficients η, κ, C_{kl}, $k,l \in S$, D_{kl}, $k,l \in S$, $\widehat{\lambda}$, θ_k, $k \in S$, λ, or $\widetilde{\chi}_k$, $k \in S$. Then, the evaluation of the tranport coefficient μ requires solving a system in the form

$$\begin{cases} G\alpha = \beta, \\ \langle \mathcal{G}, \alpha \rangle = 0, \end{cases} \tag{5.2.1}$$

where $G \in \mathbb{R}^{\mathfrak{n},\mathfrak{n}}$ is a symmetric matrix and α, β, and $\mathcal{G}$ are vectors of $\mathbb{R}^{\mathfrak{n}}$. The dimension $\mathfrak{n}$ of the linear system is exactly the dimension of the variational approximation space used for the perturbed species distribution function associated with the coefficient μ, as we have seen in Chapter 4. The transport coefficient μ is subsequently evaluated with a scalar product

$$\mu = \langle \alpha, \beta' \rangle, \tag{5.2.2}$$

where $\beta' \in \mathbb{R}^{\mathfrak{n}}$ is a vector.

The matrix G and the vectors β, $\mathcal{G}$, and β' are functions of temperature T, pressure p, and species mass fractions $Y_1, \dots, Y_n$, and they involve collisional molecular parameters. Reduced systems corresponding to each coefficient are presented in Sections 5.3–5.10. In the remaining part of Section 5.2, we examine the general structure of transport linear systems (5.2.1) and various solution strategies in order to evaluate α and μ.

5.2.2. Mathematical structure

As mentioned in Chapter 4, when the Galerkin variational approximation space associated with μ is perpendicular to the collisional invariant subspace, one can establish that

$$\begin{cases} G \ \text{ is positive semidefinite,} \\ N(G) \oplus \mathcal{G}^{\perp} = \mathbb{R}^{\mathfrak{n}}, \\ \beta \in R(G), \end{cases} \tag{5.2.3}$$

where $N(G)$ and $R(G)$ denote the nullspace and the range of G, respectively. In particular, either the matrix G is positive definite and the vector

$\mathcal{G}$ is zero or the nullspace of G is one-dimensional and $\mathcal{G}$ is nonzero. We will denote by $\mathcal{Z}$ a vector spanning the nullspace $N(G) = \mathbb{R}\mathcal{Z}$, with the convention that this vector is taken to be zero when G is invertible and $\mathcal{G}$ is zero. When $N(G)$ is one-dimensional, the condition $N(G) \oplus \mathcal{G}^\perp = \mathbb{R}^n$ is then equivalent to $\langle \mathcal{Z}, \mathcal{G} \rangle \neq 0$.

We easily deduce from (5.2.3) that the transport linear system (5.2.1) is well posed, that is, defines a unique solution α.

5.2.3. Direct inversion

A direct inversion of the natural system $G\alpha = \beta$ is generally not feasible, since G may be singular, depending on the transport coefficient μ under consideration. However, a consequence of the mathematical structure detailed in the previous section is that the transport linear systems (5.2.1) can be reformulated in the nonsingular form [Gio91] [EG94]

$$(G + \mathcal{G}\otimes\mathcal{G})\alpha = \beta. \tag{5.2.4}$$

The corresponding matrix $G + \mathcal{G}\otimes\mathcal{G}$ is indeed symmetric positive definite, so that (5.2.4) can be solved with Cholesky algorithm. This yields a first strategy for evaluating transport coefficients.

Note that various transport linear systems presented in the literature are *not* symmetric. The reason is that several authors have artificially destroyed the natural symmetry of transport matrices by using—up to scaling factors—the modified systems

$$(G + \mathcal{A}\otimes\mathcal{G})\alpha = \beta, \tag{5.2.5}$$

where $\mathcal{A}$ is a vector chosen such that the diagonal coefficients of the matrix $G + \mathcal{A}\otimes\mathcal{G}$ vanish as much as possible, following a procedure introduced by Hirschfelder et al. [HCB54] [MYM63]. Using these nonsymmetric systems and Gaussian elimination unduly complicates the formulation and nearly doubles the computational costs [EG95b] [EG96b].

5.2.4. Iterative methods

In order to minimize the costs associated with the evaluation of transport coefficients, one may solve iteratively the transport linear systems. For iterative techniques, the natural singular form (5.2.1) is more interesting than the nonsingular form, but we have to complete the properties (5.2.3). In the framework of the kinetic theory of reactive polyatomic gas mixtures, we have introduced in Chapter 4 the sparse transport matrix $db(G)$ constituted by diagonals of blocs of G. This matrix is invertible at low costs and such that, for any $n \geq 3$, we have [EG94]

$$\begin{cases} 2db(G) - G \;\text{ is symmetric positive definite,} \\ db(G) \;\text{ is symmetric positive definite.} \end{cases} \tag{5.2.6}$$

In the cases $n = 1$ or $n = 2$, the nullspace $N\big(2db(G) - G\big)$ is explicitly known, but is nonzero. Incidentally, this shows that binary mixtures are a degenerate case, which is inadequate for a general theory of multicomponent transport.

It is then possible to use projected iterative schemes in order to expand the transport coefficients. More specifically, we introduce the matrix $\mathcal{M} = db(G) + \mathrm{diag}(\sigma_1, \dots, \sigma_{\mathfrak{n}})$ with $\sigma_i \geq 0$ for $1 \leq i \leq \mathfrak{n}$, the splitting $G = \mathcal{M} - \mathcal{Z}$, and the iteration matrix $\mathcal{T} = \mathcal{M}^{-1}\mathcal{Z}$. Let P be the projector onto $\mathcal{G}^{\perp}$ parallel to $N(G)$, that is, $P = I - \mathcal{Z} \otimes \mathcal{G} / < \mathcal{Z}, \mathcal{G} >$. One can then establish that the spectral radius of the product $P\mathcal{T}$ is strictly lower than unity and that

$$\alpha = \sum_{j=0}^{\infty} (P\mathcal{T})^j P \mathcal{M}^{-1} P^t \beta, \tag{5.2.7}$$

where α is the unique solution of (5.2.1). This now implies that

$$\mu = \Big\langle \sum_{j=0}^{\infty} (P\mathcal{T})^j P \mathcal{M}^{-1} P^t \beta, \beta' \Big\rangle, \tag{5.2.8}$$

so that one can express all transport coefficients in gas mixtures in the form of convergent series. Upon truncating these series, accurate rigorous approximations are obtained [EG94] [EG96a]. One can also establish that the approximate systems obtained by using inexact collision integrals also satisfy properties (5.2.3) and (5.2.6), provided that the inexact parameters are in a suitable admissible set. One can also use projected conjugate gradient algorithms [EG94] preconditioned by the sparse transport matrix $dg(G)$. The above iterative techniques yield a second strategy for evaluating multicomponent transport coefficients.

5.2.5. Empirical expressions

There is finally a third strategy for some transport coefficients. It is indeed possible to use *empirical* formulae as, for instance, the average of order s given by

$$\mathcal{M}_s(\mu) = \big(\sum_{k \in S} X_k \mu_k^s\big)^{1/s}, \tag{5.2.9}$$

where the quantities μ_k, $k \in S$, denote the corresponding pure species transport coefficient.

5.2.6. Operational count

We have seen three strategies for evaluating transport coefficients, that is, direct inversions, a few steps of an iterative technique or an empirical formula. In order to compare these approaches, we can perform an operational count. As usual, we define an operation to be one multiplication plus one addition. We assume that the number of species n is large, keeping in mind that this condition is typically met in most multicomponent flow calculations. In addition, the transport linear systems are usually of size $\mathfrak{n} \approx rn$ with $r \in \{1, 2, 3\}$.

Gaussian elimination then requires $\mathfrak{n}^3/3 + \mathcal{O}(n^2)$ operations, whereas Cholesky decomposition only costs $\mathfrak{n}^3/6 + \mathcal{O}(n^2)$ operations. In addition, the cost of performing m steps of an iterative methods is $m\,\mathfrak{n}^2 + \mathcal{O}(mn)$ operations. It is, therefore, expected that an iterative method will be more cost-effective than a direct numerical inversion, provided that the first few iterations already yield approximate expressions with a high enough level of accuracy. We note also that the cost of evaluating the transport linear systems can be shown to be $O(n^2)$, so that it does not dominate the cost of iterative techniques.

Furthermore, it is important to note that, in a multicomponent mixture, each component interacts with all other components. Only an algorithm with a cost scaling as $O(n^2)$ is able to take into account these $n(n+1)/2$ interactions. This is why the iterative algorithms have a cost $O(n^2)$ proportional to the square of the number of species. Only *empirical* expressions can have a cost of $\mathcal{O}(n)$ operations. These low-cost expressions are not as accurate as the analytic expressions rigorously derived from the kinetic theory, but still constitute an alternative strategy, depending on the trade between accuracy and computational costs [EG94] [EG95b] [EG96b].

5.2.7. Stability for vanishing mass fractions

In practical applications, it is important, from a computational viewpoint, to understand the mathematical and numerical behavior of the iterative algorithms when some mass fractions become arbitrarily small. We have seen in Chapter 4 that zero mass fractions lead to artificial singularities in the transport linear systems, which are eliminated by considering rescaled versions of the original systems [EG94]. Provided the diffusion matrix is replaced by the flux diffusion matrix $C_{kl} = \rho Y_k D_{kl}$, $k, l \in S$, all transport coefficients are smooth rational functions of the mass fractions and admit finite limits when some mass fractions become arbitrarily small.

Moreover, the iterative algorithms obtained for positive mass fractions can be rewritten in terms of rescaled system matrices that are still defined for nonnegative mass fractions [EG94]. For positive mass fractions, all the iterative algorithms then yield the same sequence of iterates, whether applied to the original transport linear system or to the rescaled one. This result establishes rigorously the validity of a common practice in numerical calculations, which consists in evaluating transport properties of a given gas mixture by first adding to all species mass fractions a very small number, typically of the order of the machine precision.

5.3. Molecular parameters

The coefficients of the transport linear systems can be expressed in terms of the state variables p, T, $Y_k, k \in S$, of species molecular parameters, and of various collision integrals. In this section, we discuss these molecular parameters and the various temperature functions that are needed in order to formulate the transport linear systems. Throughout this chapter, we assume that $T > 0$, $p > 0$, and $Y_k > 0$, $k \in S$. Collision integrals are also estimated according to the Monchick and Mason approximations, which neglect complex collision with more than one quantum jump [MM62] [MPM65].

5.3.1. Interaction potentials

The molecular parameters are associated with the interaction potentials φ_{kl} between molecules species pairs (k, l), $k, l \in S$. Various potentials can be used to describe the underlying molecular physics.

One generally uses Lennard–Jones interaction potentials [Dix84] [FK72] [HCB54] in the form

$$\varphi_{kl} = 4\epsilon_{kl}\Big\{ \Big(\frac{\sigma_{kl}}{r}\Big)^{12} - \Big(\frac{\sigma_{kl}}{r}\Big)^{6} \Big\}, \tag{5.3.1}$$

where σ_{kl} denotes the collision diameter, ϵ_{kl} denotes the potential well depth, and r denotes the distance between the molecules.

Molecular parameters associated with species pairs ϵ_{kl} and σ_{kl} are usually expressed in terms of pure species parameters ϵ_k, σ_k by the following relations for nonpolar molecules:

$$\frac{\epsilon_{kl}}{k_B} = \sqrt{\Big(\frac{\epsilon_k}{k_B}\Big)\Big(\frac{\epsilon_l}{k_B}\Big)}, \qquad \sigma_{kl} = \frac{1}{2}(\sigma_k + \sigma_l). \tag{5.3.2}$$

When a polar molecule of the kth species interacts with a nonpolar but polarizable molecule of the lth species, these relations become

$$\frac{\epsilon_{kl}}{k_B} = \xi^2\sqrt{\Big(\frac{\epsilon_k}{k_B}\Big)\Big(\frac{\epsilon_l}{k_B}\Big)}, \qquad \sigma_{kl} = \xi^{-\frac{1}{6}}\frac{1}{2}(\sigma_k + \sigma_l), \tag{5.3.3}$$

where

$$\xi = 1 + \frac{1}{4}\alpha_l^* \mu_k^* \sqrt{\frac{\epsilon_k}{\epsilon_l}}, \tag{5.3.4}$$

where α_l^* is the reduced polarizability of the nonpolar molecule and μ_k^* is the reduced dipole moment of the polar molecule

$$\alpha_l^* = \frac{\alpha_l}{\sigma_l^3}, \qquad \mu_k^* = \frac{\mu_k}{\sqrt{\epsilon_k \sigma_k^3}}, \tag{5.3.5}$$

where α_l denotes the polarizability of molecules of the lth species. For more general potentials, including oriented averaged Stockmayer potentials for interacting polar molecules, we refer the reader to [Dix84], [HCB54], [MM61], and [EG94].

5.3.2. Collision integrals

The transport linear systems also involve collision integrals, which are functions of the state variables and of molecular parameters. These quantities are integrals along collision paths between pairs of molecules that naturally appear in the expression of tranport linear systems and depend on the dynamics of intermolecular collisions and thus on species pairs.

A first family of collision integrals is denoted by

$$\Omega_{kl}^{(i,j)*} = \Omega_{kl}^{(i,j)} \left[\left(\frac{k_B T \pi}{2 m_{kl}} \right)^{\frac{1}{2}} \frac{(j+1)!}{2} \left(1 - \frac{1+(-1)^i}{2(i+1)} \right) \sigma_{kl}^2 \right]^{-1}, \qquad k, l \in S,$$

where i, j are integers. The $*$ symbol denotes the reduced collision integral which only depend, for Lennard–Jones potentials, on the reduced temperature T_{kl}^* given by [FK72] [CC70] [HCB54] [EG94]

$$T_{kl}^* = \frac{k_B T}{\epsilon_{kl}}. \tag{5.3.6}$$

Rather than reduced collision integrals, it is often practical to tabulate the following collision integrals ratios [FK72] [CC70] [HCB54] [EG94] :

$$\bar{A}_{kl} = \frac{\Omega_{kl}^{(2,2)*}}{\Omega_{kl}^{(1,1)*}}, \tag{5.3.7}$$

$$\bar{B}_{kl} = \frac{5\Omega_{kl}^{(1,2)*} - 4\Omega_{kl}^{(1,3)*}}{\Omega_{kl}^{(1,1)*}}, \tag{5.3.8}$$

$$\bar{C}_{kl} = \frac{\Omega_{kl}^{(1,2)*}}{\Omega_{kl}^{(1,1)*}}. \tag{5.3.9}$$

5.3.3. Viscosity of pure gases and binary diffusion

Some of the collision integrals can be directly related to pure species properties, such as viscosities or species pair properties, like binary diffusion coefficients. As a consequence, it is often more convenient to tabulate these physical quantities rather than the collision integrals.

In particular, the viscosity η_k of the kth species is given by

$$\eta_k = \frac{5}{16} \frac{\sqrt{\pi \mathfrak{m}_k k_{\mathrm{B}} T}}{\pi \sigma_k^2 \Omega_{kk}^{(2,2)*}}, \tag{5.3.10}$$

where $\mathfrak{m}_k$ is the mass of the molecules of the kth species and k_{B} is the Boltzmann constant. As a consequence, the pure species viscosities can be used to fit the integral $\Omega_{kk}^{(2,2)*}$, $k \in S$.

The self-diffusion coefficients $\mathcal{D}_{kk}^{\text{bin}}$, $k \in S$, and the binary diffusion coefficients $\mathcal{D}_{kl}^{\text{bin}}$, $k, l \in S$, $k \neq l$, are given by

$$\mathcal{D}_{kl}^{\text{bin}} = \frac{3}{16} \frac{\sqrt{2\pi k_{\mathrm{B}}^3 T^3 / \mathfrak{m}_{kl}}}{p \pi \sigma_{kl}^2 \Omega_{kl}^{(1,1)*}}, \tag{5.3.11}$$

where $\mathfrak{m}_{kl}$, is the reduced mass of the species pair (k, l),

$$\mathfrak{m}_{kl} = \frac{\mathfrak{m}_k \mathfrak{m}_l}{\mathfrak{m}_k + \mathfrak{m}_l}, \tag{5.3.12}$$

σ_{kl} is the collision diameter, and $\Omega^{(1,1)*}$ is a reduced collision integral. As a consequence, the binary diffusion coefficients can be used to fit the collision integrals $\Omega_{kl}^{(1,1)*}$, $k, l \in S$, $k \neq l$.

It is thus possible to tabulate the pure species viscosities η_k, $k \in S$, binary diffusion coefficients $\mathcal{D}_{kl}^{\text{bin}}$, $k, l \in S$, $k \neq l$, and quantities $\bar{A}_{kl}$, $\bar{B}_{kl}$, and $\bar{C}_{kl}$, rather than $\Omega_{kl}^{(i,j)*}$, $k, l \in S$.

5.3.4. Relaxation and diffusion of internal energy

Some collision integrals are also related to collision numbers ξ_k^{int}, $k \in S_{\text{pol}}$, associated with internal energy relaxation, where S_{pol} denotes the polyatomic species indexing set. These collision numbers ξ_k^{int}, $k \in S_{\text{pol}}$, are associated with energy transfer either between translational degrees of freedom and internal degrees of freedom or between internal degrees of freedom. When the energy can be split into independent rotational and vibrational modes, these numbers can be estimated from

$$\frac{\mathfrak{c}_k^{\text{int}}}{\xi_k^{\text{int}}} = \frac{\mathfrak{c}_k^{\text{rot}}}{\xi_k^{\text{rot}}} + \frac{\mathfrak{c}_k^{\text{vib}}}{\xi_k^{\text{vib}}},$$

where $\mathfrak{c}_i^{\text{int}} = \mathfrak{c}_k^{\text{rot}} + \mathfrak{c}_k^{\text{vib}}$ and the vibrational contribution is often negligible. Note that a harmonic mean is obtained because the modes are constrained to have the same temperature, since only polynomials in ϵ_{iI} are considered in the functional space basis. When polynomials in each independent energy mode are considered, arithmetic means are obtained [EG94] [EG96c].

The collision numbers $\xi_k^{\rm int}$, $k \in S_{\rm pol}$, are temperature dependent with

$$\xi_k^{\rm int}(T) = \xi_k^{\rm int}(T_0)\frac{F(T_0)}{F(T)}, \tag{5.3.13}$$

and F can be estimated from [Par59] [Dix84]

$$F(T) = 1 + \frac{\pi^{\frac{3}{2}}}{2}\left(\frac{\epsilon_k/k_{\rm B}}{T}\right)^{\frac{1}{2}} + \left(\frac{\pi^2}{4}+2\right)\left(\frac{\epsilon_k/k_{\rm B}}{T}\right) + \pi^{\frac{3}{2}}\left(\frac{\epsilon_k/k_{\rm B}}{T}\right)^{\frac{3}{2}}, \tag{5.3.14}$$

where $\xi_k^{\rm int}(T_0)$ at temperature T_0 is part of the data. The specific heat of internal energy can also be estimated from

$$\mathfrak{c}_k^{\rm int} = \mathfrak{c}_{pk} - \mathfrak{c}_p^{\rm tr}, \tag{5.3.15}$$

where $\mathfrak{c}_p^{\rm tr} = \frac{5}{2}k_{\rm B}$ and the constant-pressure specific heat per molecule $\mathfrak{c}_{pk}$ is evaluated from the specific heats per unit mass c_{pk} by using $\mathfrak{c}_{pk} = \mathfrak{m}_k c_{pk}$ or from the specific heats per mole C_{pk} from $\mathfrak{c}_{pk} = C_{pk}/\mathcal{N}$, where $\mathcal{N}$ is the Avogadro number.

The diffusion of internal energy can also be characterized by species pair coefficients $\mathcal{D}_{k\,{\rm int},l}^{\rm bin}$, $k \in S_{\rm pol}$, $l \in S$. The coefficient $\mathcal{D}_{k\,{\rm int},l}^{\rm bin}$ is associated with the diffusion of internal energy of the $k^{\rm th}$ species into the $l^{\rm th}$ species. These coefficients are related to collision integrals associated with internal energy exchanges and often approximated by binary diffusion coefficients [Dix84].

5.3.5. Transport data

In summary, the parameters needed for evaluating transport coefficients are, for each species, the molecule spherical, linear, or nonlinear character, the collision diameter σ_k, the potential well depth ϵ_k, the dipole moment μ_k, the polarizability α_k, and the internal energy collision number $\xi_k^{\rm int}(T_0)$. One can find these parameters in the literature for a large number of molecules [Dix84] [KWM83]. They are generally adjusted to fit experimental viscosity measurements.

It is also necessary to know the collision integrals $\Omega_{kl}^{(m,n)*}$, which depend on the reduced temperature T_{kl}^*. These collision integrals have been computed for a variety of molecular interaction potentials. The temperature dependence of these integrals is nevertheless weak, so that simple approximation polynomials are feasible for η_k and $\mathcal{D}_{kl}$ [KWM83] [OB81] [War82] [WMD96].

Of course, these molecular parameters and collision integrals are the minimal data required to evaluate some of the transport linear systems. We can still plan that, in the future, ab initio accurate calculations of collision integrals will be available and more accurate transport coefficients will be computed, requiring more transport data.

5.4. Shear viscosity

The standard transport linear system associated with the shear viscosity is of size n. This system corresponds to an approximation of ϕ^η in (4.6.22) along the basis functions ϕ^{2000k}, $k \in S$, presented in (4.7.12). This system can be written in the form

$$\mathcal{H}\alpha^\eta = \beta^\eta, \tag{5.4.1}$$

where

$$\mathcal{H}_{kk} = \sum_{\substack{l\in S\\ l\neq k}} \frac{k_B T}{p\mathcal{D}_{kl}^{\mathrm{bin}}} \frac{X_k X_l}{\mathfrak{m}_k+\mathfrak{m}_l} \left[1 + \frac{3}{5}\frac{\mathfrak{m}_l}{\mathfrak{m}_k}\bar{A}_{kl}\right] + \frac{X_k^2}{\eta_k}, \qquad k \in S, \tag{5.4.2}$$

$$\mathcal{H}_{kl} = \frac{k_B T}{p\mathcal{D}_{kl}^{\mathrm{bin}}} \frac{X_k X_l}{\mathfrak{m}_k + \mathfrak{m}_l} \left[-1 + \frac{3}{5}\bar{A}_{kl}\right], \qquad k,l \in S, \quad k \neq l, \tag{5.4.3}$$

and

$$\beta_k^\eta = X_k, \qquad k \in S. \tag{5.4.4}$$

The shear viscosity is then given by

$$\eta = \langle \alpha^\eta, \beta^\eta \rangle = \sum_{k\in S} X_k \alpha_k^\eta. \tag{5.4.5}$$

One can prove that the matrix $\mathcal{H}$ is symmetric positive definite [EG94]. It is important to note that solving this symmetric linear system (5.4.1) is actually *faster* than using the traditional Wilke formula [Wlk50], which involves fourth-order roots [EG96b].

Using one step of the conjugate gradient procedure with a preconditioning by the diagonal yields the following formula:

$$\eta = \frac{\left(\sum_{k\in S} X_k^2/\mathcal{H}_{kk}\right)^2}{\sum_{k,l\in S} X_k X_l \mathcal{H}_{kl}/(\mathcal{H}_{kk}\mathcal{H}_{ll})}, \tag{5.4.6}$$

obtained in [EG94] and which is ten times more accurate than Wilke formula and cheaper to evaluate [EG95b] [EG96b].

It is worthwhile to point out that no simple empirical formula has been found to be generally accurate for viscosities. More specifically, empirical formulae that are accurate for methane mixtures, for instance, may do a poor job for hydrogen mixtures. One should thus be very careful when using empirical expressions for viscosities.

5.5. Volume viscosity

The traditional transport linear system associated with volume viscosity is of size $n + n^{\mathrm{p}}$, where n^{p} is the number of polyatomic species. This linear system is associated with an approximation of ϕ^{κ} in (4.6.22) along the basis functions ϕ^{0010k}, $k \in S$, and ϕ^{0001k}, $k \in S_{\mathrm{pol}}$, presented in (4.7.10) [EG94] [EG95c]. A simplified system of size n^{p}, however, was obtained in [EG94] and is associated with the basis functions $\widehat{\phi}^{0001k}$, $k \in S_{\mathrm{pol}}$, presented in (4.7.13). Within the Monchick and Mason approximations [MM62] [MPM65], neglecting complex collisions with more than one quantum jump, this system is diagonal and yields the expression

$$\kappa = \sum_{k \in S_{\mathrm{pol}}} \Big(\frac{\mathfrak{c}_{vk}}{\mathfrak{c}_v}\Big)^2 \frac{\Omega_{kk}^{(2,2)} X_k}{\sum_{l \in S} X_l \Omega_{kl}^{(2,2)}} \kappa_k, \tag{5.5.1}$$

where $\Omega_{kl}^{(2,2)} = \sigma_{kl}^2 \sqrt{2\pi k_{\mathrm{B}} T / \mathfrak{m}_{kl}}\, \Omega_{kl}^{(2,2)*}$, $k, l \in S$, and κ_k denotes the volume viscosity of the kth species

$$\kappa_k = \frac{k_{\mathrm{B}} \pi}{4} \Big(\frac{\mathfrak{c}_k^{\mathrm{int}}}{\mathfrak{c}_{vk}}\Big)^2 \frac{\xi_k^{\mathrm{int}}}{c_k^{\mathrm{int}}} \eta_k, \qquad k \in S_{\mathrm{pol}}. \tag{5.5.2}$$

The following empirical formula also has a good quality/price ratio:

$$\kappa = \Big(\sum_{k \in S_{\mathrm{pol}}} X_k \kappa_k^{3/4}\Big)^{4/3}. \tag{5.5.3}$$

The expression (5.5.2) has a cost $\mathcal{O}(n^2)$, whereas the formula (5.5.3) has a cost $\mathcal{O}(n)$, and more accurate expressions can also be found in [EG94], [EG97].

Remark 5.5.1. The volume viscosity has been generally neglected in multicomponent flow models. Kinetic theory, however, indicates that volume viscosity is associated with internal energy relaxation of polyatomic gases. Experimental measurements using acoustic wave absorptions indicate that the ratio κ/η is around unity. At ambient temperature, this ratio is $\kappa/\eta \approx 1.3$ for methane, $\kappa/\eta \approx 0.6$ for nitrogen, and $\kappa/\eta \approx 34$ for hydrogen. The approximation $\kappa/\eta = 0$ is thus a crude approximation that has strictly no justification a priori.

The term that can be considered to be negligible in a number of situations is the complete group $\boldsymbol{\partial_x}\cdot(\kappa \boldsymbol{\partial_x}\cdot \boldsymbol{v}\, \boldsymbol{I})$. For small Mach number flows, for instance, the term $\boldsymbol{\partial_x}\cdot(\kappa \boldsymbol{\partial_x}\cdot \boldsymbol{v}\, \boldsymbol{I})$ has a weak influence over the flow structure—as the term $-\frac{2}{3}\boldsymbol{\partial_x}\cdot(\eta \boldsymbol{\partial_x}\cdot \boldsymbol{v}\, \boldsymbol{I})$—except for acoustic perturbations of the correction pressure $\widetilde{p}$. For steady solutions, it only influences the pressure correction $\widetilde{p}$. For unsteady problems, however, the volume viscosity κ induces an attenuation of acoustic waves of the same order of magnitude than the one induced by η. ■

5.6. Diffusion matrix

The standard transport linear systems associated with the multicomponent diffusion coefficients are of size $2n + n^{\mathrm{p}}$ and associated with an approximation of the ϕ^{D_j}, $j \in S$, along the basis functions ϕ^{1000k}, $k \in S$, ϕ^{1010k}, $k \in S$, and ϕ^{1001k}, $k \in S_{\mathrm{pol}}$. These transport linear systems and the corresponding expressions for diffusion coefficients are beyond the scope of this book. Simplified systems of size $2n$ have also been introduced [EG94] and will be discussed in Section 5.9. In this section we investigate the first-order diffusion coefficients $D = D_{[00]}$ associated with systems of size n. These transport linear systems are associated with approximations of the ϕ^{D_j}, $j \in S$, along the basis functions ϕ^{1000k}, $k \in S$.

The corresponding transport linear systems are the n systems of order n indexed by $l \in S$ and given by

$$\begin{cases} \Delta \alpha^{D_l} = \beta^{D_l}, \\ \alpha^{D_l} \in Y^{\perp}, \end{cases} \tag{5.6.1}$$

where $\Delta \in \mathbb{R}^{n,n}$ and $\alpha^{D_l}, \beta^{D_l}, Y \in \mathbb{R}^n$. The matrix Δ is given by

$$\Delta_{kk} = \sum_{\substack{l \in S \\ l \neq k}} \frac{X_k X_l}{\mathcal{D}_{kl}^{\mathrm{bin}}}, \qquad k \in S, \tag{5.6.2}$$

$$\Delta_{kl} = -\frac{X_k X_l}{\mathcal{D}_{kl}^{\mathrm{bin}}}, \qquad k, l \in S, \quad k \neq l, \tag{5.6.3}$$

where $\mathcal{D}_{kl}^{\mathrm{bin}}$ is the binary diffusion coefficient for species pair (k, l), which only depends on temperature and pressure $\mathcal{D}_{kl}^{\mathrm{bin}} = \mathcal{D}_{kl}^{\mathrm{bin}}(T, p)$. More generally, for more accurate multicomponent diffusion coefficients, the quantities $\mathcal{D}_{kl}^{\mathrm{bin}}$, $k, l \in S$, are replaced by Schur complements from transport linear systems of size larger than n, and then depend on T, p, and Y, but have similar properties [Gio91] [EG94]. The right members β^{D_l}, $l \in S$, are given by

$$\beta_k^{D_l} = \delta_{kl} - \frac{Y_k}{\sum_{i \in S} Y_i}, \qquad k \in S, \tag{5.6.4}$$

and the vector $Y \in \mathbb{R}^n$ by $Y = (Y_1, \ldots, Y_n)^t$. Finally, the diffusion coefficients D_{kl}, $k, l \in S$, are evaluated from

$$D_{kl} = \langle \alpha^{D_l}, \beta^{D_k} \rangle = \langle \alpha^{D_k}, \beta^{D_l} \rangle = \alpha_k^{D_l} = \alpha_l^{D_k}, \qquad k, l \in S. \tag{5.6.5}$$

One can easily deduce from the preceding relations that

$$\boldsymbol{d}_k + \chi_k \boldsymbol{\partial_x} \log T = \sum_{\substack{l\in S\\ l\neq k}} \frac{X_k X_l}{\mathcal{D}_{kl}^{\text{bin}}} \boldsymbol{\mathcal{V}}_l - \Big(\sum_{\substack{l\in S\\ l\neq k}} \frac{X_k X_l}{\mathcal{D}_{kl}^{\text{bin}}}\Big) \boldsymbol{\mathcal{V}}_k, \qquad k \in S. \tag{5.6.6}$$

These equations are usually termed Stefan–Maxwell equations in the literature and must be completed by the constraint $\sum_{k\in S} Y_k \boldsymbol{\mathcal{V}}_k = 0$ associated with mass conservation in order to define uniquely the diffusion velocities $\boldsymbol{\mathcal{V}}_k$, $k \in S$. An elementary derivation of these equations has been given by Williams [Wil58]. In the following, we work directly with the transport coefficients D rather than with the vectors $\boldsymbol{\mathcal{V}}_k$, $k \in S$.

One can establish that the matrix Δ is symmetric positive semidefinite, and that $N(\Delta) = \mathbb{R}\mathcal{U}$ with $\mathcal{U} = (1,\dots,1)^t$, $R(\Delta) = \mathcal{U}^\perp$, and $\beta^{D_l} \in R(\Delta)$, $l \in S$. In addition, when $n \geq 3$, $2\text{diag}(\Delta) - \Delta$ is symmetric positive definite. Using these properties, one can then show that the n systems (5.6.1) are well posed and the matrix D is the unique matrix, such that $D\Delta D = D$, $\Delta D \Delta = \Delta$, $R(D) = Y^\perp$, and $N(D) = \mathbb{R}Y$. This matrix D is symmetric positive semidefinite, positive definite over $\mathcal{U}^\perp$, and irreducible. For any $a > 0$, one also has the relation $D = (\Delta + aY\otimes Y)^{-1} - (1/a)\mathcal{U}\otimes\mathcal{U}$ [Gio90] [Gio91]. All of these results will be carefully established in Chapter 7.

In order to avoid matrix inversions, one can expand the matrix D and define approximate diffusion coefficients by truncation. More specifically, consider the matrix $\mathcal{M} = \text{diag}(D_1^*/X_1, \dots, D_n^*/X_n)$ with

$$D_k^* = \frac{1 - Y_k / \sum_{l\in S} Y_l}{\sum_{\substack{l\in S\\ l\neq k}} X_l / \mathcal{D}_{kl}^{\text{bin}}} \tag{5.6.7}$$

and introduce the associated matrix splitting $\Delta = \mathcal{M} - \mathcal{Z}$. Let $\mathcal{T} = \mathcal{M}^{-1}\mathcal{Z}$ be the iteration matrix and P be the projector over $Y^\perp$ parallel to $\mathbb{R}\mathcal{U}$, that is, $P = I - Y\otimes\mathcal{U}/\langle Y,\mathcal{U}\rangle$. Then the spectral radius of the product $P\mathcal{T}$ is strictly lower than unity and we have the series expansion [Gio91]

$$D = \sum_{j=0}^{\infty} (P\mathcal{T})^j P\mathcal{M}^{-1}P^t.$$

As a consequence, we can introduce the approximate diffusion matrices $D^{[i]}$, $i \geq 0$, given by

$$D^{[i]} = \sum_{j=0}^{i} (P\mathcal{T})^j P\mathcal{M}^{-1}P^t, \tag{5.6.8}$$

and $D^{[i]}$, $i \geq 0$, are symmetric positive semidefinite and satisfy the properties $N(D^{[i]}) = \mathbb{R}Y$ and $R(D^{[i]}) = Y^\perp$. Each approximate diffusion matrix

$D^{[i]}$ thus satisfies the mass constraint $D^{[i]}Y = 0$ and yields a positive entropy production over the hyperplane of zero-sum diffusion driving forces $\mathcal{U}^{\perp}$.

The first term of the series expansion $D^{[0]} = P\mathcal{M}^{-1}P^t$ corresponds to the Hirschfelder–Curtiss approximation [HC49] with a mass corrector [OB81]. Indeed, it yields the following expression for the mass diffusion velocities:

$$\mathcal{V}_k = -\frac{D_k^*}{X_k}\boldsymbol{d}_k + \mathcal{V}_{\rm cor}, \tag{5.6.9}$$

where the correction velocity $\mathcal{V}_{\rm cor}$ ensures the mass conservation constraint $\sum_{k\in S} Y_k\mathcal{V}_k = 0$, which arises here as a direct result of the projection operator P [Gio91] [EG94].

The second term $D^{[1]}$ in the series expansion yields an approximation first introduced in [Gio91] and given explicitly in [EGS97] for modeling chemical vapor deposition reactors. It may be written in the form $D^{[1]} = P\mathsf{D}P^t$ with the matrix D given by

$$\mathsf{D}_{ii} = \frac{D_i^*}{X_i}(1+Y_i), \qquad i\in S, \tag{5.6.10}$$

$$\mathsf{D}_{ij} = \frac{D_i^* D_j^*}{\mathcal{D}_{ij}^{\rm bin}}, \qquad i,j\in S, \quad i\neq j. \tag{5.6.11}$$

The approximation $D^{[1]}$ is much more accurate then $D^{[0]}$ and highly recommended. The cost of this approximation still scales as $O(n^2)$ and yields the n^2 diffusion coefficients. At the next order of approximation, the cost for evaluating $D^{[i]}$, $i\geq 3$, then becomes $O(n^3)$ for the n^2 diffusion coefficients.

5.7. Thermal conductivity

The simplest transport linear system associated with thermal conductivity is of size n. This transport linear system corresponds to an approximation of $\boldsymbol{\phi}^\lambda$ along the basis functions ϕ^{10ek}, $k\in S$, defined in (4.7.14). The basis functions ϕ^{10ek}, $k\in S$, are associated with the sum of the kinetic and internal energy of the molecules [EG94] [EG95a]. The resulting system can be written in the form

$$\Lambda_{[\rm e]}\alpha_{[\rm e]}^\lambda = \beta_{[\rm e]}^\lambda, \tag{5.7.1}$$

with

$$\Lambda_{[\rm e]kk} = \sum_{\substack{l\in S\\ l\neq k}} \frac{X_k X_l}{\mathcal{D}_{kl}^{\rm bin}} \frac{\mathfrak{m}_k\mathfrak{m}_l}{(\mathfrak{m}_k+\mathfrak{m}_l)^2}\Big[\frac{15}{2}\frac{\mathfrak{m}_k}{\mathfrak{m}_l} + \frac{25}{4}\frac{\mathfrak{m}_l}{\mathfrak{m}_k} - 3\frac{\mathfrak{m}_l}{\mathfrak{m}_k}\bar{B}_{kl} + 4\bar{A}_{kl}$$

$$+\frac{4}{15}\frac{(3\mathfrak{m}_k-2\mathfrak{m}_l)^2}{\mathfrak{m}_l^2}\frac{\bar{A}_{kl}}{k_B\pi}\frac{c_k^{\text{int}}}{\xi_k^{\text{int}}}+\frac{20}{3}\frac{\bar{A}_{kl}}{k_B\pi}\frac{c_l^{\text{int}}}{\xi_l^{\text{int}}}+\frac{(\mathfrak{m}_k+\mathfrak{m}_l)^2}{\mathfrak{m}_k\mathfrak{m}_l}\frac{\mathfrak{c}_k^{\text{int}}\mathcal{D}_{kl}^{\text{bin}}}{k_B\mathcal{D}_{k\,\text{int},l}^{\text{bin}}}\Big]$$

$$+\frac{X_k^2}{\mathcal{D}_{kk}^{\text{bin}}}\Big[2\bar{A}_{kk}+\frac{16}{15}\frac{\bar{A}_{kk}}{k_B\pi}\frac{c_k^{\text{int}}}{\xi_k^{\text{int}}}+\frac{\mathfrak{c}_k^{\text{int}}\mathcal{D}_{kk}^{\text{bin}}}{k_B\mathcal{D}_{k\,\text{int},k}^{\text{bin}}}\Big],\qquad k\in S,\tag{5.7.2}$$

$$\Lambda_{[e]kl}=-\frac{X_kX_l}{\mathcal{D}_{kl}^{\text{bin}}}\frac{\mathfrak{m}_k\mathfrak{m}_l}{(\mathfrak{m}_k+\mathfrak{m}_l)^2}\Big[\frac{55}{4}-3\bar{B}_{kl}-4\bar{A}_{kl}+\frac{4}{3}\frac{\bar{A}_{kl}}{k_B\pi}\frac{3\mathfrak{m}_k-2\mathfrak{m}_l}{\mathfrak{m}_l}\frac{c_k^{\text{int}}}{\xi_k^{\text{int}}}$$
$$+\frac{4}{3}\frac{\bar{A}_{kl}}{k_B\pi}\frac{3\mathfrak{m}_l-2\mathfrak{m}_k}{\mathfrak{m}_k}\frac{c_l^{\text{int}}}{\xi_l^{\text{int}}}\Big],\qquad k,l\in S,\quad k\neq l,\tag{5.7.3}$$

and

$$\beta_{[e]}^{\lambda}=\frac{\mathfrak{c}_p^{\text{tr}}+\mathfrak{c}_k^{\text{int}}}{k_B}X_k,\qquad k\in S.\tag{5.7.4}$$

The resulting thermal conductivity $\lambda=\lambda_{[e]}$ is then given by

$$\lambda_{[e]}=\frac{p}{T}\langle\alpha_{[e]}^{\lambda},\beta_{[e]}^{\lambda}\rangle.\tag{5.7.5}$$

One can establish that this system is symmetric positive definite [EG94]. By using one step of the conjugate gradient procedure, very accurate approximations can be obtained [EG94].

The following empirical relation also yields interesting results:

$$\lambda=\Big(\sum_{k\in S}X_k\lambda_k^{1/4}\Big)^4.\tag{5.7.6}$$

This formula generally yields better results than the classical average between the arithmetic and harmonic means.

5.8. Thermal diffusion ratios

Once the linear system associated with the thermal conductivity has been solved, one can readily evaluate the thermal diffusion ratios $\chi=\chi_{[e]}$ from the relations

$$\chi_{[e]}=\mathfrak{L}_{[e]}\alpha_{[e]}^{\lambda},\tag{5.8.1}$$

where the matrix $\mathfrak{L}_{[e]}$ is given by

$$\mathfrak{L}_{[e]kk}=-\sum_{\substack{l\in S\\ l\neq k}}\frac{X_kX_l}{2\mathcal{D}_{kl}^{\text{bin}}}\frac{\mathfrak{m}_l}{\mathfrak{m}_k+\mathfrak{m}_l}(6\bar{C}_{kl}-5),\qquad k\in S,\tag{5.8.2}$$

$$\mathfrak{L}_{[e]kl}=\frac{X_kX_l}{2\mathcal{D}_{kl}^{\text{bin}}}\frac{\mathfrak{m}_k}{\mathfrak{m}_k+\mathfrak{m}_l}(6\bar{C}_{kl}-5),\qquad k,l\in S,\quad k\neq l.\tag{5.8.3}$$

The rescaled thermal diffusion ratios $\widetilde{\chi} = \widetilde{\chi}_{[e]}$ are evaluated in a similar way by omitting the X_k factor in (5.8.2) and (5.8.3). Thermal diffusion ratios and thermal diffusion coefficients are always difficult to approximate correctly and require two or three steps of the conjugate gradient procedure for $\alpha^{\lambda}_{[e]}$ [EG94].

5.9. Partial thermal conductivity

The partial thermal conductivity can also be evaluated by solving the linear system of size $2n$

$$\begin{cases} L_{[e]}\alpha^{\widehat{\lambda}}_{[e]} = \beta^{\widehat{\lambda}}_{[e]}, \\ \qquad \alpha^{\widehat{\lambda}}_{[e]} \in (Y, \mathcal{O})^{\perp}, \end{cases} \tag{5.9.1}$$

where $\mathcal{O} = (0, \ldots, 0)^t$ and $L_{[e]}$ has the block structure

$$L_{[e]} = \begin{pmatrix} \Delta & \mathfrak{L}_{[e]} \\ \mathfrak{L}^t_{[e]} & \Lambda_{[e]} \end{pmatrix} \tag{5.9.2}$$

and

$$\beta^{\widehat{\lambda}}_{[e]} = (\mathcal{O}, \beta^{\lambda}_{[e]})^t. \tag{5.9.3}$$

The partial thermal conductivity $\widehat{\lambda} = \widehat{\lambda}_{[e]}$ is then given by

$$\widehat{\lambda}_{[e]} = \frac{p}{T}\langle \alpha^{\widehat{\lambda}}_{[e]}, \beta^{\widehat{\lambda}}_{[e]} \rangle. \tag{5.9.4}$$

One can establish that this system is symmetric positive semidefinite [EG94]. By using one or two steps of the conjugate gradient procedure, very accurate approximations can be obtained [EG94]. Note, however, that evaluating the thermal conductivity λ is cheaper than evaluating the partial thermal conductivity $\widehat{\lambda}$.

Remark 5.9.1. By solving the transport linear systems of size $2n$ in the form

$$\begin{cases} L_{[e]}\alpha^{D_l}_{[e]} = \beta^{D_l}_{[e]}, \\ \qquad \alpha^{D_l}_{[e]} \in (Y, \mathcal{O})^{\perp}, \end{cases} \qquad l \in S, \tag{5.9.5}$$

where $L_{[e]}$ is given by (5.9.2) and

$$\beta^{D_l}_{[e]} = (\beta^{D_l}, \mathcal{O})^t, \tag{5.9.6}$$

and defining

$$D_{[e]kl} = \langle \alpha^{D_l}_{[e]}, \beta^{D_k}_{[e]} \rangle = \langle \alpha^{D_k}_{[e]}, \beta^{D_l}_{[e]} \rangle, \tag{5.9.7}$$

one can obtain a more accurate diffusion matrix $D = D_{[e]}$ [EG94]. This matrix can also be expanded as a convergent series as for the Stefan–Maxwell system, and we refer the reader to [EG94] for more details. However, the matrix $D_{[e]}$ does not significantly differ from the first-order diffusion matrix $D = D_{[00]}$ defined in (5.6.1)–(5.6.5).

The partial thermal conductivity $\widehat{\lambda} = \widehat{\lambda}_{[e]}$ can then be evaluated from the thermal conductivity $\lambda = \lambda_{[e]}$ and the thermal diffusion ratios $\chi = \chi_{[e]}$ by using the relation

$$\widehat{\lambda}_{[e]} = \lambda_{[e]} + \frac{p}{T} \sum_{k,l \in S} D_{[e]kl} \chi_{[e]k} \chi_{[e]l}, \tag{5.9.8}$$

which is a special case of (2.5.21) and (4.6.46). ∎

5.10. Thermal diffusion coefficients

Once the partial thermal conductivity has been evaluated, the thermal diffusion ratios $\theta = \theta_{[e]}$ are readily obtained from

$$\theta_{[e]k} = -\langle \alpha_{[e]}^{\widehat{\lambda}}, \beta_{[e]}^{D_k} \rangle, \qquad k \in S. \tag{5.10.1}$$

Note that we have the relation

$$\theta_{[e]} = D_{[e]} \chi_{[e]}, \tag{5.10.2}$$

where $D_{[e]}$ is defined as in (5.9.7). From a theoretical point of view, only the coefficients obtained with a consistent set of approximations, that is, a consistent set of basis functions, satisfy the relations (5.9.8) and (5.10.2). We have found in practice, however, that $D_{[e]}$ and the matrix $D_{[00]}$ defined as in (5.6.1)–(5.6.5) are very close.

5.11. Notes

5.1. Numerical evaluations of multicomponent transport properties using Gaussian elimination have been considered by Dixon-Lewis [Dix67] [Dix68], Coffee and Heimerl [CH81], Kee et al. [K*al*86], Lebedev [Leb91], and Warnatz [War82]. These authors, however, have used nonsymmetric formulations that are unduly complex in comparison with the symmetric positive definite formulations obtained in [EG94].

5.2. Solving transport linear systems by direct methods is computationally expensive in multicomponent flow simulations. Indeed, the size of the transport linear systems can be relatively large, and transport properties have to be evaluated at each computational cell in space and time. As a consequence, the use of iterative techniques is an interesting alternative. For diffusion velocities, which involve the solution of a constrained singular system, iterative schemes have been introduced by Jones and Boris [JB81], who have shown numerically the convergence of their algorithms. In order to select the proper diffusion velocities, a corrector term needs to be added after convergence [JB81] [OB81]. This corrector term was written explicitly by Oran and Boris [OB81] and corresponds to applying a projector matrix, as shown in [Gio91]. The convergence of the Jones–Oran–Boris algorithm has been proven rigorously by the author [Gio91], who also established that the corresponding iteration matrix has a spectral radius unity. Additional algorithms have been introduced for the multicomponent diffusion matrices in [Gio91], for which the iteration matrix has a spectral radius strictly lower than unity as in Section 5.6. As a consequence, the first-order multicomponent diffusion matrix can be written as a symmetric convergent series, for which each partial sum is symmetric, conserves mass, and yields a positive entropy production on the hyperplane of zero-sum driving forces [Gio91].

The mathematical and numerical theory of iterative algorithms for solving the transport linear systems presented in Chapters 4 and 5 has been given in [EG94]. The structure of the linear systems was deduced from the Boltzmann equation and, in particular, from the properties of the kinetic integral bracket operator. These results have also been extended to the singular case of vanishing mass fractions by introducing appropriately rescaled transport linear systems. Matrices obtained with inexact collision integrals resulting from practical approximations [MM62] [MPM65] are shown to possess the mathematical structure deduced from the kinetic theory, provided that the approximate parameters are in a suitable admissible set. Finally, numerical tests have been performed for various gas mixtures arising in combustion applications, and the optimization of transport property evaluation in multicomponent flow calculations has also been discussed [EG94] [EG95a] [EG95b] [EG95c] [EG96a] [EG96b] [EG96c] [EG96d] [EG97] [EG98].

5.3. The present transport algorithms, with many extensions, are included in a Fortran library available for academic purposes [EG96d]. This library contains not only single mixture routines but also highly optimized multiple mixtures routines that typically correspond to computational cells for vector and parallel architectures.

5.12. References

[CC70] S. Chapman and T. G. Cowling, *The Mathematical Theory of Non-Uniform Gases,* Cambridge University Press, Cambridge, (1970).

[CH81] T. P. Coffee and J. M. Heimerl, *Transport Algorithms for Premixed, Laminar Steady-State Flames,* Comb. Flame, **43**, (1981), pp. 273–289.

[Dix67] G. Dixon-Lewis, *Flame Structure and Flame Reaction Kinetics, I. Solution of Conservation Equations and Application to Rich Hydrogen-Oxygen Flames,* Proc. Roy. Soc., **A 298**, (1967), pp. 495–513.

[Dix68] G. Dixon-Lewis, *Flame Structure and Flame Reaction Kinetics, II. Transport Phenomena in Multicomponent Systems,* Proc. Roy. Soc., **A 307**, (1968), pp. 111–135.

[Dix84] G. Dixon-Lewis, *Computer Modeling of Combustion Reactions in Flowing Systems with Transport,* in W. C. Gardiner, Ed., *Combustion Chemistry.* Springer-Verlag, New York, (1984), pp. 21–125.

[EG94] A. Ern and V. Giovangigli, *Multicomponent Transport Algorithms,* Lecture Notes in Physics, New Series "Monographs," **m 24**, Springer-Verlag, Berlin, (1994).

[EG95a] A. Ern and V. Giovangigli, *Thermal Conduction and Thermal Diffusion in Dilute Polyatomic Gas Mixtures,* Physica-A, **214**, (1995), pp. 526–546.

[EG95b] A. Ern and V. Giovangigli, *Fast and Accurate Multicomponent Property Evaluations,* J. Comp. Physics, **120**, (1995), pp. 105–116.

[EG95c] A. Ern and V. Giovangigli, *Volume viscosity of Dilute Polyatomic Gas Mixtures,* Eur. J. Mech., B/Fluids, **14**, (1995), pp. 653–669.

[EG96a] A. Ern and V. Giovangigli, *The Structure of Transport Linear Systems in Dilute Isotropic Gas Mixtures,* Phys. Rev. E, **53**, (1996), pp. 485–492.

[EG96b] A. Ern and V. Giovangigli, *Optimised Transport Algorithms for Flame Codes,* Comb. Sci. Tech., **118**, (1996), pp. 387–395.

[EG96c] A. Ern and V. Giovangigli, *Kinetic Theory of Dilute Gas Mixtures with Independent Energy Modes near Equilibrium,* Physica-A, **224**, (1996), pp. 613–625.

[EG96d] A. Ern and V. Giovangigli, *EGlib server and user's manual,* http://www.cmap.polytechnique.fr/www.eglib/

[EG97] A. Ern and V. Giovangigli, *Projected Iterative Algorithms with Application to Multicomponent Transport,* Linear Algebra Appl., **250**, (1997), pp. 289–315.

[EGS97] A. Ern, V. Giovangigli, and M. Smooke, *Detailed Modeling of Three-Dimensional Chemical Vapor Deposition,* J. Crystal Growth, **180**, (1997), pp. 670–679.

[FK72] J. H. Ferziger and H. G. Kaper, *Mathematical Theory of Transport Processes in Gases,* North Holland Publishing Company, Amsterdam, (1972).

[Gio90] V. Giovangigli, *Mass Conservation and Singular Multicomponent Diffusion Algorithms*, IMPACT Comput. Sci. Eng., **2**, (1990), pp. 73–97.

[Gio91] V. Giovangigli, *Convergent Iterative Methods for Multicomponent Diffusion,* IMPACT Comput. Sci. Eng., **3**, (1991), pp. 244–276.

[HC49] J. O. Hirschfelder and C. F. Curtiss, *Flame Propagation in Explosive Gas Mixtures,* in Third Symposium (International) on Combustion, Reinhold, New York, (1949), pp. 121–127.

[HCB54] J. O. Hirschfelder, C. F. Curtiss, and R. B. Bird, *Molecular Theory of Gases and Liquids,* John Wiley & Sons, Inc., New York, (1954).

[JB81] W. W. Jones and J. P. Boris, *An Algorithm for Multispecies Diffusion Fluxes.* Comput. Chem., **5**, (1981), pp. 139–146.

[K*al*86] R. J. Kee, G. Dixon-Lewis, J. Warnatz, M. E. Coltrin, and J. A. Miller, *A Fortran Computer Code Package for the Evaluation of Gas-Phase Multicomponent Transport Properties.* SANDIA National Laboratories Report, SAND86-8246, (1986).

[KWM83] R. J. Kee, J. Warnatz, and J. A. Miller, *A Fortran Computer Code Package for the Evaluation of Gas-Phase Viscosities, Conductivities, and Diffusion Coefficients,* SANDIA National Laboratories Report, SAND83-8209, (1983).

[Leb91] V. N. Lebedev, *Numarical Calculation of the Quenching of Stretched Laminar Flames in H_2/Air and CH_4/Air Mixtures,* Fiz. Goreniya Vzryva (Comb. Explos. Shock Waves), **27**, (1991), pp. 58–62.

[MM61] L. Monchick and E. A. Mason, *Transport Properties of Polar Gases,* J. Chem. Phys., **35**, (1961), pp. 1676–1697.

[MM62] E. A. Mason and L. Monchick, *Heat Conductivity of Polyatomic and Polar Gases,* J. Chem. Phys., **36**, (1962), pp. 1622–1639.

[MYM63] L. Monchick, K. S. Yun, and E. A. Mason, *Formal Kinetic Theory of Transport Phenomena in Polyatomic Gas Mixtures,* J. Chem. Phys., **39**, (1963), pp. 654–669.

[MPM65] L. Monchick, A. N. G. Pereira, and E. A. Mason, *Heat Conductivity of Polyatomic and Polar Gases and Gas Mixtures,* J. Chem. Phys., **42**, (1965), pp. 3241–3256.

[OB81] E. S. Oran and J. P. Boris, *Detailed Modeling of Combustion Systems,* Prog. Energy Combust. Sci., **7**, (1981), pp. 1–72.

[Par59] J. G. Parker, *Rotational and Vibrational Relaxation in Diatomic Gases,* Phys. of Fluids, **2**, (1959), pp. 449–462.

[War82] J. Warnatz, *Influence of Transport Models and Boundary Conditions on Flame Structure,* In N. Peters and J. Warnatz, Eds., Numerical Methods in Laminar Flame Propagation, Vieweg-Verlag, Braunschweig, (1982), pp. 87–111.

[WMD96] J. Warnatz, U. Maas, and R. W. Dibble, *Combustion,* Springer-Verlag, Berlin, (1996).

[Wlk50] C. R. Wilke, *A Viscosity Equation for Gaz Mixtures*, J. Chem. Phys., **18**, (1950), pp. 517–519.

[Wil58] F. A. Williams, *Elementary Derivation of the Multicomponent Diffusion Equation,* Amer. J. Phys., **26**, (1958), pp. 467–469.

6

Mathematics of Thermochemistry

6.1. Introduction

In this chapter we investigate the mathematical properties of thermodynamic functions and chemistry source terms. We first specify the mathematical assumptions associated with thermodynamics and investigate smoothness, convexity and differentials of various functionals. A fundamental alternative arises, however, concerning the mathematical thermodynamic formalism to be used.

Two types of thermodynamic formalisms can indeed be obtained. More specifically, by using the state variables $(T, \rho_1, \ldots, \rho_n)$, the entropy is found to be nonhomogeneous with respect to $(\rho_1, \ldots, \rho_n)$, but entropy Hessians are found to be nonsingular. On the other hand, by using the state variables $(T, p, Y_1, \ldots, Y_n)$, homogeneity of entropy is maintained with respect to the corresponding species variables $(Y_1, \ldots, Y_n)$ and the classical Gibbs relations are recovered, but entropy Hessians are found to be singular. Both formalisms are useful—depending on the context—and can be found in the literature. As a consequence, both mathematical formalisms are presented in this chapter.

The first formalism, associated with the state variables $(T, \rho_1, \ldots, \rho_n)$, is somewhat more natural and simpler. It has been used, in particular, by Meixnier [Mei43], Krambeck [Kra70], and Giovangigli and Massot [GM98a] and will be used in Chapters 8, 9, and 10. On the other hand, the second formalism, associated with the state variables $(T, p, Y_1, \ldots, Y_n)$, is equivalent to Gibbs formalism expressed in terms of intensive variables or densities. This formalism has been used for instance by de Groot and Mazur [dGM84], Boillat [Boi95], Pousin [Pou93], and Giovangigli [Gio99]. This formalism is especially suited to problems in which the mass fractions appear as the natural variables and will be used, in particular, in Chapter 11 for investigating a plane flame model.

The origin of these multiple formalisms is that, in nonhomogeneous flows, extensivity is associated with volume integration and state variables can only be intensive variables or densities, either volumetric, massic, or molar. A direct consequence is that any set of $n+2$ intensive state variables—which are associated with independent extensive state variables after volume integration in homogeneous flows—are now *dependent* variables. Eliminating one of these density state variables then leads to $n+1$ independent intensive state variables and to positive definite Hessians for the entropy. This procedure leads to the first thermochemistry formalism investigated in Section 6.2. On the other hand, considering the a priori dependent mass fractions as formally independent lead to expressions similar to Gibbs thermodynamics and to degenerate Hessians for the entropy. This yields the second thermochemistry formalism investigated in Section 6.3.

In Section 6.4, we then specify the mathematical assumptions concerning the chemical production rates. The chemical production rates that we consider are the Maxwellian rates derived from the kinetic theory. These rates are compatible with the law of mass action. The rate constants satisfy reciprocity relations, which are consequences of reciprocity relations satisfied by transition probabilities appearing in Boltzmann chemical source terms. These reciprocity relations imply in particular that the entropy production due to chemical reactions is nonnegative.

We then investigate chemical equilibrium states in atom conservation manifolds in Section 6.5. Equilibrium states are obtained by a minimization technique and detailed balance is obtained at equilibrium. Detailed balance at equilibrium is again a consequence of the reciprocity relations satisfied by the reaction rate constants. Note that both thermodynamic formalisms can be used to investigate existence and uniqueness of equilibrium points.

In Section 6.6, we also examine the existence of equilibrium states with zero mass fractions, although there also exists a positive equilibrium point. These boundary equilibrium points depend on the actual values of the forward and backward stoichiometric coefficients, whereas positive equilibrium points only depend on the linear space spanned by the reaction vectors. We introduce, in particular, a natural assumption concerning the chemical mechanism—the species decomposition chain property—which automatically eliminates these spurious boundary equilibrium points.

We investigate various inequalities concerning thermodynamic functions that vanish at chemical equilibrium. In particular, we compare the production rates, entropy difference with equilibrium, and chemical dissipation rate. These inequalities are first established locally around equilibrium points in Section 6.7, but some inequalities are also globally valid on atom conservation manifolds, as investigated in Section 6.8.

Finally note that numerous state variables are used in this chapter. As a consequence, in order to avoid notational complexities, we will frequently commit small abuses of notation by denoting by the same symbol a given quantity as function of different state variables, unless explicitly mentioned. In addition, for any vector x with components $x = (x_1, \ldots, x_n)^t$, we write $x \geq 0$, when $x_k \geq 0$, $k \in S$, and $x > 0$, when $x_k > 0$, $k \in S$.

6.2. Thermodynamics with volume densities

In this section, we investigate thermodynamic properties as a function of the state variables $(T, \rho_1, \ldots, \rho_n)$. In agreement with the volumetric definition of the species densities $(\rho_1, \ldots, \rho_n)$, we consider thermodynamic properties per unit volume. Of course, we could have used the species concentrations $\gamma_k = \rho_k/m_k$, $k \in S$, or the species partial pressures $p_k = \rho_k RT/m_k$, $k \in S$, instead of the mass densities ρ_k, $k \in S$, but the resulting formalism would be very similar to the one presented in this section. This formalism has been used in particular by Meixnier [Mei43] and Krambeck [Kra70] and is especially interesting for its simplicity. It will be used in Chapters 8, 9, and 10 when investigating asymptotic stability of equilibrium states for multicomponent reactive flows. We point out, however, that this formalism differs from that of Gibbs [SS65].

6.2.1. State variables $(T, \rho_1, \ldots, \rho_n)$

For convenience, we will use the compact notation

$$\varrho = (\rho_1, \ldots, \rho_n)^t, \tag{6.2.1}$$

so that the state variables are (T, ϱ). The pressure p is then defined from

$$p = RT \sum_{k \in S} \frac{\rho_k}{m_k}, \tag{6.2.2}$$

and the total density ρ is defined from

$$\rho = \sum_{k \in S} \rho_k. \tag{6.2.3}$$

We also define the mean molecular weight $\overline{m}$ as

$$\frac{\rho}{\overline{m}} = \sum_{k \in S} \frac{\rho_k}{m_k}, \tag{6.2.4}$$

and the mass fractions from

$$Y_k = \frac{\rho_k}{\rho}, \qquad k \in S. \tag{6.2.5}$$

Similarly, we define the mole fraction X_k, partial pressure p_k, and concentration γ_k of the kth species from

$$X_k = Y_k \frac{\overline{m}}{m_k}, \qquad p_k = RT \frac{\rho_k}{m_k}, \qquad \gamma_k = \frac{\rho_k}{m_k}, \tag{6.2.6}$$

and the constraints $\sum_{k \in S} Y_k = 1$ and $\sum_{k \in S} X_k = 1$ are then *automatically* satisfied.

6.2.2. Energy and enthalpy per unit volume

We define the energy per unit volume $\mathcal{E}$ from

$$\mathcal{E}(T, \varrho) = \sum_{k \in S} \rho_k e_k(T) \tag{6.2.7}$$

and write the specific energy e_k of the kth species in the form

$$e_k(T) = e_k^{\text{st}} + \int_{T^{\text{st}}}^{T} c_{vk}(T')\, dT'. \tag{6.2.8}$$

It will be convenient to define the formation energy e_k^0 of the kth species at zero temperature

$$e_k^0 = e_k^{\text{st}} - \int_0^{T^{\text{st}}} c_{vk}(T')\, dT', \tag{6.2.9}$$

in such a way that

$$e_k(T) = e_k^0 + \int_0^{T} c_{vk}(T')\, dT'. \tag{6.2.10}$$

Similarly, we define the enthalpy per unit volume from

$$\mathcal{H}(T, \varrho) = \sum_{k \in S} \rho_k h_k(T), \tag{6.2.11}$$

where h_k is defined from

$$h_k(T) = e_k(T) + r_k T, \qquad T \geq 0, \tag{6.2.12}$$

and $r_k = R/m_k$. We define the constant-pressure specific heats from

$$c_{pk}(T) = c_{vk}(T) + r_k, \qquad k \in S, \tag{6.2.13}$$

so that

$$h_k(T) = h_k^{\text{st}} + \int_{T^{\text{st}}}^{T} c_{pk}(T')\, dT' = h_k^0 + \int_0^{T} c_{pk}(T')\, dT', \tag{6.2.14}$$

and $h_k(0) = h_k^0 = e_k^0$ is the formation enthalpy per unit mass at zero temperature. A straightforward calculation also yields that

$$\mathcal{H}(T,\varrho) = \mathcal{E}(T,\varrho) + p(T,\varrho). \tag{6.2.15}$$

6.2.3. Entropy and Gibbs function per unit volume

The entropy per unit volume of the mixture $\mathcal{S}$ is defined in terms of the species specific enthalpies $\mathcal{S}_k$, $k \in S$, from the relation

$$\mathcal{S}(T,\varrho) = \sum_{k\in S} \rho_k \mathcal{S}_k(T,\rho_k), \tag{6.2.16}$$

and the specific entropy $\mathcal{S}_k$ of the kth species is written in the form

$$\mathcal{S}_k(T,\rho_k) = s_k^{\mathrm{st}} + \int_{T^{\mathrm{st}}}^{T} \frac{c_{vk}(T')}{T'}\, dT' - r_k \log\left(\frac{\rho_k}{\gamma^{\mathrm{st}} m_k}\right), \tag{6.2.17}$$

where s_k^{st} is the formation entropy of the kth species at the standard temperature T^{st} and standard pressure p^{st} and the quantity $\gamma^{\mathrm{st}} = p^{\mathrm{st}}/RT^{\mathrm{st}}$ is the standard concentration. Note that we have introduced the new symbol $\mathcal{S}_k$ to denote the specific entropy of the kth species as function of (T,ρ_k). Of course, we have $\mathcal{S}_k(T,\rho_k) = s_k(T,p,Y)$, where $Y = (Y_1,\ldots,Y_n)^t$, but we have preferred to use different symbols to avoid confusion between differentials.

Similarly, we can express the Gibbs function per unit volume of the mixture $\mathcal{G}$ in terms of the species Gibbs functions $\mathcal{G}_k$, $k \in S$, from the relation

$$\mathcal{G}(T,\varrho) = \sum_{k\in S} \rho_k \mathcal{G}_k(T,\rho_k), \tag{6.2.18}$$

and the specific Gibbs function $\mathcal{G}_k$ of the kth species is given by

$$\mathcal{G}_k(T,\rho_k) = h_k(T) - T\mathcal{S}_k(T,\rho_k). \tag{6.2.19}$$

Note again that we have $\mathcal{G}_k(T,\rho_k) = g_k(T,p,Y)$, but we use different symbols since the variables differ.

We also have

$$\mathcal{G}(T,\varrho) = \mathcal{H}(T,\varrho) - T\mathcal{S}(T,\varrho). \tag{6.2.20}$$

It will be convenient to define the reduced chemical potentials

$$\mu_k(T,\rho_k) = \frac{\mathcal{G}_k(T,\rho_k)}{RT}, \qquad \mu_k^{\mathrm{u}}(T) = \frac{\mathcal{G}_k(T,m_k)}{RT}, \qquad k \in S, \tag{6.2.21}$$

which can be split into

$$\mu_k(T,\rho_k) = \mu_k^{\mathrm{u}}(T) + \frac{1}{m_k}\log \gamma_k, \qquad k \in S. \tag{6.2.22}$$

We also define the corresponding vectors

$$\mu = (\mu_1,\ldots,\mu_n)^t, \qquad \mu^{\mathrm{u}} = (\mu_1^{\mathrm{u}},\ldots,\mu_n^{\mathrm{u}})^t, \tag{6.2.23}$$

which often appear in thermochemistry expressions, in particular, in the chemical production rates.

6.2.4. Assumptions

The following mathematical assumption concerns the volumetric thermodynamic properties.

(Th_1) The species molar masses m_k, $k \in S$, and the gas constant R are positive constants. The formation energies e_k^{st}, $k \in S$, and the formation entropies s_k^{st}, $k \in S$, are constants. The specific heats c_{vk}, $k \in S$, are C^∞ functions of $T \in [0,\infty)$. Furthermore, there exist positive constants $\underline{c}_v$ and $\overline{c}_v$ with $0 < \underline{c}_v \le c_{vk}(T) \le \overline{c}_v$, for $T \ge 0$ and $k \in S$.

A direct consequence of this assumption is that $\mathcal{E}$ and $\mathcal{H}$ are smooth functions over the domain $\{\, T \ge 0,\ \varrho \ge 0 \,\}$. Note that the specific heats c_{vk}, $k \in S$, are defined for $T \ge 0$, that is, up to zero temperature. The extension up to zero temperature of specific heats and enthalpies is commonly used in thermodynamics. The specific heats that we consider remain bounded from zero since we consider gases governed by Boltzmann statistics [Woo86].

Similarly, the functions $\mathcal{S}_k$, $\mathcal{G}_k$, and μ_k of the kth species are defined and smooth over the domain $T > 0$ and $\rho_k > 0$, for $k \in S$, and the functions $\mathcal{S}$ and $\mathcal{G}$ are defined and smooth over the domain $T > 0$ and $\varrho > 0$. Moreover, $\mathcal{S}$ is easily extended by continuity to the domain $T > 0$ and $\varrho \ge 0$, making use of $0 \log 0 = 0$. Similarly, $\mathcal{G}$ is extended to the domain $T \ge 0$ and $\varrho \ge 0$, and $\mathcal{G}_k$ is extended to the domain $T \ge 0$ and $\rho_k > 0$. Note, however, that the gas entropy explodes like $\log T$ as the temperature T goes to zero. Indeed, the entropy is unbounded for small temperatures since we consider gaseous mixtures obeying Boltzmann statistics. The third law of thermodynamics only applies to gases obeying Fermi–Dirac or Bose–Einstein statistics [Woo86] or else to cold crystals [Gug57]. The explosion of $\mathcal{S}$ for small T will be of fundamental importance in the mathematical sections for estimating the temperature. In other words, gases obeying Boltzmann statistics tend to be repulsed away from the absolute zero temperature by the explosion of $\mathcal{S}$.

6.2.5. Differentials and convexity

In this section we investigate various differential relations as well as concavity properties of the entropy per unit volume $\mathcal{S}$ in terms of the state variables (T, ϱ) and $(\mathcal{E}, \varrho)$.

Lemma 6.2.1. *The entropy per unit volume $\mathcal{S}$ is a smooth function of the variables (T, ϱ) over the domain $T > 0$, $\varrho > 0$, which can be extended by continuity over the domain $T > 0$, $\varrho \geq 0$. When $T > 0$ and $\varrho > 0$, we have the relations*

$$\partial_T \mathcal{S} = \frac{\mathcal{C}_v}{T}, \qquad \partial_{\rho_k} \mathcal{S} = \mathcal{S}_k - r_k, \quad k \in S, \tag{6.2.24}$$

where $\mathcal{C}_v = \sum_{k\in S} \rho_k c_{vk}$ and

$$\partial^2_{T,T} \mathcal{S} = -\frac{\mathcal{C}_v}{T^2} + \sum_{k\in S} \rho_k \frac{\partial_T c_{vk}}{T}, \qquad \partial^2_{T,\rho_k} \mathcal{S} = \frac{c_{vk}}{T}, \quad k \in S, \tag{6.2.25}$$

$$\partial^2_{\rho_k,\rho_l} \mathcal{S} = -\delta_{kl} \frac{r_k}{\rho_k}, \qquad k, l \in S, \tag{6.2.26}$$

where δ_{kl} is the Kronecker symbol.

Lemma 6.2.2. *The map $(T, \varrho) \to (\mathcal{E}, \varrho)$ is a smooth diffeomorphism from the domain $T > 0$, $\varrho > 0$ onto the open set*

$$\Big\{ (\mathcal{E}, \varrho) \in \mathbb{R}^{1+n}, \ \varrho > 0, \ \mathcal{E} > \sum_{k\in S} \rho_k e_k^0 \Big\}.$$

Over this domain, denoting by $\overline{\partial}$ the derivation with respect to $(\mathcal{E}, \varrho)$, we have the relations

$$\overline{\partial}_{\mathcal{E}} \mathcal{S} = \frac{1}{T}, \qquad \overline{\partial}_{\rho_k} \mathcal{S} = -\frac{\mathcal{G}_k}{T}, \quad k \in S, \tag{6.2.27}$$

so that

$$T\, \mathbb{D}\mathcal{S} = \mathbb{D}\mathcal{E} - \sum_{k\in S} \mathcal{G}_k \, \mathbb{D}\rho_k, \tag{6.2.28}$$

where $\mathbb{D}$ denotes the total derivative, and we further have

$$\overline{\partial}^2_{\mathcal{E},\mathcal{E}} \mathcal{S} = -\frac{1}{T^2 \mathcal{C}_v}, \qquad \overline{\partial}^2_{\mathcal{E},\rho_k} \mathcal{S} = \frac{e_k}{T^2 \mathcal{C}_v}, \quad k \in S, \tag{6.2.29}$$

$$\overline{\partial}^2_{\rho_k,\rho_l} \mathcal{S} = -\frac{e_k e_l}{T^2 \mathcal{C}_v} - \delta_{kl} \frac{r_k}{\rho_k}, \qquad k, l \in S. \tag{6.2.30}$$

As a consequence, for $x = (x_{\mathcal{E}}, x_1, \ldots, x_n)^t$, we have

$$-\langle \overline{\partial}^2 \mathcal{S}\, x, x \rangle = \frac{(x_{\mathcal{E}} - \sum_{k\in S} x_k e_k)^2}{T^2 \mathcal{C}_v} + \sum_{k\in S} \frac{r_k x_k^2}{\rho_k}, \tag{6.2.31}$$

where $\langle x, y\rangle$ denotes the scalar product between vectors x and y, so that $\mathcal{S}$ is a strictly concave function of $(\mathcal{E}, \varrho)$.

Proof. Both lemmas are straightforward from definitions (6.2.1)–(6.2.23) and assumptions (Th_1). ∎

Remark 6.2.3. The partial Hessian matrices $\partial_\varrho^2 \mathcal{S}$ and $\overline{\partial}_\varrho^2 \mathcal{S}$ are both negative definite, and $\partial_\varrho^2 \mathcal{S}$ is diagonal. ∎

6.3. Thermodynamics with mass densities

In this section, we investigate thermodynamic properties as a function of the state variables $(T, p, Y_1, \dots, Y_n)$. In agreement with the specific definition of mass fractions Y_k, $k \in S$, we consider thermodynamic properties per unit mass. Of course, we could have used the species reduced concentrations Y_k/m_k, $k \in S$, or the species mole fractions $X_k = Y_k\overline{m}/m_k$, $k \in S$, instead of the mass fractions, but the resulting formalism would be very similar to the one presented in this section.

This formalism has been used, in particular, by de Groot and Mazur [dGM84], Boillat [Boi95], and Pousin [Pou93], and is similar to that of Gibbs in terms of intensive variables. This formalism is especially suited to problems in which the pressure is constant or when the mass fractions appear as natural variables, and will be used, in particular, in Chapter 11 to investigate a plane flame model.

6.3.1. State variables $(T, p, Y_1, \dots, Y_n)$

The main advantage of using the state variables $(T, p, Y_1, \dots, Y_n)$ is that homogeneity with respect to the species variables can be maintained and that the classical Gibbs relations are recovered in terms of intensive quantities.

It is important to note here that the mass fractions are now considered as independent variables and that the relation $\sum_{k\in S} Y_k = 1$ does not hold a priori, but has to be an a posteriori consequence of the conservation equations.

It will be convenient to use the notation

$$Y = (Y_1, \dots, Y_n)^t, \tag{6.3.1}$$

so that the state variables are (T, p, Y). Using these state variables, the density ρ is defined with

$$\frac{1}{\rho} = \frac{RT}{p} \sum_{k \in S} \frac{Y_k}{m_k}, \tag{6.3.2}$$

and the mean molecular weight $\overline{m}$ from

$$\frac{\sum_{k \in S} Y_k}{\overline{m}} = \sum_{k \in S} \frac{Y_k}{m_k}. \tag{6.3.3}$$

Note that the factor $\sum_{k \in S} Y_k$ has important consequences concerning Jacobian matrices [Gio90] [Gio91] and that $\overline{m}$ is homogeneous with respect to the species mass fractions.

We also define the partial densities from

$$\rho_k = \rho Y_k, \qquad k \in S. \tag{6.3.4}$$

The mole fraction of the kth species X_k is defined by

$$X_k = \frac{\overline{m}}{m_k} Y_k, \qquad k \in S, \tag{6.3.5}$$

and represents locally the number of mole of the kth species with respect to the number of mole of the mixture. It is interesting to note that, from (6.3.3) and (6.3.5), we obtain the relation $\sum_{k \in S} X_k = \sum_{k \in S} Y_k$. In particular, whenever the conservation equations and boundary conditions yield that $\sum_{k \in S} Y_k = 1$, we also obtain $\sum_{k \in S} X_k = 1$.

Other mole related quantities are the partial pressures p_k, $k \in S$, and the molar concentrations γ_k, $k \in S$, defined from

$$p_k = \frac{\rho Y_k RT}{m_k}, \qquad \gamma_k = \frac{p_k}{RT} = \frac{\rho_k}{m_k}, \qquad k \in S. \tag{6.3.6}$$

From the preceding relations, we deduce—after some algebra—that

$$p_k = \mathcal{X}_k \, p, \qquad k \in S, \tag{6.3.7}$$

where $\mathcal{X}_k$ is given by

$$\mathcal{X}_k = \frac{Y_k / m_k}{\sum_{l \in S} Y_l / m_l} = \frac{X_k}{\sum_{l \in S} X_l}, \qquad k \in S, \tag{6.3.8}$$

so that p_k is homogeneous with respect to the species mass fractions or the species mole fractions. These quantities $\mathcal{X}_k$, $k \in S$, essentially represent the species mole fractions X_k, $k \in S$, but have the properties of being homogeneous with respect to mass or mole fractions. These quantities will often appear in various expressions of thermodynamic functions.

Remark 6.3.1. Note that the relations obtained in Sections 6.2.1–6.2.3 are easily recovered by letting $\sum_{k\in S} Y_k = 1$, keeping in mind that, from (6.3.3) and (6.3.5), we then have

$$\sum_{k\in S} Y_k = \sum_{k\in S} X_k.$$

In other words, the algebraic expressions obtained by assuming that the mass fractions are independent are always more general than the corresponding relations obtained in 6.2.1–6.2.3. The properties of differentials, however, are different, as shown in the following sections. ∎

6.3.2. Energy and enthalpy per unit mass

The internal energy per unit mass e is decomposed into

$$e(T,Y) = \sum_{k\in S} Y_k e_k(T), \tag{6.3.9}$$

where e_k is the specific internal energy of the kth species. The quantity e_k is defined by

$$e_k(T) = e_k^{\text{st}} + \int_{T^{\text{st}}}^{T} c_{vk}(T')\, dT', \tag{6.3.10}$$

where $e_k^{\text{st}} = e_k(T^{\text{st}})$ is the formation energy per unit mass of the kth species at the positive standard temperature T^{st}. Introducing the formation energy at zero temperature, as in (6.2.9), we also have

$$e_k(T) = e_k^0 + \int_0^T c_{vk}(T')\, dT'. \tag{6.3.11}$$

Similarly, the enthalpy per unit mass is defined from

$$h(T,Y) = \sum_{k\in S} Y_k h_k(T), \tag{6.3.12}$$

with

$$h_k(T) = e_k(T) + r_k T, \qquad T \geq 0, \tag{6.3.13}$$

where $r_k = R/m_k$. We also define the constant-pressure specific heats

$$c_{pk}(T) = c_{vk}(T) + r_k, \qquad k \in S, \tag{6.3.14}$$

in such a way that

$$h_k(T) = h_k^{\mathrm{st}} + \int_{T^{\mathrm{st}}}^{T} c_{pk}(T')\, dT' = h_k^0 + \int_0^T c_{pk}(T')\, dT',$$

where $h_k(0) = h_k^0 = e_k^0$ is the formation enthalpy per unit mass at zero temperature of the kth species.

Finally, the enthalpy h can also be written in the form

$$h(T, Y) = e(T, Y) + \frac{p}{\rho}, \tag{6.3.15}$$

from the definition of species enthalpies and the state law.

6.3.3. Entropy and Gibbs function per unit mass

The mixture entropy per unit mass s is defined in terms of the species entropies per unit mass s_k, $k \in S$, from

$$s(T, p, Y) = \sum_{k \in S} Y_k s_k(T, p, Y), \tag{6.3.16}$$

and the specific entropy of the kth species s_k is given by

$$s_k(T, p, Y) = s_k^{\mathrm{st}} + \int_{T^{\mathrm{st}}}^{T} \frac{c_{pk}(T')}{T'}\, dT' - r_k \log\Big(\mathcal{X}_k \frac{p}{p^{\mathrm{st}}}\Big), \tag{6.3.17}$$

where s_k^{st} is the formation entropy of the kth species at the standard temperature T^{st} and the standard pressure p^{st} and $\mathcal{X}_k$ is defined in (6.3.8). Note that we have $s_k(T, p, Y) = \mathcal{S}_k(T, \rho_k)$, but that two symbols have been introduced to avoid confusion between thermodynamic function differentials.

Similarly, we can express the mixture Gibbs function per unit mass g in terms of the species Gibbs functions per unit mass g_k, $k \in S$, from

$$g(T, p, Y) = \sum_{k \in S} Y_k g_k(T, p, Y), \tag{6.3.18}$$

where the specific Gibbs function of the kth species g_k is given by

$$g_k(T, p, Y) = h_k(T) - T s_k(T, p, Y), \tag{6.3.19}$$

in such a way that

$$g(T, p, Y) = h(T, Y) - T s(T, p, Y). \tag{6.3.20}$$

We can also introduce the entropy $s_k^{\mathrm{atm}} = s_k^{\mathrm{atm}}(T)$ of the kth species at atmospheric pressure $p_k = p^{\mathrm{st}} = p^{\mathrm{atm}}$ and the corresponding Gibbs function $g_k^{\mathrm{atm}} = g_k^{\mathrm{atm}}(T)$

$$s_k^{\mathrm{atm}}(T) = s_k^{\mathrm{st}} + \int_{T^{\mathrm{st}}}^{T} \frac{c_{pk}(T')}{T'}\, dT', \qquad g_k^{\mathrm{atm}}(T) = h_k(T) - T s_k^{\mathrm{atm}}(T), \tag{6.3.21}$$

which are easily expressed in terms of the temperature. These quantities, already presented in Chapter 2, are tabulated in the literature in molar units [StPr71] [Cal85].

6.3.4. Assumptions

The mathematical assumption concerning thermodynamic properties is the same as the one used in Section 6.2.4.

(Th$_1$) The species molar masses m_k, $k \in S$, and the gas constant R are positive constants. The formation energies e_k^{st}, $k \in S$, and the formation entropies s_k^{st}, $k \in S$, are constants. The specific heats c_{vk}, $k \in S$, are C^∞ functions of $T \in [0, \infty)$. Furthermore, there exist positive constants $\underline{c}_v$ and $\overline{c}_v$ with $0 < \underline{c}_v \leq c_{vk}(T) \leq \overline{c}_v$, for $T \geq 0$ and $k \in S$.

A direct consequence of this assumption is that e and h are smooth functions over the domain $T \geq 0$, $Y \geq 0$. We emphasize again that the extension up to zero temperature of specific heats, energies and enthalpies is classical in thermodynamics.

Similarly, the functions s_k, g_k, and μ_k are defined and smooth over the domain $T > 0$, $p > 0$, $Y \geq 0$, and $Y_k > 0$, and functions s and g are defined and smooth over the domain $T > 0$, $p > 0$, and $Y > 0$. However, s is easily extended by continuity to the domain $T > 0$, $p > 0$, $Y \geq 0$, making use of $0 \log 0 = 0$, g is extended to the domain $T \geq 0$, $p > 0$, and $Y \geq 0$, and g_k is extended to the domain $T \geq 0$, $p > 0$, $Y \geq 0$, and $Y_k > 0$. The explosion of entropy at zero temperature has been discussed in Section 6.2.4.

6.3.5. Differentials and convexity

In this section we investigate various differential relations as well as concavity properties of the entropy per unit mass s in terms of the state variables (T, p, Y) and $(e, 1/\rho, Y)$.

Lemma 6.3.2. *The entropy per unit mass s is a smooth function of the variables (T, p, Y) over the domain $T > 0$, $p > 0$, and $Y > 0$, which can be* *extended by continuity over the domain $T > 0$, $p > 0$, $Y \geq 0$. For $T > 0$, $p > 0$, and $Y > 0$, we have the relations*

$$\partial_T s = \frac{c_p}{T}, \qquad \partial_p s = -\frac{1}{\rho T}, \qquad \partial_{Y_k} s = s_k, \quad k \in S, \tag{6.3.22}$$

where $c_p(T,Y) = \sum_{k\in S} Y_k c_{pk}(T)$ and

$$\partial_T g = -s, \qquad \partial_p g = \frac{1}{\rho}, \qquad \partial_{Y_k} g = g_k, \quad k \in S. \tag{6.3.23}$$

In addition, for T and p fixed, we have

$$\partial^2_{Y_k,Y_l} s = -\frac{R}{m_k m_l}\Big(\frac{\delta_{kl}}{Y_l/m_l} - \frac{1}{\sum_{i\in S} Y_i/m_i}\Big), \qquad k,l \in S, \tag{6.3.24}$$

so that $\partial^2_{Y,Y} s$ is negative semidefinite with nullspace $\mathbb{R}Y$.

Lemma 6.3.3. *The map $(T,p,Y) \to (e,\ 1/\rho,\ Y)$ is a smooth diffeomorphism from the domain $T > 0$, $p > 0$, $Y > 0$ onto the open set*

$$\Big\{(e,v,Y) \in \mathbb{R}^{2+n},\ v > 0, Y > 0,\ e > \sum_{k\in S} Y_k e_k^0\Big\}.$$

Over this domain, denoting by $\widehat{\partial}$ the derivation with respect to $(e,\ 1/\rho,\ Y)$, we have the relations

$$\widehat{\partial}_e s = \frac{1}{T}, \qquad \widehat{\partial}_{\frac{1}{\rho}} s = \frac{p}{T}, \qquad \widehat{\partial}_{Y_k} s = -\frac{g_k}{T}, \quad k \in S, \tag{6.3.25}$$

so that

$$\mathbb{D}s = \frac{\mathbb{D}e}{T} + \frac{p}{T}\mathbb{D}\Big(\frac{1}{\rho}\Big) - \sum_{k\in S} \frac{g_k}{T}\mathbb{D}Y_k, \tag{6.3.26}$$

which is one of the Gibbs relations applied to densities [dGM84]. *Moreover, we have*

$$\widehat{\partial}^2_{e,e} s = -\frac{1}{T^2 c_v}, \qquad \widehat{\partial}^2_{e,\frac{1}{\rho}} s = 0, \qquad \widehat{\partial}^2_{e,Y_k} s = \frac{e_k}{T^2 c_v}, \quad k \in S, \tag{6.3.27}$$

$$\widehat{\partial}^2_{\frac{1}{\rho},\frac{1}{\rho}} s = -\frac{\rho p}{T}, \qquad \widehat{\partial}^2_{\frac{1}{\rho},Y_k} s = \frac{p}{T m_k}\frac{1}{\sum_{l\in S} Y_l/m_l}, \quad k \in S, \tag{6.3.28}$$

$$\widehat{\partial}^2_{Y_k,Y_l} s = -\frac{e_k e_l}{T^2 c_v} - \frac{R}{Y_k m_k}\delta_{kl}, \qquad k,l \in S. \tag{6.3.29}$$

In particular, for $x = (x_e, x_{1/\rho}, x_1, \ldots, x_n)^t$ *we have*

$$-\langle \widehat{\partial}^2 s\, x, x\rangle = \frac{(x_e - \sum_{k\in S} x_k e_k)^2}{T^2 c_v} + \frac{\rho p}{T}\Big(x_{1/\rho} - \sum_{k\in S} \frac{x_k}{m_k \rho \sum_{l\in S} Y_l/m_l}\Big)^2, \\ + R\Big(\sum_{k\in S} \frac{x_k^2}{m_k Y_k} - \frac{(\sum_{k\in S} x_k/m_k)^2}{\sum_{l\in S} Y_l/m_l}\Big), \tag{6.3.30}$$

so that s *is a concave function of* $(e,\, 1/\rho,\, Y)$ *and* $N(\widehat{\partial}^2 s) = \mathbb{R}(e,\, 1/\rho,\, Y)$.

Remark 6.3.4. Note that the Gibbs relation (6.3.22) holds with respect to mass densities, but that $\widehat{\partial}^2 \mathfrak{s}$ is singular. ∎

6.3.6. Miscellaneous

The number of state variables that may arise in mathematical physics is very large. For any new set of state variables, the corresponding modifications of lemmas 6.3.2 and 6.3.3 are straightforward. As a typical example, we consider the state variables $(h, p, Y_1, \ldots, Y_n)$, which naturally appear in the modeling of plane premixed laminar flames.

Lemma 6.3.5. *The map* $(T, p, Y) \to (h, p, Y)$ *is a smooth diffeomorphism from the domain* $T > 0$, $p > 0$, $Y > 0$ *onto the open set*

$$\Big\{\ (h,p,Y),\ p>0,\ Y>0,\ h > \sum_{k\in S} Y_k h_k^0,\ \Big\}.$$

Over this domain, denoting by $\check{\partial}$ *the derivation with respect to* (h, p, Y), *we have relations*

$$\check{\partial}_h s = \frac{1}{T}, \qquad \check{\partial}_p s = -\frac{1}{\rho T}, \qquad \check{\partial}_{Y_k} s = -\frac{g_k}{T}, \qquad k \in S. \tag{6.3.31}$$

Furthermore, at a fixed pressure p, *the Euler relation holds*

$$s = \langle \check{\partial}_{(h,Y)} s, (h, Y)\rangle, \tag{6.3.32}$$

and, for $x = (x_h, 0, x_1, \ldots, x_n)^t$, *the Hessian* $\check{\partial}^2 s$ *is given by*

$$-\frac{\langle (\check{\partial}^2 s)x, x\rangle}{R} = \frac{(x_h - \sum_{k\in S} h_k x_k)^2}{RT^2 c_p} + \sum_{k\in S} \frac{(x_k/m_k)^2}{Y_k/m_k} - \frac{(\sum_{k\in S} x_k/m_k)^2}{\sum_{l\in S} Y_l/m_l}. \tag{6.3.33}$$

As a consequence, for a fixed pressure p, s *is concave with respect to* (h, Y) *and* $N(\check{\partial}^2_{(h,Y),(h,Y)} s) = \mathbb{R}(h, Y)^t$.

In order to relate the species mole and mass fractions gradients, we will further need the matrix E defined by

$$E = \partial_Y X. \tag{6.3.34}$$

A direct calculation yields

$$\begin{cases} E_{kk} = \dfrac{\overline{m}}{m_k} + \mathcal{X}_k\Big(1 - \dfrac{\overline{m}}{m_k}\Big), & k \in S, \\ E_{kl} = \mathcal{X}_k\Big(1 - \dfrac{\overline{m}}{m_l}\Big), & k,l \in S, \quad k \neq l. \end{cases} \tag{6.3.35}$$

The matrix $E = \partial_Y X$ is invertible, and we have [Gio90]

$$\det E = \prod_{k \in S} \frac{\overline{m}}{m_k}. \tag{6.3.36}$$

Note that the matrix E is invertible because of the factor $\sum_{k\in S} Y_k$ in (6.3.3) [Gio90]. Moreover, we have $EY = X$, where $X = (X_1, \dots, X_n)^t$, and $E^t\mathcal{U} = \mathcal{U}$, where $\mathcal{U} = (1, \dots, 1)^t$.

We will also need the matrix $\partial_Y \mathcal{X}$, which can be expressed as

$$\partial_Y \mathcal{X} = \frac{1}{\langle Y, \mathcal{U}\rangle}\Big(I - \frac{X \otimes \mathcal{U}}{\langle Y, \mathcal{U}\rangle}\Big)\, E, \tag{6.3.37}$$

so that $(\partial_Y \mathcal{X})x = Ex/\langle Y, \mathcal{U}\rangle$ for $x \in \mathcal{U}^\perp$, since $E^t\mathcal{U} = \mathcal{U}$. In other words, over the physical hyperplane of zero-sum gradients $\mathcal{U}^\perp$, the matrices $\partial_Y \mathcal{X}$ and E coincide.

6.4. Chemistry sources

In this section, we focus on the mathematical structure of chemistry source terms. We investigate, in particular, atom conservation, mass conservation, and various forms of molar production rates.

6.4.1. Chemical reactions

We consider a reaction mechanism composed of an arbitrary number n^{r} of elementary chemical reactions between n species, which can be written in the general form

$$\sum_{k\in S} \nu_{ki}^{\mathrm{f}}\, \mathfrak{M}_k \rightleftharpoons \sum_{k\in S} \nu_{ki}^{\mathrm{b}}\, \mathfrak{M}_k, \qquad i \in R, \tag{6.4.1}$$

where $\mathfrak{M}_k$ is the chemical symbol of the kth species, ν_{ki}^{f} and ν_{ki}^{b} are the forward and backward stoichiometric coefficients of the kth species in the ith reaction, respectively, and $R = \{1, \dots, n^{\text{r}}\}$ is the set of reaction indices.

Note that all chemical reactions are reversible and that the number of reactions n^{r} is arbitrary. Indeed, we are primarily interested in *elementary* chemical reactions which effectively take place in the mixture, and elementary reactions are always reversible, as already pointed out in Chapters 2 and 4. Typical examples of reaction mechanisms have also been referred to in Chapter 2, and a chemical mechanism modeling the combustion of hydrogen in air will be presented in Chapter 12.

6.4.2. Maxwellian production rates

The molar production rates that we consider are the Maxwellian production rates obtained from the kinetic theory, as summarized in Chapter 4. These rates are obtained when the chemistry characteristic times are larger than the mean free times of the molecules. They are compatible with the law of mass action. The molar production rate of the kth species reads

$$\omega_k = \sum_{i\in R} \nu_{ki}\, \tau_i, \tag{6.4.2}$$

where

$$\nu_{ki} = \nu_{ki}^{\text{b}} - \nu_{ki}^{\text{f}}, \tag{6.4.3}$$

and τ_i is the rate of progress of the ith reaction [Wil85] [EG94]. The reaction rates of progress are given by

$$\tau_i = K_i^{\text{f}} \prod_{l\in S} \gamma_l^{\nu_{li}^{\text{f}}} - K_i^{\text{b}} \prod_{l\in S} \gamma_l^{\nu_{li}^{\text{b}}}, \qquad i \in R, \tag{6.4.4}$$

where $\gamma_k = \rho_k/m_k = p_k/RT$ is the molar concentration of the kth species and K_i^{f} and K_i^{b} are the forward and backward rate constants of the ith reaction, respectively.

The quantities K_i^{f} and K_i^{b} are functions of the temperature, and their ratio is the equilibrium constant K_i^{e} of the ith reaction

$$K_i^{\text{e}}(T) = \frac{K_i^{\text{f}}(T)}{K_i^{\text{b}}(T)}, \tag{6.4.5}$$

given by

$$\log K_i^{\text{e}}(T) = -\sum_{k\in S} \frac{\nu_{ki} m_k}{RT} \mathcal{G}_k(T, m_k) = -\sum_{k\in S} \nu_{ki} m_k \mu_k^{\text{u}}(T), \tag{6.4.6}$$

where $\mathcal{G}_k(T, m_k) = RT\mu_k^{\mathrm{u}}(T) = g_k(T, RT, e^k)$ and e^k is the kth canonical basis vector of $\mathbb{R}^n$. For actual calculations in terms of standard state quantities, the equilibrium constant can also be written in the form (2.4.6)

$$\log K_i^{\mathrm{e}}(T) = -\sum_{k \in S} \frac{\nu_{ki} m_k}{RT} \Big(g_k^{\mathrm{atm}}(T) - \frac{RT}{m_k} \log\big(\frac{p^{\mathrm{atm}}}{RT}\big) \Big), \qquad (6.4.7)$$

where $g_k^{\mathrm{atm}}(T)$ is the Gibbs function at the standard pressure $p^{\mathrm{st}} = p^{\mathrm{atm}}$.

6.4.3. Assumptions

We introduce here the assumptions concerning the chemical reactions and the species production rates.

The species of the mixture are assumed to be constituted by atoms, and we denote by $\mathfrak{a}_{kl}$ the number of lth atom in the kth species. We also denote by $\mathfrak{A} = \{1, \ldots, n^{\mathrm{a}}\}$ the set of atom indices and by $n^{\mathrm{a}} \geq 1$ the number of atoms—or elements—in the mixture. The following mathematical assumptions concern molecules, atoms, and chemical reactions.

(Th$_2$) The stoichiometric coefficients ν_{ki}^{f} and ν_{ki}^{b}, $k \in S$, $i \in R$, and the atomic coefficients $\mathfrak{a}_{kl}$, $k \in S$, $l \in \mathfrak{A}$, are nonnegative integers. The atomic vectors $\mathfrak{a}_l$, $l \in \mathfrak{A}$, defined by $\mathfrak{a}_l = (\mathfrak{a}_{1l}, \ldots, \mathfrak{a}_{nl})^t$, and the reaction vectors ν_i, $i \in R$, defined by $\nu_i = (\nu_{1i}, \ldots, \nu_{ni})^t$, satisfy the atom conservation relations

$$\langle \nu_i, \mathfrak{a}_l \rangle = 0, \qquad i \in R, \quad l \in \mathfrak{A}. \qquad (6.4.8)$$

We also define the forward and backward reaction vectors $\nu_i^{\mathrm{f}} = (\nu_{1i}^{\mathrm{f}}, \ldots, \nu_{ni}^{\mathrm{f}})^t$ and $\nu_i^{\mathrm{b}} = (\nu_{1i}^{\mathrm{b}}, \ldots, \nu_{ni}^{\mathrm{b}})^t$, $i \in R$, respectively, which satisfy $\nu_i = \nu_i^{\mathrm{b}} - \nu_i^{\mathrm{f}}$ and $\langle \nu_i^{\mathrm{f}}, \mathfrak{a}_l \rangle = \langle \nu_i^{\mathrm{b}}, \mathfrak{a}_l \rangle$, $i \in R$, $l \in \mathfrak{A}$. The space spanned by the reaction vectors is denoted by

$$\mathcal{R} = \mathrm{span}\{\ \nu_i,\ i \in R\ \},$$

and the space spanned by the atom vectors is denoted by

$$\mathcal{A} = \mathrm{span}\{\ \mathfrak{a}_l,\ l \in \mathfrak{A}\ \},$$

in such a way that $\mathcal{R} \subset \mathcal{A}^{\perp}$ and $\mathcal{A} \subset \mathcal{R}^{\perp}$.

(Th$_3$) The atom masses $\widetilde{m}_l$, $l \in \mathfrak{A}$, are positive constants, and the species molar masses m_k, $k \in S$, are given by

$$m_k = \sum_{l \in \mathfrak{A}} \widetilde{m}_l\, \mathfrak{a}_{kl}. \qquad (6.4.9)$$

Denoting by m the mass vector $m = (m_1, \dots, m_n)^t$, these relations can be written in vector form

$$m = \sum_{l \in \mathfrak{A}} \widetilde{m}_l \, \mathfrak{a}_l. \tag{6.4.10}$$

(Th_4) The rate constants K_i^{f}, and K_i^{b}, $i \in R$, are C^∞ positive functions of $T > 0$ and satisfy the reciprocity relations (6.4.5) (6.4.6).

Remark 6.4.1. The atoms vectors $\mathfrak{a}_l$, $l \in \mathfrak{A}$, have to be linearly independent. When this is not the case, it is first necessary to eliminate linearly dependent atomic vectors. For realistic complex chemistry networks, the number of chemical reactions is always much larger than the number of chemical species and one usually has $\mathcal{R} = \mathcal{A}^\perp$. In other words, the chemical reactions are spanning the largest possible space. When this is not the case, one has simply to use the space $\mathcal{R}^\perp$ instead of $\mathcal{A}$ [Kra70]. ∎

6.4.4. Mass weights

In this section, we investigate the mass weighted species production rates defined by $m_k \omega_k$, $k \in S$. These rates directly appear in the species conservation equations. To this purpose, we introduce the mass weights matrix M of order n defined by

$$M = \mathrm{diag}(m_1, \dots, m_n). \tag{6.4.11}$$

The mass weighted stoichiometric coefficients are then the vectors $M\nu_i$, $i \in R$, and the specific atom compositions are the vectors $M^{-1}\mathfrak{a}_l$, $l \in \mathfrak{A}$. The atomic mass conservation now reads

$$\langle M\nu_i, M^{-1}\mathfrak{a}_l \rangle = 0, \qquad i \in R, \quad l \in \mathfrak{A}, \tag{6.4.12}$$

and the corresponding linear spaces $M\mathcal{R}$ and $M^{-1}\mathcal{A}$ are then such that $M\mathcal{R} \subset (M^{-1}\mathcal{A})^\perp$ and $M^{-1}\mathcal{A} \subset (M\mathcal{R})^\perp$ in the composition space $\mathbb{R}^n$.

6.4.5. Mass conservation

From atom conservation and the definition of species masses, we now deduce the mass conservation property.

Lemma 6.4.2. *The vector of chemical production rates $\omega = (\omega_1, \dots, \omega_n)^t$ can be written in the vector form*

$$\omega = \sum_{i \in R} \tau_i \, \nu_i, \tag{6.4.13}$$

in such a way that

$$\omega \in \mathcal{R} \qquad \text{and} \qquad M\omega \in M\mathcal{R}. \tag{6.4.14}$$

Moreover, the unity vector $\mathcal{U} \in \mathbb{R}^n$ *defined by* $\mathcal{U} = (1, \dots, 1)^t$ *satisfies*

$$\mathcal{U} \in (M\mathcal{R})^\perp, \tag{6.4.15}$$

and we have the total mass conservation relation

$$\langle \mathcal{U}, M\omega \rangle = \sum_{k \in S} m_k \omega_k = 0. \tag{6.4.16}$$

Proof. These properties are straightforward from (Th_2) and (Th_3), which imply that

$$\mathcal{U} = \sum_{l \in \mathfrak{A}} \widetilde{m}_l M^{-1} \mathfrak{a}_l, \tag{6.4.17}$$

so that $\mathcal{U} \in (M\mathcal{R})^\perp$. ∎

6.4.6. Creation and destruction rates

In the following, it will be useful to decompose the production rates between the creation and destruction rates.

Lemma 6.4.3. *The chemical production rates* ω_k, $k \in S$, *can be split into the form*

$$\omega_k = \mathcal{C}_k - \mathcal{D}_k = \mathcal{C}_k - \rho_k \widehat{\mathcal{D}}_k, \qquad k \in S, \tag{6.4.18}$$

where $\mathcal{C}_k$, $\mathcal{D}_k = \rho_k \widehat{\mathcal{D}}_k$, *and* $\widehat{\mathcal{D}}_k$, *are, respectively, the creation, destruction, and reduced destruction rates of the* k^{th} *species. These rates satisfy* $\mathcal{C}_k \geq 0$, $\mathcal{D}_k \geq 0$, *and* $\widehat{\mathcal{D}}_k \geq 0$ *and they are smooth functions of* (T, ϱ), *for* $T > 0$, $\varrho \geq 0$.

Proof. We simply use (6.4.2)–(6.4.4) and write $\omega_k = \mathcal{C}_k - \mathcal{D}_k$ with

$$\mathcal{C}_k = \sum_{i \in R} \Big(\nu_{ki}^{\mathrm{b}} K_i^{\mathrm{f}} \prod_{l \in S} \gamma_l^{\nu_{li}^{\mathrm{f}}} + \nu_{ki}^{\mathrm{f}} K_i^{\mathrm{b}} \prod_{l \in S} \gamma_l^{\nu_{li}^{\mathrm{b}}} \Big) \tag{6.4.19}$$

and

$$\mathcal{D}_k = \sum_{i \in R} \Big(\nu_{ki}^{\mathrm{f}} K_i^{\mathrm{f}} \prod_{l \in S} \gamma_l^{\nu_{li}^{\mathrm{f}}} + \nu_{ki}^{\mathrm{b}} K_i^{\mathrm{b}} \prod_{l \in S} \gamma_l^{\nu_{li}^{\mathrm{b}}} \Big). \tag{6.4.20}$$

Since the stoichiometric coefficients are nonnegative integers, we can now write that

$$\mathcal{D}_k = \gamma_k\Big(\sum_{\substack{i\in R\\ \nu_{ki}^{\mathrm{f}}\geq 1}} \nu_{ki}^{\mathrm{f}} K_i^{\mathrm{f}} \gamma_k^{\nu_{ki}^{\mathrm{f}}-1} \prod_{\substack{l\in S\\ l\neq k}} \gamma_l^{\nu_{li}^{\mathrm{f}}} + \sum_{\substack{i\in R\\ \nu_{ki}^{\mathrm{b}}\geq 1}} \nu_{ki}^{\mathrm{b}} K_i^{\mathrm{b}} \gamma_k^{\nu_{ki}^{\mathrm{b}}-1} \prod_{\substack{l\in S\\ l\neq k}} \gamma_l^{\nu_{li}^{\mathrm{b}}} \Big), \tag{6.4.21}$$

so that $\mathcal{D}_k = \rho_k \widehat{\mathcal{D}}_k$, where

$$\widehat{\mathcal{D}}_k m_k = \sum_{\substack{i\in R\\ \nu_{ki}^{\mathrm{f}}\geq 1}} \nu_{ki}^{\mathrm{f}} K_i^{\mathrm{f}} \gamma_k^{\nu_{ki}^{\mathrm{f}}-1} \prod_{\substack{l\in S\\ l\neq k}} \gamma_l^{\nu_{li}^{\mathrm{f}}} + \sum_{\substack{i\in R\\ \nu_{ki}^{\mathrm{b}}\geq 1}} \nu_{ki}^{\mathrm{b}} K_i^{\mathrm{b}} \gamma_k^{\nu_{ki}^{\mathrm{b}}-1} \prod_{\substack{l\in S\\ l\neq k}} \gamma_l^{\nu_{li}^{\mathrm{b}}}, \tag{6.4.22}$$

which is a smooth function of (T, ϱ), for $T > 0$, $\varrho \geq 0$. ∎

6.4.7. Symmetric formulation for the rates of progress

The purpose of this section is to rewrite the rates of progress into a natural form involving a single reaction constant—instead of the forward and backward reaction constants, as in (6.4.4)—making use of the expression of the equilibrium constant (6.4.5) (6.4.6). This formulation naturally arises in mathematical studies [Kra70] [Pou93] [Gio99] and has also been introduced in the context of Onsager reciprocal relations for linearized rates [Mei43]. This formulation also arises from molecular aspects, that is, from reactive Boltzman investigations, as shown in Chapter 4. The usual form (6.4.4) was first introduced for the sake of clarity and readability.

In order to simplify the expression of the rate of progress τ_i of the ith reaction, we introduce the new reaction constant K_i^{s} defined by

$$\log K_i^{\mathrm{s}} = \log K_i^{\mathrm{f}} - \langle M\nu_i^{\mathrm{f}}, \mu^{\mathrm{u}}\rangle = \log K_i^{\mathrm{b}} - \langle M\nu_i^{\mathrm{b}}, \mu^{\mathrm{u}}\rangle, \tag{6.4.23}$$

where $\mu^{\mathrm{u}} = (\mu_1^{\mathrm{u}}, \ldots, \mu_n^{\mathrm{u}})^t$ and $\mu_k^{\mathrm{u}} = \mu_k(T, m_k) = \mathcal{G}_k(T, m_k)/(RT)$, $k \in S$, is the value of μ_k when γ_k is unity in (6.2.21). Note that we have used here the expression of the equilibrium constant

$$\log K_i^{\mathrm{e}} = -\langle M\nu_i, \mu^{\mathrm{u}}\rangle = \langle M\nu_i^{\mathrm{f}}, \mu^{\mathrm{u}}\rangle - \langle M\nu_i^{\mathrm{b}}, \mu^{\mathrm{u}}\rangle, \qquad i \in R. \tag{6.4.24}$$

We can then rewrite (6.4.4) in the form

$$\tau_i = K_i^{\mathrm{s}}\Big(\exp\langle M\nu_i^{\mathrm{f}}, \mu^{\mathrm{u}}\rangle \prod_{l\in S} \gamma_l^{\nu_{li}^{\mathrm{f}}} - \exp\langle M\nu_i^{\mathrm{b}}, \mu^{\mathrm{u}}\rangle \prod_{l\in S} \gamma_l^{\nu_{li}^{\mathrm{b}}}\Big). \tag{6.4.25}$$

Using now $\mu = (\mu_1, \ldots, \mu_n)^t$ with $\mu_k = \mu_k^{\mathrm{u}}(T) + (1/m_k)\log\gamma_k$, $k \in S$, we finally obtain the natural expression

$$\tau_i = K_i^{\mathrm{s}}\Big(\exp\langle M\nu_i^{\mathrm{f}}, \mu\rangle - \exp\langle M\nu_i^{\mathrm{b}}, \mu\rangle\Big), \tag{6.4.26}$$

which coincides with the natural expression (4.6.71) obtained from the kinetic theory of gases in Chapter 4. These expression (6.4.26) also suggests a new way of fitting reaction constants. The assumptions (Th_4) concerning the rate constants $\log K_i^{\mathrm{f}}$ and $\log K_i^{\mathrm{b}}$, $i \in R$, can then be rewritten in terms of the quantities K_i^{s}, $i \in R$, as follows.

(Th_4)′ The rate constants K_i^{s}, $i \in R$, are C^∞ positive functions of the temperature T for $T > 0$.

Of course, the assumptions (Th_4) and (Th_4)′ are strictly equivalent, taking into account the smoothness of the species specific heats. Note that we do not assume that the quantities K_i^{f} and K_i^{b}, $i \in R$, or K_i^{s}, $i \in R$, are bounded functions of the temperature T.

6.5. Positive equilibrium points

In this section, we investigate existence and uniqueness of positive equilibrium points in atom conservation manifolds. We first characterize equilibrium points and then restate that detailed balance holds at equilibrium. Existence of equilibrium states is generally obtained by extremalizing a thermodynamic functional over an atom conservation affine subspace. The thermodynamic functional to be maximized or minimized depends on which thermal properties are kept fixed.

Both thermodynamic formulations investigated in the previous sections can be used for convenience. More specifically, we can use the formalism of Section 6.2 associated with strictly convex thermodynamic functionals [Kra70] or the formalism of Section 6.3, which involves thermodynamic functionals with singular Hessians, following Gibbs [SS65]. In order to illustrate both formalisms—and for future needs—we study in this section various types of equilibriums.

We first investigate equilibrium points with fixed T and ρ by using the formalism of Section 6.2 associated with the state variables $(T, \rho_1, \ldots, \rho_n)$. These results will be used in Chapter 9 when investigating global solutions for the Cauchy problem around stationary states and in Chapter 10 when investigating chemical equilibrium flows.

We then investigate the existence of equilibrium points with fixed h and p by using the formalism of Section 6.3 associated with the state variables $(h, p, Y_1, \ldots, Y_n)$. These results will be used in Chapter 11 for plane flame problems whose natural variables are the species mass fractions. Note that all of these techniques for establishing existence and uniqueness of equilibrium points are essentially similar. The choice of a set of variables and of the thermodynamic formalism essentially depends on the context under consideration.

We next establish the smoothness of positive equilibrium points with respect to the thermal variables and atom concentrations.

We will investigate in Section 6.6 another type of equilibrium points, which is different in nature, that is, boundary equilibrium points. These points are such that at least one mass fraction is vanishing, although there also exists a positive equilibrium point.

6.5.1. Definition of equilibrium points

In the next proposition, we characterize positive concentration points where chemistry source terms vanish.

Proposition 6.5.1. *The reduced entropy production due to chemical reactions*

$$\zeta(T,\varrho) = -\langle \mu, M\omega\rangle = -\frac{1}{RT}\sum_{k\in S} \mathcal{G}_k m_k \omega_k,$$

defined for $T > 0$ and $\varrho > 0$, is nonnegative and admits 0 as a minimum at any point where source terms vanish. Any point $(T^{\rm e}, \varrho^{\rm e})$ with $T^{\rm e} > 0$ and $\varrho^{\rm e} = (\rho_1^{\rm e}, \dots, \rho_n^{\rm e})^t > 0$ where the source term vanishes

$$\omega_k(T^{\rm e}, \varrho^{\rm e}) = 0, \qquad k \in S, \tag{6.5.1}$$

is also such that the rate of progress of each reaction vanishes

$$\tau_i(T^{\rm e}, \varrho^{\rm e}) = 0, \qquad i \in R, \tag{6.5.2}$$

which can also be written in the form

$$\Big\langle \mu(T^{\rm e}, \varrho^{\rm e}), M\nu_i \Big\rangle = 0, \qquad i \in R, \tag{6.5.3}$$

or, equivalently, $(\mu_1^{\rm e}, \dots, \mu_n^{\rm e})^t \in (M\mathcal{R})^\perp$ or $(\mathcal{G}_1^{\rm e}, \dots, \mathcal{G}_n^{\rm e})^t \in (M\mathcal{R})^\perp$.

Proof. Rewriting ζ in the form

$$\zeta = -\sum_{\substack{i\in R\\ k\in S}} \mu_k m_k \nu_{ki} \tau_i = \sum_{i\in R} \langle \mu, M(\nu_i^{\rm f} - \nu_i^{\rm b})\rangle \,\tau_i,$$

and using (6.4.26), we obtain

$$\zeta = \sum_{i\in R} K_i^{\rm s}\Big(\langle \mu, M\nu_i^{\rm f}\rangle - \langle \mu, M\nu_i^{\rm b}\rangle\Big)\Big(\exp\langle \mu, M\nu_i^{\rm f}\rangle - \exp\langle \mu, M\nu_i^{\rm b}\rangle\Big), \tag{6.5.4}$$

so that $\zeta(T,\varrho) \geq 0$ and $\zeta(T,\varrho) = 0$ if and only if $\langle \mu, M\nu_i\rangle = 0$, $i \in R$, that is to say, if and only if $\tau_i = 0$, $i \in R$. In addition, $\tau_i = 0$, $i \in R$, if and only if $\omega_k = 0$, $k \in S$, from the expression of ζ. ∎

Definition 6.5.2. *A point* $(T^{\mathrm{e}}, \varrho^{\mathrm{e}})$ *with* $T^{\mathrm{e}} > 0$ *and* $\varrho^{\mathrm{e}} > 0$*—or, equivalently, a point* $(T^{\mathrm{e}}, p^{\mathrm{e}}, Y^{\mathrm{e}})$ *with* $T^{\mathrm{e}} > 0$, $p^{\mathrm{e}} > 0$, *and* $Y^{\mathrm{e}} > 0$*—which satisfies the equivalent properties of Proposition* 6.5.1, *will be termed an equilibrium point.*

Note that we are only interested here in positive equilibrium states which are in the interior of the composition space. Boundary points where source terms vanish will be investigated in Section 6.6.

6.5.2. Equilibrium points with T and ρ fixed

In this section we investigate equilibrium states by using the formalism of Section 6.2. We restate existence and uniqueness of an equilibrium density vector at a fixed temperature in a given affine submanifold of atom conservation [Kra70]. These results will be used in Chapter 9 when investigating asymptotic stability of equilibrium states.

When defining chemical equilibrium states, it is necessary to use equations expressing the fact that atoms are neither created nor destroyed in chemical reactions. These atom conservation relations are typically in the form $\langle \varrho - \varrho^{\mathrm{f}}, M^{-1}\mathfrak{a} \rangle = 0$, where $\mathfrak{a}$ is an atom vector and ϱ^{f} is a given state, and more generally in the form $\langle \varrho - \varrho^{\mathrm{f}}, u \rangle = 0$, where $u \in (M\mathcal{R})^{\perp}$. As a consequence, equilibrium points have to be investigated in atom conservation affine manifolds in the form $\varrho^{\mathrm{f}} + M\mathcal{R}$. Note that the total density ρ is constant on these manifolds since $\mathcal{U} \in (M\mathcal{R})^{\perp}$ and $\rho = \langle \varrho, \mathcal{U} \rangle$.

Proposition 6.5.3. *Consider a temperature* $T^{\mathrm{e}} \in (0, \infty)$, *a point* ϱ^{f} *of* $(0,\infty)^n$, *and the associated reaction simplex* $(\varrho^{\mathrm{f}} + M\mathcal{R}) \cap (0,\infty)^n$ *and assume that Properties* (Th_1)–(Th_4) *hold. Then there exists a unique equilibrium state* ϱ^{e} *in the simplex* $(\varrho^{\mathrm{f}} + M\mathcal{R}) \cap (0,\infty)^n$ *where the chemical source term* ω *vanishes, and, therefore, where the reaction rates of progress also vanish and where* $\mu = (\mu_1^{\mathrm{e}}, \dots, \mu_n^{\mathrm{e}})^t \in (M\mathcal{R})^{\perp}$. *In addition, as a function of* ϱ, *the entropy production due to chemical reactions admits* 0 *as a strict maximum at* ϱ^{e} *over the simplex* $(\varrho^{\mathrm{f}} + M\mathcal{R}) \cap (0,\infty)^n$.

Proof. We characterize the equilibrium point as the only extremum of the reduced Helmholtz free energy function [Kra70]

$$\mathfrak{F}(T, \varrho) = \frac{\mathcal{G} - p}{RT} = \langle \varrho, \mu - M^{-1}\mathcal{U} \rangle.$$

The function $\mathfrak{F}$ is a C^{∞} strictly convex function of ϱ in $(0,\infty)^n$ at a fixed temperature $T = T^{\mathrm{e}}$, which can be written as

$$\mathfrak{F}(T^{\mathrm{e}}, \varrho) = \frac{1}{R} \sum_{k \in S} r_k \rho_k \big(\log \rho_k - 1 + a_k(T^{\mathrm{e}}) \big),$$

where $a_k(T) = \mathcal{G}_k(T,1)/r_kT$. The partial derivative with respect to the mass density vector ϱ is also given by

$$\partial_\varrho \mathfrak{F} = \mu.$$

The function $\mathfrak{F}$ is easily extended over the closure of $(\varrho^{\mathrm{f}} + M\mathcal{R}) \cap (0,\infty)^n$ into a continuous function by using $0 \log 0 = 0$, and, therefore, $\mathfrak{F}$ admits a minimum on this nonempty convex compact set. Note that the boundedness of the reaction simplex is a direct consequence of $\mathcal{U} \in (M\mathcal{R})^\perp$, that is, of mass conservation. This minimum cannot be reached at the boundaries as is easily checked by inspecting the sign of the normal derivative. As a consequence, it is reached in the interior, and, thanks to the strict convexity of $\mathfrak{F}$, this minimum is unique. Since the minimum is reached in the interior of the reaction simplex, we must have

$$\mu^{\mathrm{e}} = \mu(T^{\mathrm{e}}, \varrho^{\mathrm{e}}) \in (M\mathcal{R})^\perp. \tag{6.5.5}$$

As a consequence, $\zeta = -\langle \mu, M\omega \rangle$ vanishes at $(\varrho^{\mathrm{e}}, T^{\mathrm{e}})$, which is therefore an equilibrium point. Conversely, from (6.5.4), any equilibrium point on the reaction simplex is such that the quantities $\langle \mu, M\nu_i \rangle$, $i \in R$, vanish, so that the partial derivatives of $\mathfrak{F}$ along the reaction simplex are zero. Since $\mathfrak{F}$ is a strictly convex function of ϱ, it reaches a minimum at this point. Therefore, this point coincides with the unique minimum of $\mathfrak{F}$ and the proof is complete. ∎

Remark 6.5.4. As a typical example of conservation manifold and equilibrium point, consider a homogeneous mixture governed by the simplified equations $\partial_t \rho_k = m_k \omega_k$, $k \in S$. Then, for any vector $u \in (M\mathcal{R})^\perp$, we obtain $\langle \varrho(t) - \varrho^{\mathrm{f}}, u \rangle = 0$, where ϱ^{f} denotes the initial state $\varrho^{\mathrm{f}} = \varrho(0)$. As a consequence, the equilibrium limit point will be in the affine conservation manifold $\varrho^{\mathrm{f}} + M\mathcal{R}$. More generally, consider a closed tank occupying a domain Ω and assume Neumann boundary conditions for the temperature T and the mass fractions Y at the tank boundary $\partial\Omega$. Integrating the conservation equations over the whole tank, making use of boundary conditions, then yields that the limit equilibrium point is in $\varrho^{\mathrm{f}} + M\mathcal{R}$, where ϱ^{f} is now the initial average $(1/|\Omega|) \int_\Omega \varrho(0, \boldsymbol{x})\, d\boldsymbol{x}$. ∎

6.5.3. Equilibrium points with h and p fixed

In this section we investigate existence and uniqueness of equilibrium points with h and p fixed. These results will be used in Chapter 11 when investigating steady plane flame problems.

We use the formalism of Section 6.3 associated with the state variables $(h, p, Y_1, \ldots, Y_n)$, which is suited to the problem. Consistently, equilibrium points have to be investigated in an atom conservation manifold in the form $Y^{\mathrm{f}} + M\mathcal{R}$ with a given specific enthalpy $h^{\mathrm{e}} = h^{\mathrm{f}}$ and a given pressure $p^{\mathrm{e}} = p^{\mathrm{f}}$. Therefore, the equilibrium mass fractions Y^{e} are sought in the simplex

$$\mathfrak{Y} = (Y^{\mathrm{f}} + M\mathcal{R}) \cap (0,\infty)^n \cap \{\, Y, \sum_{k\in S} Y_k h_k^0 < h^{\mathrm{f}} \,\}. \tag{6.5.6}$$

Moreover, in the previous section, the state $(T^{\mathrm{f}}, p^{\mathrm{f}}, Y^{\mathrm{f}})$ defining the reaction simplex was taken to be such that $Y^{\mathrm{f}} > 0$. In order to generalize this assumption, we only assume in this section that $T^{\mathrm{f}} > 0$, $p^{\mathrm{f}} > 0$, $Y^{\mathrm{f}} \geq 0$, $\langle Y^{\mathrm{f}}, \mathcal{U}\rangle = 1$, that is, the initial state Y^{f} is allowed to have zero mass fractions. Therefore, in order to obtain a nonempty simplex, we also have to assume that

$$(Y^{\mathrm{f}} + M\mathcal{R}) \cap (0,\infty)^n \neq \emptyset. \tag{6.5.7}$$

When this property does not hold, one simply has to eliminate from the chemical mechanism the species that cannot be reached starting from Y^{f}. This property (6.5.7) is weaker than the reachability property needed for homogeneous reactors introduced in [VH85].

Proposition 6.5.5. *Assume that* (Th_1)–(Th_4) *hold and that* $Y^{\mathrm{f}} \geq 0$, $\langle Y^{\mathrm{f}}, \mathcal{U}\rangle = 1$, $\sum_{k\in S} Y_k^{\mathrm{f}} h_k^0 < h^{\mathrm{f}}$, *and* $(Y^{\mathrm{f}} + M\mathcal{R}) \cap (0,\infty)^n \neq \emptyset$. *Then there exists a unique equilibrium vector* Y^{e} *in the simplex* $\mathfrak{Y}$. *At* Y^{e}, *the chemical source term* ω *vanishes, the reaction rates of progress also vanish, and we have* $(g_1^{\mathrm{e}}, \ldots, g_n^{\mathrm{e}})^t \in (M\mathcal{R})^{\perp}$.

Proof. First note that $\mathfrak{Y}$ is nonempty since there exists $Y^{\mathrm{p}} \in Y^{\mathrm{f}} + M\mathcal{R}$ with $Y^{\mathrm{p}} > 0$ so that $(1-\tau)Y^{\mathrm{f}} + \tau Y^{\mathrm{p}}$ is in $\mathfrak{Y}$ for small positive τ. We characterize the equilibrium point Y^{e} as the only extremum of the function $Y \to \breve{s}(h^{\mathrm{f}}, p^{\mathrm{f}}, Y)$, where $\breve{s}(h,p,Y) = s(T,p,Y)$ denotes the entropy s as a function of (h,p,Y). The function $\breve{s}$ is a C^∞ function of Y over $\mathfrak{Y}$, and the partial derivative of $\breve{s}(h^{\mathrm{f}}, p^{\mathrm{f}}, Y)$ with respect to the mass fractions Y is

$$\partial_Y \, \breve{s}(h^{\mathrm{f}}, p^{\mathrm{f}}, Y) = -R \, \breve{\mu}(h^{\mathrm{f}}, p^{\mathrm{f}}, Y),$$

where $\breve{\mu} = (\breve{\mu}_1, \ldots, \breve{\mu}_n)^t$ and $\breve{\mu}_k(h,p,Y) = g_k(T,p,Y)/RT$. We claim that the function $\breve{s}(h^{\mathrm{f}}, p^{\mathrm{f}}, Y)$ is also strictly concave over the convex set $\mathfrak{Y}$. Indeed, we have from (6.3.33)

$$-\frac{1}{R} \sum_{k,l\in S} (\partial^2_{Y_k Y_l} \breve{s}) x_k x_l \geq \sum_{k\in S} \frac{(x_k/m_k)^2}{Y_k/m_k} - \frac{(\sum_{k\in S} x_k/m_k)^2}{\sum_{l\in S} Y_l/m_l}, \tag{6.5.8}$$

so that $\partial_Y^2 \breve{s}$ is negative semidefinite and $N(\partial_Y^2 \breve{s}) \subset \mathbb{R}Y$. However, we have $Y \notin M\mathcal{R}$ from $M\mathcal{R} \subset \mathcal{U}^{\perp}$ and $Y \notin \mathcal{U}^{\perp}$, since $Y > 0$ over $\mathfrak{Y}$, so that $\partial_Y^2 \breve{s}$ is negative definite over $M\mathcal{R}$ and the strict concavity of $\breve{s}$ over $\mathfrak{Y}$ follows.

Because we want to maximize $\check{s}(h^{\mathrm{f}}, p^{\mathrm{f}}, Y)$, we only need to consider the subset $\mathfrak{Y}^*$

$$\mathfrak{Y}^* = \{\, Y \in \mathfrak{Y},\ \check{s}(h^{\mathrm{f}}, p^{\mathrm{f}}, Y) > \check{s}(h^{\mathrm{f}}, p^{\mathrm{f}}, Y^{\mathrm{f}}) - 1 \,\}.$$

This set $\mathfrak{Y}^*$ is convex and nonempty since $\check{s}$ is concave and we have $(1-\tau)Y^{\mathrm{f}} + \tau Y^{\mathrm{p}} \in \mathfrak{Y}^*$ for τ small and positive. The mass fractions are bounded over $\mathfrak{Y}^*$ since $Y \in \mathfrak{Y}^*$ implies $Y > 0$ and $\langle Y, \mathcal{U}\rangle = \langle Y^{\mathrm{f}}, \mathcal{U}\rangle$ from the mass conservation constraint $\mathcal{U} \in (M\mathcal{R})^{\perp}$. Moreover the temperature $\check{T}(h^{\mathrm{f}}, p^{\mathrm{f}}, Y)$ is easily seen to be positively bounded from below over $\mathfrak{Y}^*$ since $\check{s}$ is bounded from below. This implies that there exists a positive α such that $\alpha \leq \check{T}(h^{\mathrm{f}}, p^{\mathrm{f}}, \mathfrak{Y}^*)$. As a consequence, the function $\check{s}(h^{\mathrm{f}}, p^{\mathrm{f}}, Y)$ can be extended by continuity over the closure $\overline{\mathfrak{Y}^*}$ of $\mathfrak{Y}^*$. Therefore, $\check{s}(h^{\mathrm{f}}, p^{\mathrm{f}}, Y)$ admits a maximum on this nonempty convex compact set. Denoting Y^{m} any point where the maximum is reached, we claim that Y^{m} is not on the boundary of $\mathfrak{Y}^*$. Of course, it cannot be reached at the boundary $\check{s} = \check{s}(h^{\mathrm{f}}, p^{\mathrm{f}}, Y^{\mathrm{f}}) - 1$. In addition, it cannot be reached at the boundaries $Y_k = 0$, $k \in S$. Indeed, arguing by contradiction, we consider a point Y^{in} in the interior of $\mathfrak{Y}^*$ and the function $u(\tau) = \check{s}\big(h^{\mathrm{f}}, p^{\mathrm{f}}, \tau Y^{\mathrm{in}} + (1-\tau) Y^{\mathrm{m}}\big)$ which is continuous over $[0,1]$, differentiable over $(0,1]$, and reaches its maximum for $\tau = 0$. Over the interval $(0,1]$, the derivative of u is given by

$$u'(\tau) = \langle \partial_Y \check{s}, Y^{\mathrm{in}} - Y^{\mathrm{m}}\rangle = \sum_{\substack{k\in S\\ Y_k^{\mathrm{m}}=0}} (\partial_{Y_k}\check{s}) Y_k^{\mathrm{in}} + \sum_{\substack{k\in S\\ Y_k^{\mathrm{m}}>0}} (\partial_{Y_k}\check{s})(Y_k^{\mathrm{in}} - Y_k^{\mathrm{m}}),$$

with $\partial_Y \check{s}$ evaluated at $\big(h^{\mathrm{f}}, p^{\mathrm{f}}, \tau Y^{\mathrm{in}} + (1-\tau)Y^{\mathrm{m}}\big)$. This implies that u' is positive in the neighborhood of $\tau = 0$, since $\partial_{Y_k}\check{s}\big(h^{\mathrm{f}}, p^{\mathrm{f}}, \tau Y^{\mathrm{in}} + (1-\tau)Y^{\mathrm{m}}\big)$ goes to $+\infty$ for $\tau \to 0$ when $Y_k^{\mathrm{m}} = 0$, whereas the sum over $Y_k^{\mathrm{m}} > 0$ remains bounded, an obvious contradiction. As a consequence, $\check{s}(h^{\mathrm{f}}, p^{\mathrm{f}}, Y)$ reaches its maximum in the interior of $\mathfrak{Y}$, and, thanks to the strict concavity of $\check{s}(h^{\mathrm{f}}, p^{\mathrm{f}}, Y)$, this maximum is unique and we denote by Y^{e} the corresponding point.

Since this maximum is reached in the interior of $\mathfrak{Y}^*$, we must have

$$\check{\mu}^{\mathrm{e}} = (1/RT^{\mathrm{e}})(g_1^{\mathrm{e}}, \ldots, g_n^{\mathrm{e}})^t \in (M\mathcal{R})^{\perp}. \tag{6.5.9}$$

As a consequence, $\zeta = -\langle \mu, M\omega\rangle$ vanishes at $(h^{\mathrm{e}}, p^{\mathrm{e}}, Y^{\mathrm{e}}) = (h^{\mathrm{f}}, p^{\mathrm{f}}, Y^{\mathrm{e}})$, which is therefore an equilibrium point. Conversely, from Proposition 6.5.1, any equilibrium point on the simplex $\mathfrak{Y}$ is such that the quantities $\langle \check{\mu}, M\nu_i\rangle$, $i \in R$, vanish, so that the partial derivatives of $\check{s}(h^{\mathrm{f}}, p^{\mathrm{f}}, Y)$ along the simplex are zero. Since $\check{s}(h^{\mathrm{f}}, p^{\mathrm{f}}, Y)$ is a strictly concave function over $\mathfrak{Y}$, it reaches a maximum at this point. Therefore, this point coincides with the unique maximum of $\check{s}(h^{\mathrm{f}}, p^{\mathrm{f}}, Y)$ and the proof is complete. ∎

Remark 6.5.6. As a typical example of conservation manifold and equilibrium point, consider a mixture governed by one-dimensional simplified equations in the form $c\partial_x Y_k = m_k\omega_k$ for $k \in S$, and $x \geq 0$, where c is a constant mass flux. Then for any vector $u \in (M\mathcal{R})^\perp$, we obtain $\langle Y(x) - Y^{\mathrm{f}}, u\rangle = 0$, where $Y^{\mathrm{f}} = Y(0)$ is the state at the origin $x = 0$. As a consequence, any limit point will be in the simplex $Y^{\mathrm{f}} + M\mathcal{R}$. ■

Remark 6.5.7. The assumption $(Y^{\mathrm{f}} + M\mathcal{R}) \cap (0, \infty)^n \neq \emptyset$ implies in particular that $\langle Y^{\mathrm{f}}, M^{-1}\mathfrak{a}_l\rangle > 0$ for $l \in \mathfrak{A}$. Indeed, there exist scalars α_i, $i \in R$, such that $Y^{\mathrm{f}} + \sum_{i\in R} \alpha_i M\nu_i$ is positive. Taking the scalar product with $M^{-1}\mathfrak{a}_l$ and using the element conservation relations (6.4.12) directly yields $\langle Y^{\mathrm{f}}, M^{-1}\mathfrak{a}_l\rangle > 0$. ■

6.5.4. Smoothness of equilibrium points

Smoothness of equilibrium points is easily obtained by using the implicit function theorem. Very different equilibrium points can exist, e.g., with various thermal properties kept fixed. For illustration, we only consider here the equilibrium points obtained with (h, p) fixed, but the method is very general and applies to all situations.

Proposition 6.5.8. *Let $\mathcal{P}$ be the orthogonal projector in $\mathbb{R}^{2+n}$ onto the linear space $(0, 0, M\mathcal{R})^\perp$. Then the equilibrium point $(h^{\mathrm{e}}, p^{\mathrm{e}}, Y^{\mathrm{e}})$ defined in Proposition 6.5.5 only depends on $\mathcal{P}(h^{\mathrm{f}}, p^{\mathrm{f}}, Y^{\mathrm{f}})$ and is a smooth function of $\mathcal{P}(h^{\mathrm{f}}, p^{\mathrm{f}}, Y^{\mathrm{f}})$.*

Proof. The equilibrium point $(h^{\mathrm{e}}, p^{\mathrm{e}}, Y^{\mathrm{e}})$ only depends on $\mathcal{P}(h^{\mathrm{f}}, p^{\mathrm{f}}, Y^{\mathrm{f}})$ by construction since $(h^{\mathrm{f}}, p^{\mathrm{f}}, Y^{\mathrm{f}}) - \mathcal{P}(h^{\mathrm{f}}, p^{\mathrm{f}}, Y^{\mathrm{f}}) \in M\mathcal{R}$. Denote by $v_1, \ldots, v_n$ an orthonormal basis of $\mathbb{R}^n$, such that $v_1, \ldots, v_d$ is a basis of $M\mathcal{R}$ and $v_{d+1}, \ldots, v_n$ a basis of $(M\mathcal{R})^\perp$, where $d = \dim(\mathcal{R})$. The vectors $(1, 0, O)^t$, $(0, 1, O)^t$—where $O = (0, \ldots, 0)^t \in \mathbb{R}^n$—and $(0, 0, v_i)^t$, for $d + 1 \leq i \leq n$, form a basis of $(0, 0, M\mathcal{R})^\perp$ and the corresponding components of the vector $\mathcal{P}(h^{\mathrm{f}}, p^{\mathrm{f}}, Y^{\mathrm{f}})$ are h^{f}, p^{f}, and $\langle Y^{\mathrm{f}}, v_i\rangle$, for $d + 1 \leq i \leq n$, respectively. The equilibrium point $(h^{\mathrm{e}}, p^{\mathrm{e}}, Y^{\mathrm{e}})$ is then the unique solution (h, p, Y) of the system

$$\begin{cases} h = h^{\mathrm{f}}, & \\ p = p^{\mathrm{f}}, & \\ \langle Y, v_i\rangle = \langle Y^{\mathrm{f}}, v_i\rangle, & d + 1 \leq i \leq n, \\ \langle \tilde{\mu}(h, p, Y), v_i\rangle = 0, & 1 \leq i \leq d. \end{cases}$$

The Jacobian matrix of this system is easily shown to have full rank at the equilibrium point $(h^{\mathrm{e}}, p^{\mathrm{e}}, Y^{\mathrm{e}})$ since $N(\partial_Y \tilde{\mu}) \cap (M\mathcal{R}) = \{0\}$, so that the implicit function theorem applies. ■

6.6. Boundary equilibrium points

In Section 6.5, we have investigated existence and uniqueness of positive equilibrium points in atom conservation manifolds. From Proposition 6.5.1, we know that any point with positive mass fractions where the chemical production terms vanish coincides with the unique equilibrium point on the proper atom conservation manifold. In this section, we further investigate the boundaries—with respect to species mass densities—of atom conservation manifolds, which may hide spurious points where the chemical production terms vanish.

Positive equilibrium points depend on the linear space spanned by the reaction vectors, whereas boundary equilibrium points depend on the actual values of the forward and backward stoichiometric coefficients. We introduce, in particular, a natural assumption concerning the chemical reaction mechanism, the species decomposition chain property, which automatically eliminates these spurious boundary equilibrium points.

6.6.1. Definition of boundary equilibrium points

Definition 6.6.1. *A point (T, ϱ) with $T > 0$ and $\varrho \geq 0$, $\varrho \neq 0$—or equivalently a point (T, p, Y) with $T > 0$, $p > 0$, and $Y \geq 0$, $Y \neq 0$—is said to be a boundary equilibrium point if $\omega(T, \varrho) = 0$ and if there exists at least a species $k \in S$ such that $\rho_k = 0$.*

Note that it is always possible to construct chemical mechanisms with boundary equilibrium points, e.g., by adding a given species both as a reactant and as a product in each reaction of a given mechanism. By repeating this operation for different species and by taking the union of the corresponding modified mechanisms, it is readily seen that any given vertex or corner can be turned into a set of boundary equilibrium points.

Proposition 6.6.2. *Assume that (T, ϱ) is a boundary equilibrium point, and define $S^+ = \{ k \in S,\ Y_k > 0 \}$ and $S^0 = \{ k \in S,\ Y_k = 0 \}$. Let us introduce the sets*

$$R^+ = \{ \ i \in R,\ \forall k \in S^0,\ \nu^{\mathrm{f}}_{ki} = 0 \text{ and } \nu^{\mathrm{b}}_{ki} = 0 \ \},$$

$$R^0 = \{ \ i \in R,\ \exists k \in S^0,\ \nu^{\mathrm{f}}_{ki} > 0 \text{ or } \nu^{\mathrm{b}}_{ki} > 0 \ \}.$$

Then, at (T, ϱ), we have $\tau_i = 0$, for all $i \in R$ and ϱ^+ is an equilibrium point of the S^+ submixture for the R^+ chemical submechanism.

Proof. For any $k \in S^0$, the destruction rate vanishes since $\mathcal{D}_k = \rho_k \widehat{\mathcal{D}}_k = 0$, so that the condition $\omega_k = 0$ implies that $\mathcal{C}_k = 0$. Since $\mathcal{C}_k$ is a sum of nonnegative terms, all of these terms must vanish. This shows that $\tau_i = 0$ for any reaction $i \in R^0$. On the other hand, considering the submixture S^+, and the chemical submechanism R^+—which may be empty—we conclude that $\varrho^+ = \{ \rho_k,\ k \in S^+ \}$ is an equilibrium point of the submixture S^+. ∎

Note that equilibrium points—which are interior to atom conservation manifolds—only depend on the linear space $\mathcal{R}$ spanned by the reaction vectors $\nu_i = \nu_i^{\mathrm{b}} - \nu_i^{\mathrm{f}}$, $i \in R$. On the other hand, boundary equilibrium points depend on the effective values of the integer stoichiometric coefficients ν_{ki}^{f} and ν_{ki}^{b}, $i \in R$, $k \in S$.

6.6.2. Decomposition chain property

In this section, we introduce a sufficient condition on stoichiometric coefficients that automatically eliminates boundary equilibrium points, provided that all atoms are present in the reactive mixture.

Definition 6.6.3. *A chemical mechanism will be said to have the decomposition chain property if, for any $T > 0$ and any $\varrho \geq 0$, $\varrho \neq 0$, we have*

$$\omega(T, \varrho) = 0 \text{ and } \exists k \in S,\ \rho_k = 0 \quad \Longrightarrow \quad \exists l \in \mathfrak{A},\ \langle \varrho, M^{-1}\mathfrak{a}_l \rangle = 0.$$

In other words, a boundary equilibrium point can only be obtained provided that one atom concentration $\langle \varrho, M^{-1}\mathfrak{a}_l \rangle$ is zero at this point. The reaction scheme describing the combustion of hydrogen in air [GS87] which will be used in Chapter 12, has the decomposition chain property for instance. Heuristically, a reaction mechanism has the decomposition chain property when sufficient three body recombination/decomposition reactions are taken into account. Three body reactions are typically in the form

$$\mathfrak{M}_k + \mathfrak{M}_l + \mathrm{M} \rightleftharpoons \mathfrak{M}_m + \mathrm{M},$$

where M denotes any species present in the mixture. These reactions yield the property that $\rho_m = 0$ if and only if $\rho_k = 0$ or $\rho_l = 0$ whenever chemical source terms vanish. In this situation, the decomposition reactions form chains that link the largest molecules to the atomic species and propagate the zero concentration property to at least one atomic species. This implies, in turn, that all species containing the corresponding atom are missing in the mixture, still making use of three body reactions, so that the corresponding atom concentration is zero.

6.7. Inequalities near equilibrium

In this section, we investigate various inequalities involving the chemical entropy production, also termed the chemical dissipation rate. We first compare the production rates squared and the chemical dissipation rate.

We then compare the difference of entropy with its value at equilibrium and the chemical dissipation rate. In this section, these inequalities are investigated in the neighborhood of equilibrium points. In Section 6.8, we will investigate global extensions over atom conservation manifolds.

6.7.1. Production rates and chemical dissipation

In this section we compare the chemical production rates squared with the entropy production due to chemical reactions. This inequality is only valid locally and has no global extension for large ϱ.

Proposition 6.7.1. *Let $(T^{\rm e}, \varrho^{\rm e})$ be an equilibrium point. Then, there exists a positive constant δ such that, in the neighborhood of $(T^{\rm e}, \varrho^{\rm e})$, we have*

$$\delta\,|\omega|^2 \le -\sum_{k\in S} \frac{g_k m_k \omega_k}{RT}. \tag{6.7.1}$$

Proof. We have from (6.5.4)

$$\zeta = \sum_{i\in R} K_i^{\rm s}\Big(\langle M\nu_i^{\rm f},\mu\rangle - \langle M\nu_i^{\rm b},\mu\rangle\Big)\Big(\exp\langle M\nu_i^{\rm f},\mu\rangle - \exp\langle M\nu_i^{\rm b},\mu\rangle\Big).$$

Denoting by $\mathfrak{V}$ any closed ball centered on $(T^{\rm e}, \varrho^{\rm e})$ where the temperature and species remain positive, this implies

$$\alpha \sum_{i\in R} \langle \mu, M\nu_i\rangle^2 \le -\langle \mu, M\omega\rangle \le \beta \sum_{i\in R} \langle \mu, M\nu_i\rangle^2, \tag{6.7.2}$$

for positive constants α and β depending on $\mathfrak{V}$. On the other hand, the relations $M\omega = \sum_{i\in R} \tau_i M\nu_i$ and (6.4.26) also imply that

$$|M\omega|^2 \le \beta' \sum_{i\in R} \tau_i^2 \le \beta'' \sum_{i\in R} \langle \mu, M\nu_i\rangle^2, \tag{6.7.3}$$

where β' and β'' are positive constants. Combining inequalities (6.7.2) and (6.7.3) completes the proof. ■

Corollary 6.7.2. *Let $(T^{\rm e}, \varrho^{\rm e})$ be an equilibrium point. Then, there exists a positive constant δ such that, in the neighborhood of $(T^{\rm e}, \varrho^{\rm e})$, we have*

$$\delta\,|\omega|^2 \le -\big\langle \mu - \mu^{\rm e}, M\omega\big\rangle. \tag{6.7.4}$$

Proof. We indeed have $\mu^{\rm e} \in (M\mathcal{R})^\perp$, so that

$$\zeta = -\big\langle \mu, M\omega\big\rangle = -\big\langle \mu - \mu^{\rm e}, M\omega\big\rangle, \tag{6.7.5}$$

and inequality (6.7.4) follows directly from Proposition 6.7.1.

6.7.2. Entropy difference and chemical dissipation

In this section we obtain a stability inequality first derived by Boillat and Pousin under similar assumptions [Boi95] [Pou93]. This inequality compares the entropy production due to chemical reactions ζ to the difference of entropy with its value at equilibrium, on a given atom conservation manifold. Both quantities indeed behave as quadratics near the equilibrium point.

Various types of equilibrium could be considered, and we investigate, as a typical example, equilibrium with h and p fixed. For convenience, we use the second formalism of Section 6.5.3 in terms of mass fractions. These results will be extended in Section 6.8 and used in Chapter 11 when investigating plane flames.

First, we remark that the manifold $(h^{\mathrm{f}}, p^{\mathrm{f}}, Y^{\mathrm{f}}) + (0, 0, M\mathcal{R})$ can be rewritten in the form $\mathcal{P}(x) = \mathcal{P}(x^{\mathrm{f}})$, where $x = (h, p, Y)$, $x^{\mathrm{f}} = (h^{\mathrm{f}}, p^{\mathrm{f}}, Y^{\mathrm{f}})$, and $\mathcal{P}$ is the orthogonal projector onto $(0, 0, M\mathcal{R})^{\perp}$. Equivalently, this manifold can be rewritten in the form $\mathcal{P}(x) = \mathcal{P}(x^{\mathrm{e}})$ with $x^{\mathrm{e}} = (h^{\mathrm{e}}, p^{\mathrm{e}}, Y^{\mathrm{e}})$.

Proposition 6.7.3. *There exists a neigborhood $\mathfrak{V}$ of the equilibrium point $x^{\mathrm{e}} = (h^{\mathrm{e}}, p^{\mathrm{e}}, Y^{\mathrm{e}})$ and a constant δ such that*

$$\forall x \in \mathfrak{V} \cap \{\, \mathcal{P}(x - x^{\mathrm{e}}) = 0 \,\}, \qquad \delta\big(s^{\mathrm{e}} - s(x)\big) \leq -\sum_{k \in S} \frac{g_k m_k \omega_k}{RT}.$$

Proof. We introduce the notation $\breve{\mu}_k(h, p, Y) = g_k(T, p, Y)/RT$, $k \in S$, $\breve{s}_k(h, p, Y) = s_k(T, p, Y)$, $k \in S$, and $\breve{\mu} = (\breve{\mu}_1, \ldots, \breve{\mu}_n)^t$. We have seen in (6.5.4) that

$$\zeta = \sum_{i \in R} K_i^{\mathrm{s}} \Big(\langle M\nu_i^{\mathrm{f}}, \breve{\mu} \rangle - \langle M\nu_i^{\mathrm{b}}, \breve{\mu} \rangle \Big) \Big(\exp\langle M\nu_i^{\mathrm{f}}, \breve{\mu} \rangle - \exp\langle M\nu_i^{\mathrm{b}}, \breve{\mu} \rangle \Big), \quad (6.7.6)$$

where K_i^{s}, $i \in R$, are positive functions. As a consequence, there exists a neighborhood of $x^{\mathrm{e}} = (h^{\mathrm{e}}, p^{\mathrm{e}}, Y^{\mathrm{e}})$ such that

$$\delta \sum_{i \in R} \langle M\nu_i, \breve{\mu} \rangle^2 \leq \zeta, \quad (6.7.7)$$

where δ is a positive constant. On the other hand, we note that

$$\breve{\mu} - \breve{\mu}^{\mathrm{e}} = \int_0^1 (\partial_Y \breve{\mu})\big(h^{\mathrm{e}}, p^{\mathrm{e}}, Y^{\mathrm{e}} + \tau(Y - Y^{\mathrm{e}})\big)\, d\tau \, (Y - Y^{\mathrm{e}}).$$

Using the positive definiteness of $R\partial_Y \breve{\mu} = -\partial_Y^2 \breve{s}$ over $M\mathcal{R}$, established in the proof of Proposition 6.5.5, we obtain that, in a neighborhood of x^{e} and on the simplex $\mathcal{P}(x - x^{\mathrm{e}}) = 0$, we have

$$\big\| Y - Y^{\mathrm{e}} \big\|^2 \leq \beta \big| \langle \breve{\mu} - \breve{\mu}^{\mathrm{e}}, Y - Y^{\mathrm{e}} \rangle \big|.$$

Denoting by Π the orthogonal projector onto $M\mathcal{R}$, this implies that

$$\|Y - Y^{\mathrm{e}}\| \leq \beta \left\|\Pi(\breve{\mu} - \breve{\mu}^{\mathrm{e}})\right\|,$$

since $Y - Y^{\mathrm{e}} = \Pi(Y - Y^{\mathrm{e}})$ on $\mathcal{P}(x - x^{\mathrm{e}}) = 0$. Moreover, we also have

$$\breve{s} - \breve{s}^{\mathrm{e}} = \Big\langle \int_0^1 \int_0^\tau (\partial_Y^2 \breve{s})\big(h^{\mathrm{e}}, p^{\mathrm{e}}, Y^{\mathrm{e}} + t(Y - Y^{\mathrm{e}})\big)\, dt\, d\tau\, (Y - Y^{\mathrm{e}}),\ (Y - Y^{\mathrm{e}})\Big\rangle,$$

so that, in the neighborhood of Y^{e}, we have

$$\breve{s}^{\mathrm{e}} - \breve{s} \ \leq\ \beta\, \|Y - Y^{\mathrm{e}}\|^2, \tag{6.7.8}$$

since we stay on the simplex $x^{\mathrm{e}} + (0, 0, M\mathcal{R})$, where $\partial_Y^2 \breve{s}$ is negative definite. Combining the above inequalities yields

$$\breve{s}^{\mathrm{e}} - \breve{s} \ \leq\ \beta \left\|\Pi(\breve{\mu} - \breve{\mu}^{\mathrm{e}})\right\|^2, \tag{6.7.9}$$

and using now

$$\left\|\Pi(\breve{\mu} - \breve{\mu}^{\mathrm{e}})\right\|^2 \ \leq\ \beta \sum_{i \in R} \langle M\nu_i, \breve{\mu} - \breve{\mu}^{\mathrm{e}}\rangle^2 = \beta \sum_{i \in R} \langle M\nu_i, \breve{\mu}\rangle^2,$$

since $\breve{\mu}^{\mathrm{e}} \in (M\mathcal{R})^{\perp}$, and inequality (6.7.7) completes the proof. ∎

A straightforward extension of the preceding proposition yields the following inequality.

Corollary 6.7.4. *There exists a neighborhood* $\mathfrak{V}$ *of the equilibrium point* $x^{\mathrm{e}} = (h^{\mathrm{e}}, p^{\mathrm{e}}, Y^{\mathrm{e}})$ *and a positive constant* δ *such that*

$$\forall x \in \mathfrak{V}, \qquad \delta\Big(s\big(\pi^{\mathrm{e}}(x)\big) - s(x)\Big) \ \leq\ -\sum_{k \in S} \frac{g_k m_k \omega_k}{RT}, \tag{6.7.10}$$

where $\pi^{\mathrm{e}}(x)$ *denotes the unique equilibrium point on* $x + (0, 0, M\mathcal{R})$, *that is, the unique equilibrium point* π^{e}, *such that* $\mathcal{P}(\pi^{\mathrm{e}}) = \mathcal{P}(x)$.

6.8. A global stability inequality

In this section, we investigate a global version of the inequality established in Proposition 6.7.3 and Corollary 6.7.4. More specifically, we extend this inequality to the whole atom conservation manifold, provided that the reaction mechanism and the atom conservation simplex are such that no boundary equilibrium point exists. This assumption is easily shown to be a necessary requirement.

In order to obtain a global stability inequality, we also assume that the temperature is bounded, that is, we investigate a global inequality in the set

$$\mathcal{M}_\epsilon = \{ \ x = (h, p, Y), \ \alpha \leq T \leq \beta, \ Y > 0, \ \|\mathcal{P}(x - x^{\mathrm{e}})\| \leq \epsilon \ \}. \tag{6.8.1}$$

Proposition 6.8.1. *Let α and β be positive constants with $\alpha < T^{\mathrm{e}} < \beta$, and assume that there are no boundary equilibrium points in the set*

$$(Y^{\mathrm{f}} + M\mathcal{R}) \cap [0, \infty)^n \cap \{ Y, \sum_{k \in S} Y_k h_k^0 < h^{\mathrm{f}} \}.$$

Then, there exist $\epsilon_0 > 0$ and $\delta > 0$ such that

$$\forall x \in \mathcal{M}_{\epsilon_0} \qquad \delta\Big(s\big(\pi^{\mathrm{e}}(x)\big) - s(x) \Big) \ \leq \ - \sum_{k \in S} \frac{g_k m_k \omega_k}{RT},$$

where $\pi^{\mathrm{e}}(x)$ denotes the unique equilibrium point on $x + (0, 0, M\mathcal{R})$, that is, the unique equilibrium point π^{e} such that $\mathcal{P}(\pi^{\mathrm{e}}) = \mathcal{P}(x)$.

This proposition is proved by using a compacity argument. We already know from Corollary 6.7.4 that such an inequality holds near equilibrium points. We also know that the dissipation rate $\zeta(x)$ is positive for $Y > 0$ when x is not an equilibrium point. In addition, $s\big(\pi^{\mathrm{e}}(x)\big) - s(x)$ is globally bounded over the set $\mathcal{M}_\epsilon$. As a consequence, we only need to investigate the behavior of the chemical dissipation rate ζ near the boundary—with respect to zero mass fractions—of the set $\mathcal{M}_\epsilon$.

Lemma 6.8.2. *Consider the set $\mathcal{M}_\epsilon$, where α and β are positive constants such that $\alpha < T^{\mathrm{e}} < \beta$. Then, there exists $\epsilon > 0$ such that the functional ζ is bounded from below by a positive constant near the zero mass fraction boundary.*

Proof. Assume that ϵ is small enough so that there are no boundary equilibrium points in the set $\|\mathcal{P}(x - x^{\mathrm{e}})\| \leq \epsilon$. Consider then a boundary point $(T^{\mathrm{b}}, p^{\mathrm{b}}, Y^{\mathrm{b}})$, such that Y^{b} is nonpositive. We can introduce the sets S^+, S^0, R^+, and R^0, associated with Y^{b}, as in Proposition 6.6.2. For Y in the neighborhood of Y^{b} and Y positive, we now write

$$\zeta = \frac{1}{RT}\Big(\sum_{k \in S^0} g_k m_k (\rho Y_k \widehat{\mathcal{D}}_k - \mathcal{C}_k) - \sum_{k \in S^+} g_k m_k \omega_k \Big).$$

We have $-g_l m_l \mathcal{C}_l \geq 0$, $g_l Y_l \to 0$, and $-g_l \to +\infty$, for $l \in S^0$, when $Y \to Y^{\mathrm{b}}$. On the other hand, the quantities $g_l m_l \omega_l$, $l \in S^+$, remain bounded in the neighborhood of Y^{b}. As a consequence, if there exists $k \in S^0$ with $\mathcal{C}_k^{\mathrm{b}} > 0$, then ζ goes to $+\infty$ as $Y \to Y^{\mathrm{b}}$, since $-g_k m_k \mathcal{C}_k \to +\infty$. On the other hand, if $\mathcal{C}_k^{\mathrm{b}} = 0$ for all $k \in S^0$, then the term $-\sum_{k\in S^0} g_k m_k \mathcal{C}_k$ is nonnegative if Y is sufficiently close to Y^{b}—although it may be arbitrary large,—whereas the last term goes to

$$-\frac{1}{RT^{\mathrm{b}}} \sum_{k\in S^+} g_k m_k \omega_k^+,$$

where ω^+ only involves the chemical reactions of R^+ associated with the S^+ submixture. This limit is then positive, otherwise Y^{b} would be a boundary equilibrium point from Proposition 6.5.1 applied to the submixture S^+. ∎

6.9. Notes

6.1. The two alternative thermodynamic formulations can easily be understood by considering the limiting situation of homogeneous mixtures. For uniform mixtures, a natural choice of state variables indeed exists. Denoting by V the volume of gas under consideration, M_k the mass of species k in V, the classical natural variables are $(T, p, M_1, \dots, M_n)$. Denoting by Σ the entropy of the gas in V, we have

$$\Sigma = \sum_{k\in S} M_k s_k,$$

where s_k can be written as

$$s_k = s_k^{\mathrm{st}} + \int_{T^{\mathrm{st}}}^{T} \frac{c_{vk}(T')}{T'}\, dT' - r_k \log\left(\frac{M_k}{V\gamma^{\mathrm{st}} m_k}\right). \qquad (6.9.1)$$

Similarly, denoting by Θ the enthalpy of the gas in volume V, we have

$$\Theta = \sum_{k\in S} M_k h_k,$$

and similar relations hold for all extensive thermodynamic functions. However, considering now intensive *densities* per unit volume instead of extensive quantities, the functions Θ, Σ, and $M_1, \dots, M_n$ have to be replaced by Θ/V, Σ/V, and $M_1/V, \dots, M_n/V$, that is, by $\mathcal{H}$, $\mathcal{S}$, and $\rho_1, \dots, \rho_n$. In this situation, the variables $(T, V, M_1, \dots, M_n)$ are transformed into $(T, 1, M_1/V, \dots, M_n/V)$, which are no longer independent. Similarly, starting from the variables $(T, p, M_1, \dots, M_n)$, we obtain—after dividing by the volume V—the reduced variables $(T, p, \rho_1, \dots, \rho_n)$, which again are not independent. This is consistent with the classical Gibbs–Duhem relation $s\,\mathbb{D}T - (1/\rho)\,\mathbb{D}p + \sum_{k\in S} Y_k\,\mathbb{D}g_k = 0$, which states that the natural set of *intensive* variables is a dependent set of variables. In summary, only

$n+1$ intensive variables are independent, whereas $n+2$ independent variables were previously available by making use of extensivity. In the first formalism, we simply *eliminate* one of the dependent intensive variables. More specifically, we either eliminate 1 from $(T, 1, M_1/V, \dots, M_n/V)$ or we eliminate the pressure from $(T, p, \rho_1, \dots, \rho_n)$ and we recover the variables $(T, \rho_1, \dots, \rho_n)$. Of course, we could have eliminated another variable instead of the pressure, but it would lead to the same type of formalism and be somewhat less elegant. Therefore, it appears that the first thermochemistry formalism is associated with *elimination* of one dependent intensive variable.

If we consider densities per unit mass instead of densities per unit volume, the quantities Θ, Σ, and $M_1, \dots, M_n$ have to be replaced by Θ/M, Σ/M, and $M_1/M, \dots, M_n/M$, that is, by h, s, and $Y_1, \dots, Y_n$, and the variables $(T, p, M_1, \dots, M_n)$ become $(T, p, M_1/M, \dots, M_n/M)$, that is to say, $(T, p, Y_1, \dots, Y_n)$. We again obtain a set of dependent variables since the mass fractions are supposed to satisfy a priori the constraint $\sum_{k \in S} Y_k = 1$. By eliminating one of the mass fractions, we would obtain a formalism similar to the preceding one, that is, we would obtain positive definite Hessians for the entropy. Instead of eliminating one of the mass fractions, however, we alternatively can consider the mass fractions as independent unknowns. In this situation, an a priori constraint between the mass fractions no longer exists, and the relation $\sum_{k \in S} Y_k = 1$ must be a posteriori *deduced* from the governing equations. This yields the second thermodynamics formalism, which involves singular Hessians, but is formally identical to the classical Gibbs thermodynamics in terms of intensive variables. Note that we could have also started from the variables $(T, V, M_1, \dots, M_n)$ and obtained the variables $\big(T, (1/\rho), Y_1, \dots, Y_n\big)$, which leads to a similar formalism.

Finally, we note that *both* formalisms have been used in the literature. We mention, in particular, Meixnier and Krambeck who have used the first formalism, and de Groot and Mazur, and Boillat and Pousin, who have used the second formalism.

6.2. The explosion of entropy for small or large temperature is of fundamental importance in the modeling. The explosion of entropy for small temperatures is associated with the fact that we consider gaseous mixtures governed by Boltzmann statistics. On the contrary, gaseous mixures governed by Fermi–Dirac or Bose–Einstein statistics [Woo86] or cold crystals [Gug57] satisfy the third principle of thermodynamics, that is, their entropy goes to zero when the absolute temperature goes to zero .

The explosion of entropy is very useful in order to estimate the temperature in reactive flows. In plane flame problems, in particular, once the species are estimated, one only has to show that the entropy is bounded to obtain a priori bounds for the absolute temperature T.

6.3. Formalisms involving generalized mass action kinetics have been introduced, in particular, by Feinberg [Fei95a] [Fei95b], Horn, and Jackson [Hor72]

[HJ72]. With these formalisms, it is possible to associate an integer, named the default, to each reaction network. In this formalism, only "zero-default" reaction networks have a satisfactory behavior for *all* possible values of the rate constants. With these formalisms, however, forward and backward reaction rate constants are assumed to be independent. Negative results thus concern networks for which the ratio $K_i^{\mathrm{f}}(T)/K_i^{\mathrm{b}}(T)$ is arbitrary. This is not the case according to the kinetic theory, which yields the fundamental constraint $K_i^{\mathrm{f}}(T) = K_i^{\mathrm{b}}(T)/K_i^{\mathrm{e}}(T)$. All of these formalisms thus appear to be *fundamentally* contradictory with kinetic theory. Note also that the relation $K_i^{\mathrm{f}}(T) = K_i^{\mathrm{b}}(T)/K_i^{\mathrm{e}}(T)$ can be interpreted as a natural *compatibility condition* between chemistry and thermodynamics and imply detailed balance at equilibrium. When this fundamental relation is assumed to hold, all classical results from thermochemistry are valid [SS65] [Kra70] and there are strictly no restrictions on the number of chemical reactions—other than obvious combinatorics—or on their linear dependency. In this situation, Feinberg's default can be *arbitrarily large*, but the corresponding networks have a very satisfactory behavior and can be naturally investigated by using Gibbs classical thermochemistry [SS65] [Kra70] [GM98a].

In the framework of generalized mass action kinetics, it has also been shown by Horn [Hor72] that certain relations have to be satisfied between reaction constants in order to obtain a satisfactory behavior. We point out, however, that these relations are automatically satisfied from the kinetic theory of gases. Open homogeneous systems are also often advocated as a reason for using nonconservative chemical reactions, which are artificially introduced to take care of inflow or outflow conditions [Fei95a]. We note, however, that with these formal procedures—restricted to homogeneous flows—the source terms are simply confused with the boundary conditions. Confusing boundary conditions with chemical reactions appears to be of limited interest for nonhomogeneous flows.

6.4. The case of infinitely fast chemistry, that is, the case of equilibrium flows, will be considered in Chapter 8. Such flows are governed by equations expressing the conservation of atom concentrations, momentum, and energy.

6.10. References

[Boi95] E. Boillat, *Existence and Uniqueness of the Solution to the Edge Problem in a Two-Dimensional Reactive Boundary Layer,* Math. Mod. Meth. Appl. Sci., **5**, (1995), pp. 1–27.

[*Cal*85] M. W. Chase, Jr., C. A. Davies, J. R. Downey, Jr., D. J. Frurip, R. A. McDonald, and A. N. Syverud, *JANAF Thermochemical Tables, Third Edition,* J. Phys. Chem. Ref. Data, **14**, Suppl. 1, (1985), pp. 1–1856.

[dGM84] S. R. de Groot and P. Mazur, *Non-Equilibrium Thermodynamics,* Dover, New York, (1984).

[EG94] A. Ern and V. Giovangigli, *Multicomponent Transport Algorithms,* Lecture Notes in Physics, New Series "Monographs", **m 24**, Springer-Verlag, Berlin, 1994.

[Fei95a] M. Feinberg, *The Existence and Uniqueness of Steady States for a Class of Chemical Reaction Networks,* Arkiv Rat. Mech. Anal., **132**, (1995), pp. 311–370.

[Fei95b] M. Feinberg, *Multiple Steady States for Chemical Reaction Networks of Deficiency One,* Arkiv Rat. Mech. Anal., **132**, (1995), pp. 371–406.

[Gio99] V. Giovangigli, *Plane Flames with Multicomponent Transport and Complex Chemistry,* Math. Mod. Meth. Appl. Sci., **9**, (1999), pp. 337–378.

[GM98a] V. Giovangigli and M. Massot, *Asymptotic Stability of Equilibrium States for Multicomponent Reactive Flows,* Math. Mod. Meth. Appl. Sci., **8**, (1998), pp. 251–297.

[GS87] V. Giovangigli and M. Smooke, *Extinction Limits of Strained Premixed Laminar Flames with Complex Chemistry,* Comb. Sci. Tech., **53**, (1987), pp. 23–49.

[Gug57] E. A. Guggenheim, *Thermodynamics,* Interscience, New York, (1957).

[Hor72] F. J. M. Horn, *Necessary and Sufficient Conditions for Complex Balancing in Chemical Kinetics,* Arch. Rational Mech. Anal., **49**, (1972), pp. 172–186.

[HJ72] F. J. M. Horn and R. Jackson, *General Mass Action Kinetics,* Arch. Rational Mech. Anal., **47**, (1972), pp. 81–116.

[Kra70] F. J. Krambeck, *The Mathematical Structure of Chemical Kinetics,* Arch. Rational Mech. Anal., **38**, (1970), pp. 317–347.

[Mei43] J. Meixnier, *Zur Thermodynamik der Irreversiblen Prozesse in Gases mit Chemish Reagierenden, Dissoziierenden und Anregbaren Komponent,* Annalen der Physik, **5**, (1943), pp. 244–270.

[Pou93] J. Pousin, *Modélisation et Analyse Numérique de Couches Limites Réactives d'Air,* Doctorat es Sciences, Ecole Polytechnique Fédérale de Lausanne, 1112, (1993).

[SS65] N. Z. Shapiro and L. S. Shapley, *Mass Action Law and the Gibbs Free Energy Function,* SIAM J. Appl. Math., **13**, (1965), pp. 353–375.

[StPr71] D. R. Stull and H. Prophet, *JANAF Thermochemical Tables, Second ed.,* Washington, NBS NSRDS-NBS37, (1971).

[VH85] A. Vol'pert and S Hudjaev, *Analysis in Classes of Discontinuous Functions and Equations of Mathematical Physics,* Mechanics: Analysis 8, Martinus Nijhoff Publishers, Dordrecht, (1985).

[Wil85] F. A. Williams, *Combustion Theory,* Second ed., The Benjamin/Cummings Publishing Company, Inc., Menlo Park, (1985).

[Woo86] L. C. Woods, *The Thermodynamics of Fluid Systems,* Oxford Eng. Sci. Series, **2**, (1986).

7

Mathematics of Transport Coefficients

7.1. Introduction

The expressions for transport fluxes have been presented in various forms in Chapter 2 and Chapter 4. These transport fluxes have been written in terms of macroscopic variable gradients and transport coefficients. In this chapter we now investigate the mathematical structure and properties of these transport coefficients.

We have also seen in Chapter 4 that the transport coefficients are not given explicitly. Their evaluation indeed requires solving transport linear systems associated with Galerkin approximate solutions of linearized Boltzmann equations. As a consequence, the mathematical assumptions on transport coefficients used in this book have been deduced from previous studies concerning the transport linear systems [Gio91] [EG94].

These assumptions on the transport coefficients are generally written in the situation of nonnegative mass fractions, that is, when $T > 0$, $p > 0$, $Y \geq 0$, and $Y \neq 0$, but we also have to consider the diffusion coefficients in the case of positive mass fractions $Y > 0$, where for any vector x with components $x = (x_1, \ldots, x_n)^t$, we write $x \geq 0$ when $x_k \geq 0$, $k \in S$, and $x > 0$ when $x_k > 0$, $k \in S$. Such a distinction between positive and nonnegative mass fractions naturally arises in the kinetic theory of dilute polyatomic gas mixtures [Gio91] [EG94].

From these assumptions on transport coefficients, we first derive the mathematical properties of λ, $\widetilde{\chi}$, D, and C and then define χ, $\widehat{\lambda}$, and θ in terms of λ, $\widetilde{\chi}$, D, and C. We obtain, in particular, the asymptotic behavior of diffusion coefficients and diffusion fluxes for small species concentrations.

We then investigate the diagonal diffusion problem, that is, we investigate the situations in which the diffusion process can be represented by a diagonal matrix, either with respect to diffusion driving forces or to mass fraction gradients. Since the diffusion matrices are generally irreducible, these representations are investigated on the zero sum gradient hyperplane $\mathcal{U}^\perp$, either for all species or after discarding one species.

We next establish the fundamental diffusion inequality associated with entropy production. This inequality will be used, in particular, in Chapter 11 and shows that the natural entropy production norm is a solution weighted norm involving mass fractions at the denominator of mass fraction gradients squared.

Finally, we investigate Stefan–Maxwell equations and justify in this context the assumptions about diffusion matrices.

7.1.1. Definition of transport fluxes

In order to investigate the mathematical structure associated with the transport fluxes $\boldsymbol{\Pi}$, $\boldsymbol{\mathcal{F}}_k$, $k \in S$, and $\boldsymbol{\mathcal{Q}}$, we write these fluxes in the form

$$\boldsymbol{\Pi} = -(\kappa - \tfrac{2}{3}\eta)(\boldsymbol{\partial_x} \cdot \boldsymbol{v})\boldsymbol{I} - \eta\big(\boldsymbol{\partial_x} \boldsymbol{v} + (\boldsymbol{\partial_x} \boldsymbol{v})^t\big), \tag{7.1.1}$$

$$\boldsymbol{\mathcal{F}}_k = -\sum_{l \in S} C_{kl}(\boldsymbol{d}_l + \frac{p_l}{p}\widetilde{\chi}_l \boldsymbol{\partial_x} \log T), \qquad k \in S, \tag{7.1.2}$$

$$\boldsymbol{\mathcal{Q}} = \sum_{k \in S} h_k \boldsymbol{\mathcal{F}}_k - \lambda \boldsymbol{\partial_x} T + RT \sum_{k \in S} \frac{\widetilde{\chi}_k}{m_k} \boldsymbol{\mathcal{F}}_k, \tag{7.1.3}$$

where κ is the volume viscosity, η is the shear viscosity, ^{t} is the transposition operator, C_{kl}, $k, l \in S$, are the multicomponent flux diffusion coefficients, $\boldsymbol{d}_k$ is the diffusion driving force of the kth species, $\widetilde{\chi} = (\widetilde{\chi}_1, \ldots, \widetilde{\chi}_n)^t$ are the rescaled thermal diffusion ratios, λ is the thermal conductivity, and h_k is the enthalpy per unit mass of the kth species. The diffusion driving forces $\boldsymbol{d}_k$, $k \in S$, take into account the contribution of several macroscopic variable gradients and are taken in the form

$$\boldsymbol{d}_k = \boldsymbol{\partial_x}\Big(\frac{p_k}{p}\Big) + \Big(\frac{p_k}{p} - \frac{\rho_k}{\sum_{l \in S} \rho_l}\Big)\boldsymbol{\partial_x} \log p + \frac{\rho_k}{p}\Big(\frac{\sum_{l \in S} \rho_l \boldsymbol{b}_l}{\sum_{l \in S} \rho_l} - \boldsymbol{b}_k\Big), \quad k \in S. \tag{7.1.4}$$

Note that, in this chapter, all species mass fractions are treated as formally independent. The resulting formulae apply to all situations, as discussed in Remark 6.3.1.

7.1.2. Diffusion velocities

For positive mass fractions, the species diffusion velocities $\boldsymbol{\mathcal{V}}_k$, $k \in S$, are defined from $\boldsymbol{\mathcal{V}}_k = \boldsymbol{\mathcal{F}}_k / \rho Y_k$, $k \in S$. Therefore, these velocities can be expressed as

$$\boldsymbol{\mathcal{V}}_k = -\sum_{l \in S} D_{kl}(\boldsymbol{d}_l + \frac{p_l}{p}\widetilde{\chi}_l \boldsymbol{\partial_x} \log T), \qquad k \in S, \tag{7.1.5}$$

where the diffusion matrix $D = (D_{kl})_{k,l\in S}$ is correspondingly defined by $D_{kl} = C_{kl}/(\rho Y_k)$, $k,l \in S$, when $Y_k > 0$. The diffusion matrix D is of fundamental importance since this matrix is symmetric positive semidefinite and associated with entropy production, as it has been shown in Section 2.6. Note that the multicomponent fluxes (7.1.1)–(7.1.3) have natural symmetry properties [WT62] [CC70] [FK72] [Gio91] [EG94], which have artificially been destroyed in [HCB54].

7.1.3. Alternative formulations

By introducing the proper definition of the partial thermal conductivity $\widehat{\lambda}$ and of the thermal diffusion coefficients $\theta = (\theta_1, \ldots, \theta_n)^t$, we will later recover the alternative expressions

$$\mathcal{F}_k = -\sum_{l\in S} C_{kl}\boldsymbol{d}_l - \rho Y_k \theta_k \boldsymbol{\partial_x} \log T, \qquad k \in S, \tag{7.1.6}$$

$$\mathcal{Q} = \sum_{k\in S} h_k \mathcal{F}_k - \widehat{\lambda}\boldsymbol{\partial_x} T - p\sum_{k\in S} \theta_k \boldsymbol{d}_k. \tag{7.1.7}$$

The kinetic theory specifically yields (7.1.1)–(7.1.7) with $(\boldsymbol{\partial_x} p_k - \rho_k \boldsymbol{b}_k)/p$, $k \in S$, in place of $\boldsymbol{d}_k$, $k \in S$, [EG94]. However, both expressions are easily shown to be equivalent because of the mass constraints satisfied by the transport coefficients. In addition, the constraint $\sum_{k\in S} \boldsymbol{d}_k = 0$ satisfied by the normalized diffusion driving forces $\boldsymbol{d}_k$, $k \in S$, is important when considering generalized inverses of transport matrices and for coercivity properties associated with entropy production.

7.2. Assumptions on transport coefficients

In this section, we introduce a set of assumptions concerning the transport coefficients. These properties have been deduced from previous work on diffusion matrices [Gio91] and from kinetic theory investigations of polyatomic reactive gas mixtures [EG94].

(Tr$_1$) The flux diffusion coefficients C_{kl}, $k,l \in S$, rescaled thermal diffusion ratios $\widetilde{\chi}_k$, $k \in S$, volume vicosity κ, shear vicosity η, and thermal conductivity λ are C^∞ functions of (T, Y) for $T > 0$ and $Y \geq 0$, $Y \neq 0$.

(Tr$_2$) The flux diffusion matrix $C = (C_{kl})_{k,l \in S}$ and the rescaled thermal diffusion ratios $\widetilde{\chi} = (\widetilde{\chi}_1, \dots, \widetilde{\chi}_n)^t$ satisfy the mass constraints $N(C) = \mathbb{R}Y$, $R(C) = \mathcal{U}^\perp$, and $\widetilde{\chi} \in X^\perp$, where $\mathcal{U} = (1, \dots, 1)^t$, $Y = (Y_1, \dots, Y_n)^t$, and $X = (X_1, \dots, X_n)^t$.

(Tr$_3$) The thermal conductivity λ and the shear viscosity η are positive functions. The volume viscosity κ is a nonnegative function.

(Tr$_{4a}$) For $Y > 0$, the matrix $D = (1/\rho)\mathcal{Y}^{-1}C$ is symmetric positive semidefinite and its nullspace is given by $N(D) = \mathbb{R}Y$, where $\mathcal{Y} = \mathrm{diag}(Y_1, \dots, Y_n)$ is the mass fraction diagonal matrix.

(Tr$_{4b}$) For $Y \geq 0$, $Y \neq 0$, we define the sets $S^+ = \{\, k \in S,\ Y_k > 0 \,\}$ and $S^0 = \{\, k \in S,\ Y_k = 0 \,\}$ and denote by Υ the permutation matrix associated with the reordering of S into (S^+, S^0). We then have the block structure

$$\Upsilon^t C \Upsilon = \begin{pmatrix} C^{++} & C^{+0} \\ 0 & C^{00} \end{pmatrix},$$

where C^{00} is diagonal with positive entries and the matrix D^{++} defined by $\rho D_{kl}^{++} = C_{kl}^{++}/Y_k$, $k, l \in S^+$, is symmetric positive semidefinite and has nullspace $\mathbb{R}Y^+$, where Y^+ corresponds to the S^+ mixture, that is, $Y = \Upsilon(Y^+, 0)^t$.

(Tr$_5$) There exists a positive function $\varphi(T)$ defined for $T > 0$ such that the rescaled transport properties $\eta^0(T,Y) = \eta(T,Y)/\varphi(T)$, $\kappa^0(T,Y) = \kappa(T,Y)/\varphi(T)$, $C^0(T,Y) = C(T,Y)/\varphi(T)$, $\lambda^0(T,Y) = \lambda(T,Y)/\varphi(T)$, and $\widetilde{\chi}^0(T,Y) = \widetilde{\chi}(T,Y)$ admit a continuous extension to $T \in [0, \infty]$, $Y \geq 0$, $Y \neq 0$, satisfying (Tr$_1$)–(Tr$_4$).

Note that the matrix C, which appears in the governing equations, is singular since $CY = 0$ and is not symmetric in general. The transport coefficients C, ρD, $\widetilde{\chi}$, κ, η, and λ only depend on (T, Y) and do not depend on pressure. Only the coefficients D and θ depend on pressure, but are simply inversely proportional to p. For nonnegative mass fractions, the matrix C^{++} is also the flux diffusion matrix of the submixture S^+. The assumptions (Tr$_1$)–(Tr$_4$) are mainly taken from [Gio91], [EG94], and [GM98a]. They are more general than those considered in [GM98a], in which all mass fractions were assumed to be bounded away from zero.

The assumption (Tr$_5$) was first introduced in [Gio97], [Gio98], and [Gio99] and concerns the global temperature dependency of transport coefficients. This assumption means that there exists a common temperature scaling for all transport coefficients, that is, for all collision integrals appearing in the transport linear systems. The scaling function $\varphi = 1$ is generally used in mathematics, whereas various functions φ are suggested by the kinetic theory of gases, depending on the interaction potentials between pairs of molecules. For a rigid sphere interaction potential, we have, for instance, $\varphi = T^{1/2}$. For point centers of repulsion, we have $\varphi = T^{(\nu-4)/2\nu}$, where ν varies from $\nu = 4$—Maxwell molecules—to $+\infty$ and the temperature dependence varies, respectively, between $\varphi = 1$ and $\varphi = T^{1/2}$ [FK72]. Small values of ν correspond to soft molecules, whereas large values of ν correspond to hard molecules, and, for $\nu \to \infty$, we recover the hard sphere model. For Lennard–Jones or Stockmayer potentials the existence of φ is a consequence of the Mason and Monchick tables for collision integral ratios [FK72].

7.3. Properties of diffusion matrices

In this section we investigate various properties of the flux diffusion matrix C and of the diffusion matrix D as consequences of (Tr_1)–(Tr_4) and occasionally of (Tr_5).

7.3.1. *First properties of the diffusion matrix D*

The smoothness of the diffusion coefficients D_{kl}, $k, l \in S$, is investigated in the following lemma.

Lemma 7.3.1. *For $k, l \in S$, and $k \neq l$, the coefficient D_{kl} admits a smooth extension to the domain $T > 0$, $Y \geq 0$, $Y \neq 0$. For $k \in S$, the diagonal coefficient D_{kk} admits a smooth extension to the domain $T > 0$, $Y \geq 0$, $Y_k > 0$. In addition, the diagonal coefficient D_{kk} is nonnegative $D_{kk} \geq 0$, and $D_{kk} = 0$ if and only if $Y_l = 0$ for $l \neq k$ and $Y_k > 0$. Finally, the diagonal coefficient D_{kk} explodes like $1/Y_k$ as $Y_k \to 0$.*

Proof. Consider a point Z, such that $Z \geq 0$, $Z \neq 0$, and let Y be in the neighborhood of Z. When $Z_k > 0$, we have $Y_k > 0$ and, from the definition $D_{kl}(T,Y) = C_{kl}(T,Y)/\big(\rho(T,Y)Y_k\big)$, we obtain that $D_{kl}(T,Y)$ is smooth. Assume now that Z has at least a component such that $Z_k = 0$. We then consider $Y > 0$ in the neighborhood of Z and define

$$\pi(Y) = \frac{Y - Y_k e^k}{1 - Y_k},$$

where e^l, $l \in S$, is the canonical basis of $\mathbb{R}^n$. We have, of course, $\pi(Y) \geq 0$ and $\pi(Y)_k = 0$. As a consequence, we obtain for $k, l \in S$ and $k \neq l$ that $C_{kl}(T,Y) - C_{kl}\big(T,\pi(Y)\big) = C_{kl}(T,Y)$, since $C_{kl}\big(T,\pi(Y)\big) = 0$ for $k \neq l$, from (Tr_{4b}). This now yields

$$C_{kl}(T,Y) = \int_0^1 \Big\langle \partial_Z C_{kl}\big(T, \tau Y + (1-\tau)\pi(Y)\big), Y - \pi(Y) \Big\rangle \, d\tau.$$

However, since $Y - \pi(Y) = Y_k(e^k - Y)/(1 - Y_k) = Y_k\big(e^k - \pi(Y)\big)$ we now obtain

$$D_{kl}(T,Y) = \frac{C_{kl}(T,Y)}{\rho Y_k} = \frac{1}{\rho}\int_0^1 \Big\langle \partial_Z C_{kl}\big(T, \tau Y + (1-\tau)\pi(Y)\big), e^k - \pi(Y)\Big\rangle \, d\tau,$$

so that $D_{kl}(T,Y)$ is a smooth function of (T,Y) for $T > 0$ and $Y \geq 0$, $Y \neq 0$, keeping in mind that C and π are smooth.

The smooth extension of D_{kk} to $T > 0$, $Y \geq 0$, $Y_k > 0$ is also obvious from $D_{kk} = C_{kk}/\rho Y_k$. Moreover, the fact that D_{kk} explodes when $Y_k \to 0$ is a direct consequence of $C_{kk} > 0$ for $Y_k = 0$. Furthermore, for $Y_k > 0$, we obtain from (Tr_4) that $D_{kk} \geq 0$, since D^{++} is symmetric positive semidefinite, and, for $Y_k = 0$, we have $D_{kk} > 0$ from (Tr_{4b}). Finally, when $D_{kk} = 0$, we have necessarily $k \in S^+$ from (Tr_{4b}) and $e^{k+} \in N(D^{++})$. Since $N(D^{++}) = \mathbb{R}Y^+$, however, we deduce that Y^+ is reduced to one element and conversely. ■

We have presented several properties (2.5.15)–(2.5.18) of the diffusion matrix D in Chapter 2. These properties are direct consequences of assumptions (Tr_1)–(Tr_{4a}), and Property (2.5.18) will be strengthened in Section 7.6.

Lemma 7.3.2. *Assume that $Y > 0$. Then the diffusion matrix D satisfies $D = D^t$, $N(D) = \mathbb{R}Y$, $R(D) = Y^\perp$, and D is positive definite over $\mathcal{U}^\perp$.*

Proof. We simply note that $\mathcal{U}^\perp \cap N(D) = \mathcal{U}^\perp \cap \mathbb{R}Y = \{0\}$. ■

7.3.2. First properties of the flux diffusion matrix C

The properties (2.5.5)–(2.5.7) of the flux diffusion matrix C stated in Chapter 2 are direct consequences of assumptions (Tr_1)–(Tr_4).

Lemma 7.3.3. *Assume that $Y \geq 0$, $Y \neq 0$. Then the flux diffusion matrix C satisfies $C\mathcal{Y} = \mathcal{Y}C^t = (C\mathcal{Y})^t$, where $\mathcal{Y} = \mathrm{diag}(Y_1, \ldots, Y_n)$, $N(C) = \mathbb{R}Y$, and $R(C) = \mathcal{U}^\perp$. In addition, the diagonal coefficient C_{kk} is nonnegative $C_{kk} \geq 0$, and $C_{kk} = 0$ if and only if $Y_l = 0$ for $l \neq k$ and $Y_k > 0$. Finally, the diagonal coefficient C_{kk} has a positive limit as $Y_k \to 0$.*

Proof. The properties (2.5.6) and (2.5.7) are direct consequences of assumptions (Tr_1)–(Tr_4). Therefore, we only have to establish that $C_{kl}Y_l = C_{lk}Y_k$, $k, l \in S$. When Y_k and Y_l are both positive, we can write that $C_{kl}/Y_k = \rho D_{kl}$ and $C_{lk}/Y_l = \rho D_{lk}$, so that we conclude thanks to the symmetry of D. On the other hand, when $Y_k = 0$ and $Y_l > 0$, we have $C_{kl} = 0$ by assumption, so that $C_{kl}Y_l = C_{lk}Y_k$, and the relation is trivial when $Y_k = Y_l = 0$.

In addition, the fact that C_{kk} has a positive limit as $Y_k \to 0$ is a direct consequence of ($\mathrm{Tr}_4 b$). Furthermore, for $Y_k > 0$, we obtain from ($\mathrm{Tr}_4 b$) that $D_{kk} \geq 0$, since D^{++} is symmetric positive semidefinite, so that $C_{kk} = \rho D_{kk} \geq 0$. Finally, when $C_{kk} = 0$, we have necessarily $k \in S^+$ from (Tr_{4b}) and then $e^{k+} \in N(C^{++}) = N(D^{++})$. Since $N(D^{++}) = \mathbb{R}Y^+$, however, we deduce that Y^+ is reduced to one element and conversely. ∎

7.3.3. Flux splitting

In this section we investigate how the diffusion flux of the kth species $\mathcal{F}_k$ behaves when Y_k goes to zero. More specifically, we obtain a decomposition of the flux $\mathcal{F}_k$ into a diagonal term $-\mathfrak{d}_k \boldsymbol{\partial_x} Y_k$ and a residual term $Y_k \mathfrak{f}_k$ proportional to Y_k. This decomposition is a direct consequence of the smoothness properties of the diffusion matrix D, which have been established in Lemma 7.3.1.

Lemma 7.3.4. *The diffusion flux of the kth species $\mathcal{F}_k$ can be split into the form*

$$\mathcal{F}_k = -\mathfrak{d}_k \boldsymbol{\partial_x} Y_k - Y_k \mathfrak{f}_k, \tag{7.3.1}$$

where the scalar $\mathfrak{d}_k$ is given by

$$\mathfrak{d}_k = \frac{C_{kk}}{m_k} \frac{\overline{m}}{\langle Y, \mathcal{U} \rangle} \tag{7.3.2}$$

and the correction vector $\mathfrak{f}_k$ is given by

$$\begin{aligned} \mathfrak{f}_k = {} & \frac{C_{kk}}{m_k} \Big(-\frac{\overline{m}^2}{\langle Y, \mathcal{U} \rangle^2} \sum_{l \in S} \frac{\boldsymbol{\partial_x} Y_l}{m_l} + \frac{\overline{m} - m_k}{\langle Y, \mathcal{U} \rangle} \frac{\boldsymbol{\partial_x} p}{p} + \widetilde{\chi}_k \frac{\overline{m}}{\langle Y, \mathcal{U} \rangle} \frac{\boldsymbol{\partial_x} T}{T} \Big) \\ & + C_{kk} \frac{\rho}{p} \Big(\frac{\sum_{l \in S} \rho_l \boldsymbol{b}_l}{\sum_{l \in S} \rho_l} - \boldsymbol{b}_k \Big) + \sum_{\substack{l \in S \\ l \neq k}} \rho D_{kl} \big(\boldsymbol{d}_l + \frac{p_l}{p} \widetilde{\chi}_l \frac{\boldsymbol{\partial_x} T}{T} \big), \end{aligned} \tag{7.3.3}$$

so that the nondiagonal part of the multicomponent flux $\mathcal{F}_k$ vanishes for $Y_k = 0$. Moreover, we have $\mathfrak{d}_k \geq 0$, and $\mathfrak{d}_k = 0$ if and only if $Y_l = 0$ for $l \neq k$ and $Y_k > 0$. In particular, whenever $\langle Y, \mathcal{U} \rangle = 1$, $\mathfrak{d}_k / \varphi$ is positive and bounded away from zero when $0 \leq Y_k \leq \delta$ for $\delta < 1$.

This lemma is very useful in practice for showing that the mass fractions are strictly positive or for estimating various expressions involving mass fluxes. Note that the decomposition (7.3.1) is not unique because of various terms proportional to $Y_k \boldsymbol{\partial_x} Y_k$ in the expression of $\mathcal{F}_k$.

7.3.4. Generalized inverses of C and D

The flux diffusion matrix C and the diffusion matrix D are both singular since $CY = 0$ and $DY = 0$. However, these matrices admit generalized inverses that naturally arise in the framework of the kinetic theory of gases or when investigating Stefan–Maxwell equations [Gio90] [Gio91]. We first restate the existence of generalized inverses with prescribed range and nullspace [BG74] [BP79] and then investigate the case of diffusion matrices [Gio90] [Gio91] [EG97].

Proposition 7.3.5. *Let $G \in \mathbb{R}^{n,n}$ be a matrix, and let $\mathfrak{p}$ and $\mathfrak{q}$ be two subspaces of $\mathbb{R}^n$ such that $N(G) \oplus \mathfrak{p} = \mathbb{R}^n$ and $R(G) \oplus \mathfrak{q} = \mathbb{R}^n$. Then, there exists a unique matrix Z such that $GZG = G$, $ZGZ = Z$, $N(Z) = \mathfrak{q}$, and $R(Z) = \mathfrak{p}$. The matrix Z—termed the generalized inverse of G with prescribed range $\mathfrak{p}$ and nullspace $\mathfrak{q}$—satisfies $GZ = P_{R(G),\mathfrak{q}}$ and $ZG = P_{\mathfrak{p},N(G)}$, where $P_{\mathbb{A},\mathbb{B}}$ is defined for linear spaces $\mathbb{A}$ and $\mathbb{B}$, such that $\mathbb{A}\oplus\mathbb{B} = \mathbb{R}^n$, and denotes the projector onto $\mathbb{A}$ along $\mathbb{B}$.*

Proof. We first show that there exists a matrix $\mathcal{J} \in \mathbb{R}^{n,n}$ such that $G\mathcal{J}G = G$ and $\mathcal{J}G\mathcal{J} = \mathcal{J}$. Let a^i, $i = 1,\ldots,n$, be a basis of $\mathbb{R}^n$, such that a^i, $i = 1,\ldots,r$, is a basis of $N(G)$. Then, by construction, the vectors $b^i = Ga^i$, $i = r{+}1,\ldots,n$ are linearly independent and may be completed to form a basis b^i, $i = 1,\ldots,n$ of $\mathbb{R}^n$. Define then $\mathcal{J}$ such that $\mathcal{J}b^i = 0$, $1 \le i \le r$, and $\mathcal{J}b^i = a^i$, $r{+}1 \le i \le n$. One may then easily check that $G\mathcal{J}Ga^i = Ga^i$, $1 \le i \le n$, and $\mathcal{J}G\mathcal{J}b^i = \mathcal{J}b^i$, $1 \le i \le n$, so that $G\mathcal{J}G = G$ and $\mathcal{J}G\mathcal{J} = \mathcal{J}$. Defining now $Z = P_{\mathfrak{p},N(G)}\mathcal{J}P_{R(G),\mathfrak{q}}$, we have $GZG = GP_{\mathfrak{p},N(G)}\mathcal{J}P_{R(G),\mathfrak{q}}G = G\mathcal{J}G$, since $G = GP_{\mathfrak{p},N(G)}$ and $P_{R(G),\mathfrak{q}}G = G$, so that $GZG = G$. Similarly, from $ZGZ = P_{\mathfrak{p},N(G)}\mathcal{J}P_{R(G),\mathfrak{q}}GP_{\mathfrak{p},N(G)}\mathcal{J}P_{R(G),\mathfrak{q}}$, we obtain that $ZGZ = P_{\mathfrak{p},N(G)}\mathcal{J}G\mathcal{J}P_{R(G),\mathfrak{q}}$ and thus that $ZGZ = Z$. By construction, we also have $R(Z) \subset \mathfrak{p}$ and $\mathfrak{q} \subset N(Z)$, and, from $GZG = G$ and $ZGZ = Z$, we obtain that $\operatorname{rank}(G) = \operatorname{rank}(Z)$. Since $\dim(\mathfrak{p}) = \operatorname{rank}(G)$ and $\dim(\mathfrak{q}) = n - \operatorname{rank}(G)$ by assumption, we conclude that $R(Z) = \mathfrak{p}$ and $N(Z) = \mathfrak{q}$. From the relations $GZG = G$ and $ZGZ = Z$, we also deduce that GZ and ZG are projectors and that $\operatorname{rank}(GZ) = \operatorname{rank}(G) = \operatorname{rank}(Z) = \operatorname{rank}(ZG)$. Since we also have $R(ZG) \subset R(Z) = \mathfrak{p}$ and $N(G) \subset N(ZG)$, we obtain that $R(ZG) = \mathfrak{p}$ and $N(ZG) = N(G)$ and similarly that $R(GZ) = R(G)$ and $N(GZ) = \mathfrak{q}$, so that $GZ = P_{R(G),\mathfrak{q}}$ and $ZG = P_{\mathfrak{p},N(G)}$. Finally, if there are two such generalized inverses Z_1 and Z_2, we have $Z_i = Z_iGZ_i$, $GZ_i = P_{R(G),\mathfrak{q}}$ and $Z_iG = P_{\mathfrak{p},N(G)}$, $i = 1,2$, so that $Z_1 = Z_1GZ_1 = P_{\mathfrak{p},N(G)}Z_1 = Z_2GZ_1 = Z_2P_{R(G),\mathfrak{q}} = Z_2GZ_2 = Z_2$. ∎

Similarly, in the case where the nullspace and range of the matrix G are complementary spaces, G admits a group inverse which is characterized by the following result.

Proposition 7.3.6. *Let $G \in \mathbb{R}^{n,n}$ be a matrix, such that $N(G) \oplus R(G) = \mathbb{R}^n$. Then there exists a unique matrix Z such that $GZG = G$, $ZGZ = Z$, and $GZ = ZG$. The matrix Z is called the group inverse of G and denoted by $G^\sharp$. The group inverse is also the generalized inverse with prescribed range $R(G)$ and nullspace $N(G)$ and such that $GZ = ZG = P_{R(G),N(G)}$.*

Proof. From the relations $GZG = G$, $ZGZ = Z$, and $GZ = ZG$, we easily deduce that $R(G) = R(Z)$ and $N(G) = N(Z)$. Therefore, Z is necessarily the generalized inverse of G with prescribed range $R(G)$ and nullspace $N(G)$. Since $N(G) \oplus R(G) = \mathbb{R}^n$, we deduce from Proposition 7.3.5 that the matrix Z exists and is unique. Hence, we only have to show that $GZ = ZG$, but this directly follows from $GZ = P_{R(G),N(G)}$ and $ZG = P_{R(G),N(G)}$ established in 7.3.5 with $\mathfrak{p} = R(G)$ and $\mathfrak{q} = N(G)$. ∎

We can now apply these results to the diffusion matrices D and C.

Lemma 7.3.7. *Let $T > 0$, $Y \geq 0$, and $Y \neq 0$ and consider the flux diffusion matrix C. Then, there exists a unique matrix Γ such that $C\Gamma C = C$, $\Gamma C \Gamma = \Gamma$, $N(\Gamma) = \mathbb{R}Y$, and $R(\Gamma) = \mathcal{U}^\perp$. The matrix Γ is the group inverse of C and we have the relations $\Gamma C = C\Gamma = I - Y \otimes \mathcal{U}/\langle Y, \mathcal{U}\rangle$. In addition, the matrix Γ is a smooth function of (T, Y) over the domain $T > 0$ and $Y \geq 0$, $Y \neq 0$.*

Lemma 7.3.8. *Let $T > 0$, $p > 0$, and $Y > 0$, and consider the diffusion matrix D. Then, there exists a unique matrix Δ such that $D\Delta D = D$, $\Delta D \Delta = \Delta$, $N(\Delta) = \mathbb{R}\mathcal{U}$, and $R(\Delta) = \mathcal{U}^\perp$. The matrix Δ is the generalized inverse of D with prescribed range $\mathcal{U}^\perp$ and nullspace $\mathbb{R}\mathcal{U}$, and we have $\Delta D = I - Y \otimes \mathcal{U}/\langle Y, \mathcal{U}\rangle$ and $D\Delta = I - \mathcal{U} \otimes Y/\langle Y, \mathcal{U}\rangle$. In addition, the matrix Δ is symmetric positive semidefinite and a smooth function of T, p, Y over the domain $T > 0$, $p > 0$, and $Y > 0$.*

Proof. The existence and first properties of Γ and Δ are direct consequences of Propositions 7.3.5 and 7.3.6. The symmetry of Δ is easily obtained from that of D and the uniqueness of generalized inverses, and positive semidefiniteness results from $\langle \Delta x, x\rangle = \langle D\Delta x, \Delta x\rangle$, $x \in \mathbb{R}^n$. Finally, the smoothness of Γ and Δ is a consequence of the inversion formulae $\Gamma = (C + \alpha Y \otimes \mathcal{U})^{-1} - \beta Y \otimes \mathcal{U}$ and $\Delta = (D + \alpha \mathcal{U} \otimes \mathcal{U})^{-1} - \beta Y \otimes Y$, valid for positive α and β with $\alpha\beta\langle Y, \mathcal{U}\rangle^2 = 1$. ∎

The interest of these generalized inverses is that for first-order diffusion coefficients, that is, when only one term is retained in the Sonine polynomial expansions of the species perturbed distribution functions, the matrices Δ and Γ are known explicitly. The corresponding equations will be investigated in detail in Section 7.7. More generally, the generalized inverses Δ and Γ can be evaluated as Schur complements from larger transport linear systems [EG94]. The corresponding linear systems are generalized Stefan–Maxwell type equations. The generalized inverses Γ and Δ are very useful when manipulating the multicomponent transport fluxes.

7.3.5. Modified diffusion coefficients

It is also possible to define modified diffusion matrices $\widetilde{D}$ that are nonsingular, but that yield the proper fluxes on the physical hyperplane $\mathcal{U}^{\perp}$ [Gio90] [Gio91].

Proposition 7.3.9. *Assume that $T > 0$, $p > 0$, and $Y > 0$, let α be positive scalar and define $\tilde{\alpha} = \alpha\langle Y, \mathcal{U}\rangle$. Then the modified diffusion matrix*

$$\widetilde{D} = D + \alpha\, \mathcal{U} \otimes \mathcal{U} \tag{7.3.4}$$

is symmetric positive definite and coincides with D on the physical hyperplane $\mathcal{U}^{\perp}$. Further replace $\boldsymbol{\partial_x} \mathcal{X}_k$ by $\boldsymbol{\partial_x} X_k$ in (7.1.4) and denote by $\widetilde{\boldsymbol{d}}_k$ the corresponding diffusion driving force. Then the corresponding modified species diffusion velocities $\widetilde{\boldsymbol{\mathcal{V}}}_k = -\sum_{l\in S} \widetilde{D}_{kl}(\widetilde{\boldsymbol{d}}_l + \chi_l \boldsymbol{\partial_x} \log T)$, $k \in S$, satisfy

$$\sum_{k\in S} Y_k \widetilde{\boldsymbol{\mathcal{V}}}_k = -\tilde{\alpha}\Big(\sum_{k\in S} \widetilde{\boldsymbol{d}}_l\Big), = -\tilde{\alpha}\boldsymbol{\partial_x}\Big(\sum_{k\in S} Y_k\Big). \tag{7.3.5}$$

These modified formulations have important practical applications. More specifically, consider the equation (2.2.13) governing $\mathsf{Y} = \sum_{k\in S} Y_k$, presented in Section 2.2.3, when all species mass fractions are considered as independent unknowns. In this situation, by using the modified diffusion velocities $\widetilde{\boldsymbol{\mathcal{V}}}_k$, $k \in S$, this equation is transformed into

$$\rho\partial_t \mathsf{Y} + \rho \boldsymbol{v}\cdot\boldsymbol{\partial_x}\mathsf{Y} = \boldsymbol{\partial_x}\cdot\big(\tilde{\alpha}\boldsymbol{\partial_x}\mathsf{Y}\big) \tag{7.3.6}$$

and the new extra diffusive term $\boldsymbol{\partial_x}\cdot\big(\tilde{\alpha}\boldsymbol{\partial_x}\mathsf{Y}\big)$ has a fundamental stabilizing effect for discretized equations and boundary conditions [Gio90].

7.4. Properties of other coefficients

In this section we define the transport coefficients $\widehat{\lambda}$, θ, and χ from λ, C, D, and $\widetilde{\chi}$, and we recover the alternative formulation (7.1.6)–(7.1.7).

7.4.1. Alternative coefficients

We introduce the following mathematical definitions of the partial thermal conductivity $\widehat{\lambda}$ and the thermal diffusion coefficients θ.

Definition 7.4.1. *The coefficients θ_k, $k \in S$, and $\widehat{\lambda}$ are defined by the expressions*

$$\theta_k = \frac{RT}{p} \sum_{l \in S} \frac{\widetilde{\chi}_l}{m_l} C_{lk}, \qquad k \in S, \tag{7.4.1}$$

$$\widehat{\lambda} = \lambda + \frac{p}{T} \sum_{l \in S} \chi_l \theta_l. \tag{7.4.2}$$

The coefficients $\rho\theta_k$, $k \in S$, and $\widehat{\lambda}$ are smooth functions of $T > 0$, $Y \geq 0$, and $Y \neq 0$, and the partial thermal conductivity $\widehat{\lambda}$ is positive. In addition, from (7.4.1)–(7.4.2), the quantities $\widehat{\lambda}/\varphi(T)$ and $\rho\theta/\varphi(T)$ admit a continuous extension to $0 \leq T \leq \infty$, $Y \geq 0$, and $Y \neq 0$.

From this definition of the thermal diffusion coefficients, we now deduce several properties.

Lemma 7.4.2. *Assume that $T > 0$ and $Y \geq 0$, $Y \neq 0$. Then we have the identities*

$$\rho Y_k \theta_k = \sum_{l \in S} C_{kl} \frac{p_l}{p} \widetilde{\chi}_l, \qquad k \in S. \tag{7.4.3}$$

Proof. We deduce from the definition of θ_k, $k \in S$, that

$$\rho Y_k \theta_k = \frac{\rho RT}{p} \sum_{l \in S} Y_k \frac{\widetilde{\chi}_l}{m_l} C_{lk}, \qquad k \in S. \tag{7.4.4}$$

However, we have seen in Lemma 7.1.5 that $Y_k C_{lk} = Y_l C_{kl}$, $k, l \in S$. Substituting this relation into (7.4.4), we obtain

$$\rho Y_k \theta_k = \frac{\rho RT}{p} \sum_{l \in S} Y_l \frac{\widetilde{\chi}_l}{m_l} C_{kl}. \tag{7.4.5}$$

We now use the relation $p_k = \rho Y_k RT/m_k$ and obtain (7.4.3). ∎

Lemma 7.4.2 and Definition 7.4.1 imply, in particular, that the alternative relations (7.1.6) (7.1.7) hold.

7.4.2. Waldmann coefficients

The thermal diffusion ratios χ were first introduced by Waldmann for multicomponent mixtures [Wal47] [Wal58] [WT62]. For mathematical convenience, we have used the rescaled thermal diffusion ratios $\widetilde{\chi}$. In this section, we define the classical thermal diffusion ratios χ from the rescaled quantities $\widetilde{\chi}$ and recover classical relations associated with the coefficients χ.

Definition 7.4.3. *The thermal diffusion ratios χ_k, $k \in S$, are defined by*

$$\chi_k = \frac{p_k}{p}\widetilde{\chi}_k, \qquad k \in S, \tag{7.4.6}$$

so that $\chi \in \mathcal{U}^\perp$. Moreover, when Y is positive, we have

$$\theta_k = \sum_{l \in S} D_{kl}\chi_l \tag{7.4.7}$$

and

$$\chi_k = \sum_{l \in S} \Delta_{kl}\theta_l. \tag{7.4.8}$$

Proof. The constraint $\chi \in \mathcal{U}^\perp$ is a direct consequence of $\widetilde{\chi} \in X^\perp$. On the other hand, when Y is positive, we can write that $C = \rho\mathcal{Y}D$, so that, by using definition 7.4.1, we obtain

$$\theta_k = \frac{RT}{p}\sum_{l \in S}\frac{\widetilde{\chi}_l}{m_l}\rho Y_l D_{lk} = \sum_{l \in S}\widetilde{\chi}_l\frac{p_l}{p}D_{lk} = \sum_{l \in S} D_{kl}\chi_l, \qquad k \in S.$$

This shows that $\theta = D\chi$, and, since $\chi \in \mathcal{U}^\perp$, we obtain from Lemma 7.3.6

$$\Delta\theta = \Delta D\chi = \Big(I - \frac{Y\otimes\mathcal{U}}{\langle Y,\mathcal{U}\rangle}\Big)\chi = \chi,$$

which completes the proof. ∎

7.5. Diagonal diffusion

In this section, we first investigate the irreducibility of multicomponent diffusion matrices. Irreducibility is shown to be a direct consequence of the structure properties (Tr_1)–(Tr_4).

Because of the considerable simplifications that may result, we then investigate the situations in which the diffusion process can be represented by a diagonal matrix. We investigate diagonal representations with respect to the diffusion driving forces $\boldsymbol{d}_k$, $k \in S$, and with respect to the mass fraction gradients $\boldsymbol{\partial_x} Y_k$, $k \in S$, which is the most interesting case in practice. In both situations, we consider diagonal representations over $\mathcal{U}^\perp$, that is to say, valid for all species, or diagonal representations after discarding the last species.

7.5.1. Irreducibility of C and D

From Assumption (Tr_{4b}), we already know that, when some mass fractions are zero, the matrix C is reducible. In the next proposition, we establish that the converse is true.

Proposition 7.5.1. *Assume that $T > 0$, $p > 0$, $Y \geq 0$, and $Y_k \neq 0$, $k \in S$. Then, the flux diffusion matrix C is irreducible. In particular, the diffusion matrix D is always irreducible.*

Proof. Assuming that C is reducible, there exists a permutation matrix Π such that

$$\Pi^t C \Pi = \begin{pmatrix} C^{aa} & C^{ab} \\ 0 & C^{bb} \end{pmatrix},$$

where the superscripts ^{a} and ^{b} are used to denote the reordering associated with Π and where C^{aa} and C^{bb} are thus square matrices of size greater than or equal to unity. Since $R(C) = \mathcal{U}^\perp$, we deduce that $R(\Pi^t C \Pi) = \mathcal{U}^\perp$, so that $R(C^{aa}) \subset (\mathcal{U}^a)^\perp$. This shows that C^{aa} is singular, so that there exists $x^a \neq 0$ such that $C^{aa} x^a = 0$. Introducing the vector $x = (x^a, 0)^t$, we obtain that $\Pi^t C \Pi\, x = 0$, which yields $\Pi x \in N(C) = \mathbb{R}Y$. We have thus shown that $\Pi(x^a, 0)^t = \alpha Y$ and α cannot be zero, since $x^a \neq 0$, so that finally $Y = (1/\alpha)\Pi(x^a, 0)^t$ and some mass fractions are vanishing. ∎

This result shows that no uncoupling is generally expected between groups of species for evaluating transport fluxes in terms of species diffusion driving forces.

7.5.2. Matrix E and mass fraction gradients

Since the mass fractions appear as natural variables in the species conservation equations, diagonal representation of diffusive processes are especially interesting in terms of the mass fraction gradients $\boldsymbol{\partial_x} Y_k$, $k \in S$. Before investigating such diagonal representations, we first have to express the diffusion velocities in terms of these gradients $\boldsymbol{\partial_x} Y_k$, $k \in S$.

As a consequence, we now focus on the diffusion fluxes and velocities arising from the concentration gradients, or, equivalently, we assume that $\boldsymbol{d}_k = \boldsymbol{\partial_x} X_k$, $k \in S$. Denoting then by E the matrix $E = \partial_Y X$, we have the relations $\boldsymbol{\partial_x} X = E \boldsymbol{\partial_x} Y$, where E is given by (6.3.35), so that

$$\boldsymbol{\mathcal{F}} = -CE\, \boldsymbol{\partial_x} Y,$$

where $\boldsymbol{\mathcal{F}} = (\boldsymbol{\mathcal{F}}_1, \dots, \boldsymbol{\mathcal{F}}_n)^t$ and $\boldsymbol{\partial_x} Y = (\boldsymbol{\partial_x} Y_1, \dots, \boldsymbol{\partial_x} Y_n)^t$. As a consequence, in order to investigate the species fluxes $\boldsymbol{\mathcal{F}}_k$ in terms of the mass fraction gradients $\boldsymbol{\partial_x} Y_k$, $k \in S$, we only have to examine the product matrix CE. Similarly, for the diffusion velocities $\boldsymbol{\mathcal{V}}_k$, $k \in S$, we only have to investigate the product matrix DE.

More generally, the same conclusion holds if we use for convenience the relations $\boldsymbol{d}_k = \boldsymbol{\partial_x} \mathcal{X}_k$, $k \in S$, that is, if we replace $\boldsymbol{\partial_x}(p_k/p)$ by $\boldsymbol{\partial_x} \mathcal{X}_k$ instead of simply $\boldsymbol{\partial_x} X_k$, for $k \in S$. Indeed, we then have

$$\partial_Y \mathcal{X} = \frac{1}{\langle Y,\mathcal{U}\rangle}\Big(I - \frac{X \otimes \mathcal{U}}{\langle Y,\mathcal{U}\rangle}\Big)\, E,$$

so that $(\partial_Y \mathcal{X})x = Ex/\langle Y,\mathcal{U}\rangle$ for $x \in \mathcal{U}^\perp$, since $E^t\mathcal{U} = \mathcal{U}$. As a consequence, $C\partial_Y \mathcal{X}$ and CE coincide over $\mathcal{U}^\perp$, and, similarly, $D\partial_Y \mathcal{X}$ and DE coincide over $\mathcal{U}^\perp$, so that it is sufficient to investigate the products CE and DE.

As a first property of the product CE, we have the following result.

Lemma 7.5.2. *The nullspace and range of CE are given by $N(CE) = \mathbb{R}Z$ and $R(CE) = \mathcal{U}^\perp$, where Z is defined by*

$$EZ = Y. \tag{7.5.1}$$

The vector Z is given by

$$Z_k = Y_k\Big(\frac{m_k}{\overline{m}} + \sum_{l\in S}(1 - \frac{m_l}{\overline{m}})\frac{Y_l}{\langle Y,\mathcal{U}\rangle}\Big), \qquad k \in S,$$

and we have $\langle Y,\mathcal{U}\rangle = \langle Z,\mathcal{U}\rangle$. Finally, when $Y > 0$, the nullspace and range of DE are given by $N(DE) = \mathbb{R}Z$ and $R(DE) = Y^\perp$.

Proof. These properties are easy consequences of those of C, D, and E, keeping in mind that E is invertible and $E^t\mathcal{U} = \mathcal{U}$. ∎

In the following sections, we investigate the irreducibility of the products CE and DE. We later investigate when CE and DE can be represented by a diagonal matrix.

7.5.3. Irreducibility of CE and DE

Proposition 7.5.3. *Assume that $T > 0$, $p > 0$, $Y \geq 0$, and $Z_k \neq 0$, $k \in S$. Then, the flux diffusion matrix with respect to mass fraction gradients CE is irreducible. In this situation, we also have $Y_k \neq 0$, $k \in S$, so that D is well defined and the product DE is irreducible.*

Proof. We assume that CE is reducible, so that there exists a permutation matrix Π such that

$$\Pi^t CE \Pi = \begin{pmatrix} (CE)^{aa} & (CE)^{ab} \\ 0 & (CE)^{bb} \end{pmatrix},$$

where the superscripts ^{a} and ^{b} are used to denote the reordering associated with Π. Since $R(CE) = \mathcal{U}^\perp$, we deduce that $R(\Pi^t(CE)\Pi) = \mathcal{U}^\perp$ so that $R((CE)^{aa}) \subset (\mathcal{U}^a)^\perp$. This shows that $(CE)^{aa}$ is singular so that there exists $x^a \neq 0$ such that $(CE)^{aa}x^a = 0$. Introducing the vector $x = (x^a, 0)^t$, we obtain that $\Pi^t(CE)\Pi\, x = 0$ which yields $\Pi x \in N(CE) = \mathbb{R}Z$. We have thus shown that $\Pi(x^a, 0)^t = \alpha Z$ and α cannot be zero since $x^a \neq 0$ so that finally $Z = (1/\alpha)\Pi(x^a, 0)^t$ which completes the proof. ∎

Proposition 7.5.4. *Let $T > 0$, $p > 0$, $Y \geq 0$, and $Y \neq 0$, and assume that there exists $k \in S$ such that $Y_k = 0$. Then the matrix CE is reducible.*

Proof. This results from a straightforward calculation, making use of the explicit expression of the matrix E and of (Tr_{4b}). ∎

Using the expression of Z and Proposition (7.5.3), we deduce that, when

$$m_k + \overline{m} - \sum_{l \in S} m_l Y_l / \langle Y, U \rangle \neq 0, \qquad k \in S,$$

then CE is irreducible if and only if $Y > 0$. The pathological cases where $m_k + \overline{m} = \sum_{l \in S} m_l Y_l / \langle Y, U \rangle$ for some $k \in S$ further require numerous constraints on the structure of the flux diffusion matrix C in order to obtain reducibility of CE. Such constraints are not likely to happen in practice. These results show that no uncoupling is generally expected between groups of species for evaluating transport fluxes in terms of mass fractions gradients.

7.5.4. Diagonal diffusion of C and D over $\mathcal{U}^\perp$

We have shown in Proposition 7.5.1 that the matrices C and D are irreducible when $Y > 0$ and that these matrices cannot a fortiori be diagonal. However, we may ask for a weaker property since the vectors of physical interest—the diffusion driving forces—always lie in the hyperplane $\mathcal{U}^\perp$. As a consequence, we investigate, in this section, the situation in which C and D coincide with a diagonal matrix on the physical hyperplane $\mathcal{U}^\perp$ [Gio91].

Proposition 7.5.5. *The flux diffusion matrix C coincides with a diagonal matrix on the hyperplane $\mathcal{U}^\perp$ if and only if the matrix Γ may be written as*

$$\Gamma = \frac{1}{\mathcal{D}}\Big(I - \frac{Y \otimes \mathcal{U}}{\langle Y, \mathcal{U} \rangle}\Big), \tag{7.5.2}$$

where the quantity $\mathcal{D}$ is a positive scalar independent of the species. In this situation, we have

$$C = \mathcal{D}\Big(I - \frac{Y \otimes \mathcal{U}}{\langle Y, \mathcal{U} \rangle}\Big), \tag{7.5.3}$$

so that C coincides with $\mathcal{D}I$ over $\mathcal{U}^\perp$.

Proof. Assume that the matrix C coincides with a diagonal matrix on the hyperplane $\mathcal{U}^\perp$. We then have $\mathcal{U}^\perp \subset N(C - \Phi)$, where Φ is a diagonal matrix $\Phi = \mathrm{diag}(\phi_1, \ldots, \phi_n)$, and this yields $C = \Phi + \psi\otimes\mathcal{U}$, where $\psi \in \mathbb{R}^n$ is a vector. Since $N(C) = \mathbb{R}Y$ and $R(C) = \mathcal{U}^\perp$, we deduce that $\phi_k Y_k = -\psi_k\langle Y,\mathcal{U}\rangle$ and $\phi_k = -\langle Y, \psi\rangle$, $k \in S$. Denoting by $\mathcal{D}$ the scalar $\langle Y, \psi\rangle$, we deduce, after some algebra, that (7.5.3) holds.

In order to establish that $\mathcal{D}$ is positive, we first exclude the trivial case $n = 1$, for which $\Gamma = C = 0$ are zero matrices, so that $\mathcal{D}$ can then be chosen arbitrarily. When $n \geq 2$, we first note that $\mathcal{D}$ is nonzero from (7.5.3) and since $R(C) = \mathcal{U}^\perp$ is nonzero. We further note that $C_{kk} = \mathcal{D}(1 - y_k) \geq 0$, where $y_k = Y_k/\langle Y, U\rangle$, $k \in S$, so that, if there exists $k \in S$ with $Y_k = 0$, we have $C_{kk} = \mathcal{D} \geq 0$, whereas, if $Y > 0$, we have $0 < y_k < 1$, $k \in S$, and again $\mathcal{D} \geq 0$. We have thus shown that $\mathcal{D} \geq 0$ and $\mathcal{D} \neq 0$, so that $\mathcal{D}$ is positive.

Conversely, assuming that (7.5.3) holds, the matrix C obviously coincides with the diagonal matrix $\mathcal{D}I$ on the hyperplane $\mathcal{U}^\perp$. Finally, we note that the relations (7.5.3) and (7.5.2) are equivalent since $\Gamma C = C\Gamma = I - Y\otimes\mathcal{U}/\langle Y,\mathcal{U}\rangle$ and $\Gamma Y = CY = 0$. ■

Corollary 7.5.6. *The diffusion matrix D coincides with a diagonal matrix on the hyperplane $\mathcal{U}^\perp$ if and only if the matrix Δ may be written as*

$$\Delta = \frac{1}{\mathcal{D}}\Big(\mathcal{Y} - \frac{Y\otimes Y}{\langle Y,\mathcal{U}\rangle}\Big), \tag{7.5.4}$$

where $\mathcal{Y} = \mathrm{diag}(Y_1, \ldots, Y_n)$ and the quantity $\mathcal{D}$ is a positive scalar independent of the species. In this situation, we have

$$D = \mathcal{D}\Big(\mathcal{Y}^{-1} - \frac{\mathcal{U}\otimes\mathcal{U}}{\langle Y,\mathcal{U}\rangle}\Big). \tag{7.5.5}$$

Note that more precise results will be obtained in Section 7.7 for first-order diffusion coefficients for which the matrices Γ and Δ are known explicitly in terms of binary diffusion coefficients, species mass fractions, and molar masses.

7.5.5. Diagonal diffusion of CE and DE over $\mathcal{U}^\perp$

We have shown in Proposition 7.5.3 that the matrix CE is irreducible when $Z_k \neq 0$, $k \in S$. In this situation, we also know that $Y > 0$, so that the product DE is also irreducible. A fortiori, none of these products can be a diagonal matrix. However, since the mass fractions gradients $\boldsymbol{\partial_x} Y$ lie in the hyperplane $\mathcal{U}^\perp$, we now investigate when the products CE and DE can be represented by a diagonal matrix on $\mathcal{U}^\perp$ [Gio91].

Proposition 7.5.7. *The matrix CE coincides with a diagonal matrix on the hyperplane $\mathcal{U}^\perp$ if and only if the matrix Γ may be written as*

$$\Gamma = \frac{\overline{m}}{\mathcal{D}}\Big(M^{-1} - \frac{X\otimes M^{-1}\mathcal{U}}{\langle Y,\mathcal{U}\rangle}\Big), \tag{7.5.6}$$

where $M = \mathrm{diag}(m_1,\dots m_n)$ and the quantity $\mathcal{D}$ is a positive scalar independent of the species. Equivalently, (7.5.6) holds if and only if

$$E^{-1}\Gamma = \frac{1}{\mathcal{D}}\Big(I - \frac{Y\otimes\mathcal{U}}{\langle Y,\mathcal{U}\rangle}\Big). \tag{7.5.7}$$

In this situation, we have

$$CE = \mathcal{D}\Big(I - \frac{Z\otimes\mathcal{U}}{\langle Y,\mathcal{U}\rangle}\Big), \tag{7.5.8}$$

so that CE coincides with $\mathcal{D}I$ over $\mathcal{U}^\perp$.

Proof. By assumptions there exists a diagonal matrix $\Phi = \mathrm{diag}(\phi_1,\dots,\phi_n)$ such that $\mathcal{U}^\perp \subset N(CE - \Phi)$. As a consequence, there exists a vector ψ such that $CE = \Phi + \psi\otimes\mathcal{U}$. By transposing this relation, we also obtain that $E^tC^t = \Phi + \mathcal{U}\otimes\psi$. Using now $CEZ = 0$ and $(CE)^t\mathcal{U} = 0$, we further obtain that $\phi_k Z_k = -\psi_k\,\langle Y,\mathcal{U}\rangle$ and $\phi_k = -\langle\psi,\mathcal{U}\rangle$ and $k \in S$. Therefore, defining $\mathcal{D} = \langle\psi, Y\rangle$, we can write that $CE = \mathcal{D}\big(I - Z\otimes\mathcal{U}/\langle\mathcal{U},Y\rangle\big)$, so that (7.5.8) holds.

In order to establish that $\mathcal{D}$ is positive, we first exclude the trivial case $n = 1$, where $\Gamma = C = 0$ are zero matrices and E is the identity, so that $\mathcal{D}$ can then be chosen arbitrarily. When $n \geq 2$, we first note that $\mathcal{D}$ is nonzero from (7.5.8) and since $R(C) = \mathcal{U}^\perp$ is nonzero. In addition, multiplying on the right by E^{-1}, we also get $C = \mathcal{D}\big(E^{-1} - Z\otimes\mathcal{U}/\langle\mathcal{U},Y\rangle\big)$ since $E^t\mathcal{U} = \mathcal{U}$, whereas after some algebra, one can establish that

$$\Upsilon_k = E^{-1}_{kk} - \frac{Z_k}{\langle\mathcal{U},Y\rangle} = \frac{m_k}{\overline{m}}(1-y_k)^2 + \sum_{\substack{l\in S\\ l\neq k}} y_k y_l \frac{m_l}{\overline{m}}, \qquad k \in S,$$

where $y_k = Y_k/\langle\mathcal{U},Y\rangle$, $k \in S$. As a consequence, the diagonal coefficients Υ_k of $E^{-1} - Z\otimes\mathcal{U}/\langle\mathcal{U},Y\rangle$ are nonnegative and Υ_k vanishes only if $Y_k > 0$ and $Y_l = 0$ for $l \neq k$. Now if $n \geq 2$, two such diagonal coefficients cannot vanish simultaneously, so that, from $C_{kk} = \mathcal{D}\Upsilon_k \geq 0$, $k \in S$, we obtain that $\mathcal{D}$ is nonnegative. As a consequence, since $\mathcal{D}$ is nonzero, we deduce that $\mathcal{D}$ is positive.

Conversely, assuming (7.5.8), we obviously have that CE coincides with $\mathcal{D}I$ over $\mathcal{U}^\perp$. Furthermore, multiplying $C = \mathcal{D}\big(E^{-1} - Z\otimes\mathcal{U}/\langle\mathcal{U},Y\rangle\big)$ on the right by Γ, we then obtain (7.5.7) since we have $\Gamma^t\mathcal{U} = 0$ and $C\Gamma = I - Y\otimes\mathcal{U}/\langle\mathcal{U},Y\rangle$. Conversely, multiplying (7.5.7) on the left by E and then on the right by C yields $\Gamma C = (1/\mathcal{D})\,EC$ since $C^t\mathcal{U} = 0$, so that (7.5.7) and (7.5.8) are equivalent. Similarly, by using $EY = X$ and the expression $E = \overline{m}M^{-1} + X\otimes(\mathcal{U} - \overline{m}M^{-1}\mathcal{U})/\langle\mathcal{U},Y\rangle$, the relations (7.5.6) and (7.5.7) are easily shown to be equivalent. ∎

Corollary 7.5.8. *The matrix DE coincides with a diagonal matrix on the subspace $\mathcal{U}^\perp$ if and only if the matrix Δ may be written as*

$$\Delta = \frac{1}{\mathcal{D}}\Big(\mathfrak{X} - \frac{X\otimes X}{\langle Y,\mathcal{U}\rangle}\Big), \tag{7.5.9}$$

where $\mathfrak{X} = \mathrm{diag}(X_1,\dots,X_n)$ and the quantity $\mathcal{D}$ is a positive scalar independent of the species. In this situation, we have

$$DE = \mathcal{D}\Big(\mathcal{Y}^{-1} - \frac{\mathcal{Y}^{-1}Z\otimes\mathcal{U}}{\langle Y,\mathcal{U}\rangle}\Big). \tag{7.5.10}$$

Remark 7.5.9. Proposition (7.5.7) and Corollary (7.5.8) show that generalized Fick's laws of the type $\boldsymbol{\mathcal{F}}_k = \rho Y_k \boldsymbol{\mathcal{V}}_k = -\alpha_k \boldsymbol{\partial_x} Y_k$, where α_k is a scalar, cannot be used for all species, $k \in S$, unless all proportionality coefficients α_k, $k \in S$, are equal, i.e., $\alpha_1 = \cdots = \alpha_n$. ∎

Note that we have considered here diffusion coefficients of arbitrary order. In Section 7.7, we will focus on the first-order diffusion coefficients for which the results can be made more accurate since the matrices Δ and Γ are then known explicitly.

7.5.6. Diagonal diffusion of C and D for $n-1$ species

In this section, we investigate a more general situation than that of Section 7.5.4. More specifically, we investigate when the diffusion process can be represented by a diagonal matrix for $n-1$ species only. In other words, we first discard one species, chosen to be the last species for convenience, by using the mass constraint, and we then investigate the situations where resulting diffusion process can be represented by a diagonal matrix for the remaining $n-1$ species. In terms of the matrix C, this means that $n \geq 2$ and that there exist scalars α_k, $k \in \{1,\dots,n-1\}$, such that

$$\forall x \in \mathcal{U}^\perp, \quad \forall k \in \{1,\dots,n-1\}, \qquad (Cx)_k = \sum_{l\in S} C_{kl}x_l = \alpha_k x_k, \tag{7.5.11}$$

and this is a generalization of the property investigated in Section 7.5.4, in which diagonality was assumed to hold for all species.

Lemma 7.5.10. *Assume that Property* (7.5.11) *holds. Then the matrix* C *is in the form*

$$C = \begin{pmatrix} \alpha_1(1-y_1) & -\alpha_1 y_1 & \cdots & -\alpha_1 y_1 \\ & \ddots & & \vdots \\ -\alpha_{n-1}y_{n-1} & & \alpha_{n-1}(1-y_{n-1}) & -\alpha_{n-1}y_{n-1} \\ -\alpha_1+\sigma & \cdots & -\alpha_{n-1}+\sigma & \sigma \end{pmatrix}, \qquad (7.5.12)$$

where $y_k = Y_k/\langle Y,\mathcal{U}\rangle$, $k \in S$, *and* $\sigma = \sum_{k\in\{1,\ldots,n-1\}} \alpha_k y_k$. *In addition, we have* $\alpha_i > 0$, *for* $i \in \{1,\ldots,n-1\}$, *and whenever* $\alpha_i \neq \alpha_j$, *we must have* $Y_iY_j = 0$.

Proof. For $i \in \{1,\ldots,n-1\}$, we deduce from $e^i - e^n \in \mathcal{U}^\perp$ the property $C(e^i - e^n) - \alpha_i e^i \in \mathbb{R}e^n$. As a consequence, letting $u = Ce^n$, there exists scalars w_i, $i \in \{1,\ldots,n-1\}$, with $Ce^i = \alpha_i e^i + u + w_i e^n$. Using now $R(C) = \mathcal{U}^\perp$, we obtain $w_k = -\alpha_k$, $k \in \{1,\ldots,n-1\}$, and $\langle u,\mathcal{U}\rangle = 0$. Upon using $CY = 0$, we next obtain that $u_i = -\alpha_i Y_i/\langle Y,\mathcal{U}\rangle = -\alpha_i y_i$, $i \in \{1,\ldots,n-1\}$, and (7.5.12) is established with $\sigma = u_n$.

We now use the property that $C\mathcal{Y}$ is a symmetrix matrix established in Lemma 7.3.3. Upon calculating this matrix, we directly obtain by symmetry that $Y_iY_j(\alpha_i - \alpha_j) = 0$, for $i,j \in \{1,\ldots,n-1\}$. In addition, for $i \in \{1,\ldots,n-1\}$, we cannot have $\alpha_i = 0$, otherwise, from (7.5.12), we would get $R(C) \subset (e^i)^\perp$ contradicting $R(C) = \mathcal{U}^\perp$, since $n \geq 2$ in this section. From (7.5.12) and Lemma 7.3.3, we now obtain that $\alpha_i \geq 0$ for $i \in \{1,\ldots,n-1\}$, since either $0 \leq y_i < 1$ and $\alpha_i = C_{ii}/(1-y_i)$ or $y_i = 1$, but then $y_j = 0$ for $j \neq i$, so that $C_{nn} = \sigma = \alpha_i \geq 0$, and we then conclude that $\alpha_i > 0$ for $i \in \{1,\ldots,n-1\}$ since α_i cannot vanish. ∎

In the following, we denote by α either the common value of the coefficients α_k for $k \in \{1,\ldots,n-1\}$ and $Y_k > 0$ or the value $\alpha = 0$ in the situation where $Y_k = 0$ for all $k \in \{1,\ldots,n-1\}$. We also define the vector $\widehat{\alpha}$ by $\widehat{\alpha} = (\alpha_1,\ldots,\alpha_{n-1},\alpha)^t$.

Proposition 7.5.11. *The matrix* C *satisfies property* (7.5.11) *if and only if it can be written in the form*

$$C = \operatorname{diag}(\widehat{\alpha}) - \alpha\frac{Y\otimes\mathcal{U}}{\langle Y,\mathcal{U}\rangle} + e^n\otimes(\alpha\mathcal{U} - \widehat{\alpha}), \qquad (7.5.13)$$

where $\widehat{\alpha} = (\alpha_1,\ldots,\alpha_{n-1},\alpha)^t$, $\alpha_k > 0$, $k \in \{1,\ldots,n-1\}$, $\alpha_k = \alpha$ *whenever* $Y_k > 0$ *for* $k \in \{1,\ldots,n-1\}$, *and* $\alpha = 0$ *whenever* $Y_k = 0$ *for all* $k \in \{1,\ldots,n-1\}$.

Proof. The structure (7.5.13) is a direct consequence of Proposition 7.5.10, and the converse is obvious. ■

Defining $\beta_k = 1/\alpha_k$ whenever $k \in \{1, \ldots, n-1\}$ and either $\beta = 1/\alpha$ when $\alpha > 0$ or $\beta = 0$ when $\alpha = 0$, we now have the following result.

Proposition 7.5.12. *The matrix C satisfies property* (7.5.11) *if and only if Γ can be written in the form*

$$\Gamma = \mathrm{diag}(\widehat{\beta}) - \beta \frac{Y \otimes \mathcal{U}}{\langle Y, \mathcal{U} \rangle} + e^n \otimes (\beta \mathcal{U} - \widehat{\beta}), \tag{7.5.14}$$

where $\widehat{\beta} = (\beta_1, \ldots, \beta_{n-1}, \beta)^t$, $\beta_k > 0$, $k \in \{1, \ldots, n-1\}$, $\beta_k = \beta$ whenever $Y_k > 0$ for $k \in \{1, \ldots, n-1\}$, and $\beta = 0$ whenever $Y_k = 0$ for all $k \in \{1, \ldots, n-1\}$.

Proof. This result can be established after lengthy calculations or, more elegantly, by establishing directly that $C\Gamma = I - y \otimes \mathcal{U}$, $\Gamma C = I - y \otimes \mathcal{U}$, $\Gamma Y = 0$, and $\Gamma^t \mathcal{U} = 0$, and by using the uniqueness of the group inverse. ■

In conclusion, in order to obtain a diagonal diffusion process for $n-1$ species, part of the species can be zero and the diffusion process between the remaining positive species has to be uniform. In particular, the two limiting cases are those of positive mass fractions, where C coincides with αI on the complete hyperplane $\mathcal{U}^\perp$, and the situation $Y_1 = \cdots = Y_{n-1} = 0$ where C is exactly diagonal with respect to the $n-1$ first zero mass fraction species. This shows that only the equidiffusive limit, the dilution limit, or combinations of both yield diagonal diffusive processes. More accurate characterizations will be obtained in Section 7.7 for first order-diffusion coefficients, since the matrices Δ and Γ are then known explicitly.

7.5.7. Diagonal diffusion of CE and DE for $n-1$ species

In this section, we investigate the situations in which the diffusion process can be represented by a diagonal matrix for $n-1$ species, only but with respect to the mass fractions gradients. More specifically, we assume that $n \geq 2$ and that there exist scalars α_k, $k \in \{1, \ldots, n-1\}$, such that

$$\forall x \in \mathcal{U}^\perp, \quad \forall k \in \{1, \ldots, n-1\}, \qquad (CEx)_k = \sum_{i,j \in S} C_{ki} E_{ij} x_j = \alpha_k x_k, \tag{7.5.15}$$

which is more general than the property investigated in Section 7.5.5, where diagonality was assumed to hold for all species. These diagonal representations with respect to mass fraction gradients for $n-1$ species are often advocated in practice.

Lemma 7.5.13. *Assume that Property* (7.5.15) *holds. Then the matrix* CE *is in the form*

$$CE = \begin{pmatrix} \alpha_1(1-z_1) & -\alpha_1 z_1 & \cdots & -\alpha_1 z_1 \\ & \ddots & & \vdots \\ -\alpha_{n-1}z_{n-1} & & \alpha_{n-1}(1-z_{n-1}) & -\alpha_{n-1}z_{n-1} \\ -\alpha_1+\sigma & \cdots & -\alpha_{n-1}+\sigma & \sigma \end{pmatrix}, \tag{7.5.16}$$

where $z_k = Z_k/\langle Y,\mathcal{U}\rangle$, $k \in S$, *and* $\sigma = \sum_{k\in\{1,\dots,n-1\}} \alpha_k z_k$. *In addition, we have* $\alpha_i > 0$ *for* $k \in \{1,\dots,n-1\}$, *and whenever* $\alpha_i \neq \alpha_j$, *we must have* $Y_iY_j\big(m_i + m_j - \sum_{k\in S} m_k y_k\big) = 0$.

Proof. We proceed as in Lemma 7.5.10. For $i \in \{1,\dots,n-1\}$, we deduce from $e^i - e^n \in \mathcal{U}^\perp$ that $CE(e^i - e^n) - \alpha_i e^i \in \mathbb{R}e^n$. Letting $u = CEe^n$, there exist scalars β_i, $i \in \{1,\dots,n-1\}$, with $CEe^i = \alpha_i e^i + u + \beta_i e^n$. Using now $R(CE) = \mathcal{U}^\perp$, we obtain $\beta_k = -\alpha_k$, $k \in \{1,\dots,n-1\}$, and $\langle u,\mathcal{U}\rangle = 0$. Upon using $CEZ = 0$, we next obtain that $u_i = -\alpha_i Z_i/\langle Y,\mathcal{U}\rangle = -\alpha_i z_i$, $i \in \{1,\dots,n-1\}$, keeping in mind that $\langle Z,\mathcal{U}\rangle = \langle Y,\mathcal{U}\rangle$, and (7.5.16) is established with $\sigma = u_n$.

We now use the property that $C\mathcal{Y} = (CE)E^{-1}\mathcal{Y}$ is a symmetric matrix. Upon calculating this matrix, we obtain by symmetry that

$$Y_iY_j\big(m_i + m_j - \sum_{k\in S} m_k y_k\big)(\alpha_i - \alpha_j) = 0, \qquad i,j \in \{1,\dots,n-1\}.$$

In addition, for $i \in \{1,\dots,n-1\}$, we cannot have $\alpha_i = 0$, otherwise, from (7.5.16), we would get $R(CE) \subset (e^i)^\perp$ contradicting $R(CE) = \mathcal{U}^\perp$, since $n \geq 2$. On the other hand, multiplying (7.5.16) on the left by E^{-1}, we obtain, after some algebra, that $C_{kk} = \alpha_k \Upsilon_k$ where Υ_k is explicited in the proof of Proposition 7.5.7. When $0 \leq y_k < 1$, then $\Upsilon_k > 0$ and $\alpha_k = C_{kk}/\Upsilon_k \geq 0$, whereas, if $y_k = 1$, we have $y_j = 0$ for $j \in S$ and $j \neq k$ and we then have $C_{nn} = (m_n/m_k)\alpha_k \geq 0$, so that again $\alpha_k \geq 0$. We have thus shown that $\alpha_k \geq 0$ and $\alpha_k \neq 0$ for $k \in \{1,\dots,n-1\}$, and we can conclude that $\alpha_k > 0$ for $k \in \{1,\dots,n-1\}$. ∎

In the following, we exclude the pathological cases by assuming that

$$m_i + m_j - \sum_{k\in S} m_k y_k \neq 0, \qquad i,j \in S, \tag{7.5.17}$$

and we denote by α either the common values of the coefficients α_k for $k \in \{1,\dots,n-1\}$ and $Y_k > 0$ or the value $\alpha = 0$ in the situation where $Y_k = 0$ for all $k \in \{1,\dots,n-1\}$.

Proposition 7.5.14. *Assuming that* (7.5.17) *holds, the matrix* C *satisfies property* (7.5.15) *if and only if* CE *can be written in the form*

$$CE = \operatorname{diag}(\widehat{\alpha}) - \alpha\frac{Z\otimes\mathcal{U}}{\langle Y,\mathcal{U}\rangle} + e^n\otimes(\alpha\mathcal{U} - \widehat{\alpha}), \tag{7.5.18}$$

where $\widehat{\alpha} = (\alpha_1, \ldots, \alpha_{n-1}, \alpha)^t$, $\alpha_k > 0$, $k \in \{1, \ldots, n-1\}$, $\alpha_k = \alpha$ *whenever* $Y_k > 0$ *for* $k \in \{1, \ldots, n-1\}$, *and* $\alpha = 0$ *whenever* $Y_k = 0$ *for all* $k \in \{1, \ldots, n-1\}$.

Proof. The structure (7.5.18) is a direct consequence of Proposition 7.5.13, and the converse trivially holds. ∎

As in the preceding section, we can now obtain the structure of the matrix Γ.

Proposition 7.5.15. *Assuming that* (7.5.17) *holds, the matrix* C *satisfies property* (7.5.15) *if and only if* $E^{-1}\Gamma$ *can be written in the form*

$$E^{-1}\Gamma = \mathrm{diag}(\widehat{\beta}) - \beta \frac{Y \otimes \mathcal{U}}{\langle Y, \mathcal{U} \rangle} + e^n \otimes (\beta \mathcal{U} - \widehat{\beta}), \tag{7.5.19}$$

or if and only if the matrix Γ *can be written in the form*

$$\Gamma = \overline{m}\Big(M^{-1}\mathrm{diag}(\widehat{\beta}) - \frac{X}{\langle Y, \mathcal{U} \rangle} \otimes \big(M^{-1}\widehat{\beta} + \frac{1}{m_n}(\beta \mathcal{U} - \widehat{\beta}) \big) + \frac{e^n}{m_n} \otimes (\beta \mathcal{U} - \widehat{\beta}) \Big), \tag{7.5.20}$$

where $\widehat{\beta} = (\beta_1, \ldots, \beta_{n-1}, \beta)^t$, $\beta_k > 0$, $k \in \{1, \ldots, n-1\}$, $\beta_k = \beta$ *whenever* $Y_k > 0$ *for* $k \in \{1, \ldots, n-1\}$, *and* $\beta = 0$ *whenever* $Y_k = 0$ *for all* $k \in \{1, \ldots, n-1\}$.

In conclusion, in order to obtain a diagonal diffusion process for $n-1$ species with respect to the mass fraction gradients, part of these species can be zero and diffusion between the remaining positive species has to be uniform. In particular, the two limiting cases are that of positive mass fractions, where CE coincides with αI on the complete hyperplane $\mathcal{U}^{\perp}$, and the situation $Y_1 = \cdots = Y_{n-1} = 0$, where CE is exactly diagonal with respect to the $n-1$ first zero mass fraction species, excluding the pathological cases, as in Corollary 7.5.15. This shows that only the equidiffusive limit, the dilution limit, or combinations of both yield diagonal diffusive processes. More accurate characterizations will be obtained in Section 7.7 for first-order diffusion coefficients, since the matrices Δ and Γ are then known explicitly.

7.6. Diffusion inequalities

In this section, we investigate the positivity properties associated with the diffusion matrix D. These properties are of fundamental interest since the matrix D directly appears in the expression of the entropy production term, as shown in Section 2.6.

7.6.1. Fundamental diffusion inequality

The following inequality was first written in [Gio97], but is essentially a consequence of [Gio91].

Lemma 7.6.1. *Let T be a fixed temperature, and let $\mathcal{Y}$ denote the matrix $\mathcal{Y} = \mathrm{diag}(Y_1, \dots, Y_n)$. Then there exists a positive constant δ such that*

$$\forall Y > 0 \text{ with } \langle Y, \mathcal{U}\rangle = 1, \quad \forall x \in \mathcal{U}^\perp \qquad \delta\langle \mathcal{Y}^{-1}x, x\rangle \le \langle \rho D x, x\rangle. \tag{7.6.1}$$

Proof. Consider the quadratic form associated with the rescaled matrix

$$\begin{cases} \widehat{D}_{kk} = C_{kk}, & k \in S, \\ \widehat{D}_{kl} = \rho D_{kl}\sqrt{Y_k Y_l}, & k, l \in S, \quad k \neq l. \end{cases}$$

Then $\widehat{D}$ is a continuous function over $Y \ge 0$, $\langle Y, \mathcal{U}\rangle = 1$, from Lemma 7.3.1. Consider now the functional $(Y, x) \to \langle \widehat{D}(T, Y)x, x\rangle$ defined for $Y \ge 0$, $\langle Y, \mathcal{U}\rangle = 1$, $\|x\| = 1$, and $\sum_{k\in S} x_k\sqrt{Y_k} = 0$. This continuous functional reaches a minimum over this compact set. Denoting by $(Y^{\mathrm{m}}, x^{\mathrm{m}})$ any point where the minimum is reached, by S^+ the set of positive mass fraction species at Y^{m} and by S^0 the set of zero mass fraction species at Y^{m}, we have

$$\langle \widehat{D}(T, Y^{\mathrm{m}})x^{\mathrm{m}}, x^{\mathrm{m}}\rangle = \sum_{k,l\in S^+} \widehat{D}_{kl}(T, Y^{\mathrm{m}})x_k^{\mathrm{m}}x_l^{\mathrm{m}} + \sum_{k\in S^0} C_{kk}(T, Y^{\mathrm{m}})(x_k^{\mathrm{m}})^2.$$

Letting $z_k^{\mathrm{m}} = x_k^{\mathrm{m}}\sqrt{Y_k}$, $k \in S^+$, and using $\widehat{D}_{kl} = \rho D_{kl}\sqrt{Y_k Y_l}$ for $k, l \in S^+$, the first term can be rewritten in the form

$$\sum_{k,l\in S^+} \widehat{D}_{kl}(T, Y^{\mathrm{m}})x_k^{\mathrm{m}}x_l^{\mathrm{m}} = \sum_{k,l\in S^+} \rho D_{kl} z_k^{\mathrm{m}} z_l^{\mathrm{m}}.$$

As a consequence, this term is nonnegative from (Tr_4), and we further have $\sum_{k\in S^+} z_k^{\mathrm{m}} = 0$. Similarly, we also know that C_{kk}, $k \in S^0$, are positive, so that the sum $\sum_{k\in S^0} C_{kk}(T, Y^{\mathrm{m}})(x_k^{\mathrm{m}})^2$ is also nonnegative. Arguing now by contradiction, $\langle \widehat{D}(T, Y^{\mathrm{m}})x^{\mathrm{m}}, x^{\mathrm{m}}\rangle = 0$ implies that $x_k^{\mathrm{m}} = 0$, $k \in S^0$, and $z_k^{\mathrm{m}} = 0$, $k \in S^+$, since D is positive definite over $\mathcal{U}^\perp$, so that $x_k^{\mathrm{m}} = 0$, $k \in S^+$. We have thus shown that $x^{\mathrm{m}} = 0$, which contradicts $\|x^{\mathrm{m}}\| = 1$, so that $\langle \widehat{D}(T, Y^{\mathrm{m}})x^{\mathrm{m}}, x^{\mathrm{m}}\rangle$ is positive. Assuming finally that $Y > 0$ and using the change of variable $z_k = x_k\sqrt{Y_k}$, $k \in S$ completes the proof. ∎

Focusing then on the temperature dependency of the coercivity constant in (7.6.1), we now use the property (Tr_5). A straightforward adaptation of Lemma 7.6.1 yields the following result.

Corollary 7.6.2. *There exists a positive constant δ such that*

$$\begin{aligned}\forall T > 0,\ \forall Y > 0,\ \text{with}\ \langle Y,\mathcal{U}\rangle = 1, \qquad \forall x \in \mathcal{U}^{\perp} \\ \delta\,\varphi(T)\,\langle \mathcal{Y}^{-1}x, x\rangle \ \le\ \langle \rho D x, x\rangle.\end{aligned} \tag{7.6.2}$$

Remark 7.6.3. An assumption weaker than (Tr_5), which still allows various a priori estimates for multicomponent flow equations, is simply to assume that there exist a positive constant δ and a positive function of temperature φ such that, for any $T > 0$ and any $Y \ge 0$, $\langle Y,\mathcal{U}\rangle = 1$, we have

$$\delta \le \lambda/\varphi \le (1/\delta), \qquad \delta \le \mu/\varphi \le (1/\delta), \qquad 0 \le \kappa/\varphi \le (1/\delta),$$

whereas, for any $T > 0$ and any $Y > 0$ with $\sum_{k\in S} Y_k = 1$, we have,

$$\forall x \in \mathcal{U}^{\perp} \quad \delta\langle \mathcal{Y}^{-1}x, x\rangle \le \langle \rho D x, x\rangle/\varphi \le (1/\delta)\langle \mathcal{Y}^{-1}x, x\rangle.$$

In other words, all we need is to control the coercivity constants associated with entropy production in terms of T and Y. ∎

7.6.2. Positivity properties of C

The properties of the matrix ρD—including the fundamental dissipation inequalities (7.6.1) and (7.6.2)—can also be rewritten in terms of the flux diffusion matrix C.

Corollary 7.6.4. *The flux diffusion matrix C is such that $C\mathcal{Y}$ is symmetric positive semidefinite. Moreover, there exists a positive constant δ such that*

$$\begin{aligned}\forall T > 0,\ \forall Y \ge 0\ \text{with}\ \langle Y,\mathcal{U}\rangle = 1, \\ \forall x \in \mathbb{R}^n\ \text{with}\ \mathcal{Y}x \in \mathcal{U}^{\perp},\ \delta\varphi(T)\,\langle \mathcal{Y}x, x\rangle \ \le\ \langle C\mathcal{Y}x, x\rangle.\end{aligned} \tag{7.6.3}$$

It is important to note that the matrix $C\mathcal{Y}$—which naturally appears in symmetrized systems, as will be shown in Chapter 8—is of rank $n-k-1$ whenever k mass fractions are zero.

7.7. Stefan–Maxwell equations

We have seen in Chapters 2, 4, and 5 that the kinetic theory of gases does not yield explicit expressions for the transport coefficients. Transport linear systems have to be solved in order to evaluate these coefficients in gaseous mixtures. In this section, we focus on the mathematical structure of the simplest linear systems associated with the multicomponent diffusion coefficients presented in Chapter 5. We justify in this context the assumptions concerning the diffusion matrices C and D presented in Section 7.2.

7.7.1. Matrices Δ and D

The transport linear systems associated with the first-order multicomponent diffusion coefficients $D = D_{[00]}$ are the n linear systems of order n indexed by $l \in S$, in the form

$$\begin{cases} \Delta \alpha^{D_l} = \beta^{D_l}, \\ \alpha^{D_l} \in Y^{\perp}, \end{cases} \tag{7.7.1}$$

where Δ is a matrix of order n and α^{D_l}, β^{D_l}, and Y are vectors of $\mathbb{R}^n$. The matrix Δ is given by

$$\begin{cases} \Delta_{kk} = \sum_{\substack{l \in S \\ l \neq k}} \frac{X_k X_l}{\mathcal{D}_{kl}^{\text{bin}}}, & k \in S, \\ \\ \Delta_{kl} = -\frac{X_k X_l}{\mathcal{D}_{kl}^{\text{bin}}}, & k, l \in S, \quad k \neq l, \end{cases} \tag{7.7.2}$$

where $\mathcal{D}_{kl}^{\text{bin}}$ is the binary diffusion coefficient for the species pair (k, l), which only depends on pressure and temperature $\mathcal{D}_{kl}^{\text{bin}} = \mathcal{D}_{kl}^{\text{bin}}(T, p)$. More generally, for more accurate multicomponent diffusion coefficients, the quantities $\mathcal{D}_{kl}^{\text{bin}}$, $k, l \in S$, have to be replaced by Schur complements from transport linear systems of size larger than n, and then they are functions of T, p, and Y, but have similar properties [Gio91] [EG94].

The right members β^{D_l}, $l \in S$, are given by

$$\beta_k^{D_l} = \delta_{kl} - \frac{Y_k}{\sum_{i \in S} Y_i}, \qquad k \in S, \tag{7.7.3}$$

where δ_{kl} is the Kronecker symbol. Finally, the diffusion coefficients D_{kl}, $k, l \in S$, are evaluated by

$$D_{kl} = \alpha_k^{D_l}, \qquad k, l \in S. \tag{7.7.4}$$

The relations (7.7.4) can easily be rewritten in terms of the species diffusion velocities $\boldsymbol{\mathcal{V}}_k$, $k \in S$, and yield the relations

$$\boldsymbol{d}_k + \chi_k \boldsymbol{\partial_x} \log T = -\sum_{l \in S} \Delta_{kl} \boldsymbol{\mathcal{V}}_l, \qquad k \in S, \tag{7.7.5}$$

termed Stefan–Maxwell equations in the literature. An elementary derivation of these equations has been given by Williams [Wil58]. These equations can also be written in the compact form $\boldsymbol{d} + \chi \boldsymbol{\partial_x} \log T = -\Delta \boldsymbol{\mathcal{V}}$ and have to be completed by the mass constraint

$$\sum_{k\in S} Y_k \boldsymbol{\mathcal{V}}_k = 0, \tag{7.7.6}$$

as in the transport linear systems (7.7.1). However, although one may work with the diffusion velocities, it is more elegant to work directly with the transport coefficients.

The matrix Δ is easily shown to have the following properties [Gio91] [EG94] [EG97].

Lemma 7.7.1. *Assume that the molar masses m_k, $k \in S$, are positive constants, that the coefficients $\mathcal{D}_{kl}^{\mathrm{bin}}$, $k,l \in S$, $k \neq l$, are positive and symmetric, and that $Y > 0$. Then Δ is symmetric positive semidefinite, $N(\Delta) = \mathbb{R}\mathcal{U}$ where $\mathcal{U} = (1,\ldots,1)^t$, $R(\Delta) = \mathcal{U}^\perp$, and $\beta^{D_l} \in R(\Delta)$, $l \in S$. Moreover, Δ is irreducible and a singular M matrix.*

Proof. The symmetry of Δ is obvious from its definition and the properties of the binary diffusion coefficients $\mathcal{D}_{kl}^{\mathrm{bin}}$. Similarly, the relation $\Delta\mathcal{U} = 0$ is a direct consequence of (7.7.2). A straightforward computation also yields that, for $x \in \mathbb{R}^n$, we have

$$\langle \Delta x, x \rangle = \sum_{\substack{k,l\in S\\ k\neq l}} \frac{X_k X_l}{2\mathcal{D}_{kl}^{\mathrm{bin}}} (x_k - x_l)^2,$$

so that $\langle \Delta x, x\rangle \geq 0$ and $\langle \Delta x, x\rangle = 0$ implies $x \in \mathbb{R}\mathcal{U}$ since $X > 0$. The graph of Δ is also strongly connected since $\Delta_{kl} \neq 0$ for all $k, l \in S$, so that Δ is irreducible and the diagonal dominance is obvious [Var62]. Now let $s \in \mathbb{R}$ be such that $s \geq \Delta_{kk}$, for $k \in S$, and define $B = sI - \Delta$. Then $B \geq 0$ from the definitions of Δ and s. Applying then the Gershgorin theorem to B yields that $|\lambda| \leq s$ for $\lambda \in \mathrm{Sp}(B)$, where $\mathrm{Sp}(B)$ is the spectrum of B, since $\sum_{l\in S} B_{kl} = s$ for $l \in S$, so that Δ is a singular M matrix [BG74]. ∎

The matrix D is then related to a generalized inverse of Δ with prescribed range and nullspace [Gio90] [Gio91] [EG94] [EG97].

Proposition 7.7.2. *Keep the assumptions of Lemma 7.7.1. Then the n linear systems are well posed and the matrix D is the generalized inverse of Δ with prescribed range $Y^\perp$ and nullspace $\mathbb{R}Y$. More specifically, D is the unique matrix, such that $D\Delta D = D$, $\Delta D \Delta = \Delta$, $R(D) = Y^\perp$, and*

$N(D) = \mathbb{R}Y$. As a consequence, the matrix D is positive semidefinite, we have $\Delta D = I - Y{\otimes}\mathcal{U}/\langle\mathcal{U},Y\rangle$, and $D\Delta = I - \mathcal{U}{\otimes}Y/\langle\mathcal{U},Y\rangle$, and, for a,b positive with $ab\langle\mathcal{U},Y\rangle^2 = 1$, we have $D = (\Delta + aY{\otimes}Y)^{-1} - b\mathcal{U}{\otimes}\mathcal{U}$. Finally, the coefficients of D are smooth functions of (T,p,Y) for $T > 0$, $p > 0$, $Y > 0$, provided that the binary diffusion coefficients are smooth functions of (T,p,Y).

Proof. Let Z be the generalized inverse of Δ with prescribed range $Y^\perp$ and nullspace $\mathbb{R}Y$. By definition, we have $Z\Delta Z = Z$ and $\Delta Z\Delta = \Delta$. Transposing these relations and using the symmetry of Δ we deduce that $Z^t\Delta Z^t = Z^t$ and $\Delta Z^t\Delta = \Delta$. Moreover, $R(Z^t) = \big(N(Z)\big)^\perp$ implies that $R(Z^t) = Y^\perp$ and $N(Z^t) = \big(R(Z)\big)^\perp$ yields $N(Z^t) = \mathbb{R}Y$. From the uniqueness of the generalized inverse with prescribed range and nullspace, we deduce that $Z = Z^t$. Moreover, Z is positive semidefinite, since, for $y \in \mathbb{R}^n$, we have $\langle y, Zy\rangle = \langle Zy, \Delta Zy\rangle \geq 0$ because $Z = Z\Delta Z$, Z is symmetric, and Δ is positive semidefinite. The relations $\Delta Z = I - Y{\otimes}\mathcal{U}/\langle\mathcal{U},Y\rangle$ and $Z\Delta = I - \mathcal{U}{\otimes}Y/\langle\mathcal{U},Y\rangle$ are direct consequences from the definition of Z. A straightforward calculation then yields that

$$(\Delta + aY{\otimes}Y)(Z + b\mathcal{U}{\otimes}\mathcal{U}) = I,$$

whenever $ab\langle\mathcal{U},Y\rangle^2 = 1$, so that $Z = (\Delta + aY{\otimes}Y)^{-1} - b\mathcal{U}{\otimes}\mathcal{U}$ and Z is a smooth function of Δ since the set of invertible matrices is an open set in which the application $A \longrightarrow A^{-1}$ is smooth.

On the other hand, the n linear systems (7.7.1) are easily seen to be well posed since $N(\Delta) \cap Y^\perp = \{0\}$. Finally, the column vectors $Z_{\bullet l} = (Z_{1l},\ldots,Z_{nl})^t$, $l \in S$, of Z are easily seen to satisfy the system (7.7.1) since $\Delta Z = I - Y{\otimes}\mathcal{U}/\langle\mathcal{U},Y\rangle$. As a consequence, we have $Z_{\bullet l} = \alpha^{D_l}$, $l \in S$, and, thus, $Z_{kl} = \alpha_k^{D_l}$, $k,l \in S$, so that $D_{kl} = Z_{kl} = Z_{lk}$, $k,l \in S$, and the proof is complete. ■

7.7.2. Matrices Γ and C

When some mass fractions vanish, the diffusion velocities are no longer defined and we have to consider the mass fluxes [Gio91]. To this purpose, we introduce the matrix Γ defined by

$$\begin{cases} \Gamma_{kk} = \dfrac{1}{\rho}\dfrac{\overline{m}}{m_k} \displaystyle\sum_{\substack{l\in S\\ l\neq k}} \dfrac{X_l}{\mathcal{D}_{kl}^{\text{bin}}}, & k \in S, \\[2ex] \Gamma_{kl} = -\dfrac{1}{\rho}\dfrac{\overline{m}}{m_l}\dfrac{X_k}{\mathcal{D}_{kl}^{\text{bin}}}, & k,l \in S, \quad k \neq l, \end{cases} \tag{7.7.7}$$

which satisfies $\Delta = \rho\Gamma\mathcal{Y}$. The transport linear systems associated with the flux diffusion coefficients C are the n linear systems of order n indexed by $l \in S$,

$$\begin{cases} \Gamma\widetilde{\alpha}^{D_l} = \beta^{D_l}, \\ \quad \widetilde{\alpha}^{D_l} \in \mathcal{U}^{\perp}, \end{cases} \tag{7.7.8}$$

and the flux diffusion coefficients C_{kl}, $k, l \in S$, are evaluated by

$$C_{kl} = \widetilde{\alpha}_k^{D_l}, \qquad k, l \in S. \tag{7.7.9}$$

One can also write the analog of equations (7.7.5) for multicomponent fluxes

$$\boldsymbol{d}_k + \chi_k \boldsymbol{\partial_x} \log T = -\sum_{l\in S} \Gamma_{kl}\boldsymbol{\mathcal{F}}_l, \qquad k \in S, \tag{7.7.10}$$

with the constraint

$$\sum_{k\in S} \boldsymbol{\mathcal{F}}_k = 0. \tag{7.7.11}$$

Proposition 7.7.3. *Assume that the molar masses m_k, $k \in S$, are positive constants. Let $\mathcal{D}_{kl}^{\text{bin}}$ be positive numbers defined for $k, l \in S$, $k \neq l$, and symmetric, i.e., $\mathcal{D}_{kl}^{\text{bin}} = \mathcal{D}_{lk}^{\text{bin}}$, for $k \neq l$, and assume that $Y \geq 0$ and $Y \neq 0$. Then the matrix Γ is such that $N(\Gamma) = \mathbb{R}Y$ and $R(\Gamma) = \mathcal{U}^{\perp}$ and is a singular M matrix.*

Proof. When all mass fractions are positive, the matrix $\mathcal{Y}$ is invertible, and, from $\Delta = \rho\Gamma\,\mathcal{Y}$, we deduce that $\Delta\mathcal{Y}^{-1} = \rho\Gamma$ and thus that $N(\Gamma) = \mathcal{Y}(N(\Delta)) = \mathbb{R}Y$ and $R(\Gamma) = R(\Delta) = \mathcal{U}^{\perp}$.

In the case of vanishing mass fractions, we introduce the permutation Υ associated with the reordering of S into (S^+, S^0), so that $\Upsilon^t Y = (Y^+, 0)^t$. Using the partitioning associated with S^+ and S^0, we may decompose each $x \in \mathbb{R}^n$ into $\Upsilon^t x = (x^+, x^0)^t$, where $x^+ \in \mathbb{R}^p$ and $x^0 \in \mathbb{R}^{n-p}$. Correspondingly, for each matrix A, we can decompose the product $\Upsilon^t A \Upsilon$ into the blocks $A^{++} \in \mathbb{R}^{p,p}$, $A^{+0} \in \mathbb{R}^{p,(n-p)}$, $A^{0+} \in \mathbb{R}^{(n-p),p}$, and $A^{00} \in \mathbb{R}^{(n-p),(n-p)}$ in such a way that $y = Ax$ if and only if $y^+ = A^{++}x^+ + A^{+0}x^0$ and $y^0 = A^{0+}x^+ + A^{00}x^0$. A straightforward calculation using (7.7.7) then shows that the matrix $\Upsilon^t\Gamma\Upsilon$ admits block decomposition

$$\Upsilon^t\Gamma\Upsilon = \begin{bmatrix} \Gamma^{++} & \Gamma^{+0} \\ 0 & \Gamma^{00} \end{bmatrix}, \tag{7.7.12}$$

where $\Gamma^{0+} = 0$ and Γ^{00} is diagonal with positive elements.

Let then $x = \Upsilon(x^+, x^0)^t \in \mathbb{R}^n$ such that $\Gamma x = 0$. From (7.7.12) we get $\Gamma^{00}x^0 = 0$, so that $x^0 = 0$ and $\Gamma^{++}x^+ = 0$. In addition, the matrices Γ^{++} and Δ^{++} are exactly the matrices that would be obtained by considering the submixture S^+ with positive mass fractions Y^+, and we have with obvious notation, $\Gamma^{++}\mathcal{Y}^{++} = \Delta^{++}$. As a consequence, $N(\Gamma^{++}) = \mathcal{Y}^{++}(\mathbb{R}\mathcal{U}^+) = \mathbb{R}Y^+$, so that $\Gamma x = 0$ if and only if for some $t \in \mathbb{R}$ we have $(x^+, x^0) = (tY^+, 0) = t(Y^+, 0)$ and thus $N(\Gamma) = \mathbb{R}Y$. Moreover, we have $R(\Gamma) \subset \mathcal{U}^\perp$ and thus $R(\Gamma) = \mathcal{U}^\perp$.

Let then s such that $s \geq \Gamma_{kk}$, for $k \in S$, and define $B = sI - \Gamma$. Then $B \geq 0$, and applying the Gershgorin theorem to B^t yields $|\lambda| \leq s$ for $\lambda \in \mathrm{Sp}(B) = \mathrm{Sp}(B^t)$ since $\sum_{k\in S} B_{kl} = s$ for $l \in S$, so that Γ is a singular M matrix. ∎

Proposition 7.7.4. *Keep the assumptions of Lemma 7.7.3. Then the n linear systems 7.7.8 are well posed and C is the group inverse of Γ, that is, C is the unique matrix, such that $C\Gamma C = C$, $\Gamma C\Gamma = \Gamma$, $R(C) = \mathcal{U}^\perp$, and $N(C) = \mathbb{R}Y$. We also have $C\Gamma = \Gamma C = I - Y\otimes\mathcal{U}/\langle\mathcal{U}, Y\rangle$, and, for a, b positive with $ab\langle\mathcal{U}, Y\rangle^2 = 1$, we have $C = (\Gamma + aY\otimes\mathcal{U})^{-1} - bY\otimes\mathcal{U}$. In addition, the matrix C is a smooth function of (T, Y) for $T > 0$ and $Y \geq 0$, $Y \neq 0$, provided that the products $\widetilde{\mathcal{D}}^{\rm bin}_{kl}(T, Y) = \rho\mathcal{D}^{\rm bin}_{kl}$, $k, l \in S$, $k \neq l$, are smooth. Finally, when all mass fractions are positive, we have the identity*

$$C = \rho\mathcal{Y}D. \tag{7.7.13}$$

Proof. From Proposition 7.7.3, we know that $N(\Gamma) = \mathbb{R}Y$ and $R(\Gamma) = \mathcal{U}^\perp$, so that $N(\Gamma) \oplus R(\Gamma) = \mathbb{R}^n$ because $Y \notin \mathcal{U}^\perp$. As a consequence, the n linear systems (7.7.1) are easily seen to be well posed.

In addition, Γ admits a group inverse Z that satisfies $Z\Gamma = \Gamma Z = P_{R(\Gamma),N(\Gamma)}$, which can be written as $Z\Gamma = \Gamma Z = I - Y\otimes\mathcal{U}/\langle\mathcal{U}, Y\rangle$. Now the column vectors $Z_{\bullet l} = (Z_{1l}, \ldots, Z_{nl})^t$, $l \in S$, of Z are easily seen to satisfy the system (7.7.8) since $\Gamma Z = I - Y\otimes\mathcal{U}/\langle\mathcal{U}, Y\rangle$. As a consequence, we have $Z_{\bullet l} = \widetilde{\alpha}^{D_l}$, $l \in S$, and, thus, $Z_{kl} = \alpha_k^{D_l} = C_{kl}$, $k, l \in S$, and C coincides with Z. The relation $C = (\Gamma + aY\otimes\mathcal{U})^{-1} - bY\otimes\mathcal{U}$ for a, b positive with $ab\langle\mathcal{U}, Y\rangle^2 = 1$ is then easily established, and C is a smooth function of Γ since the set of invertible matrices is an open set in which the application $A \longrightarrow A^{-1}$ is smooth.

Finally, when all mass fractions are positive, using the fact that D is the generalized inverse of Δ with $N(D) = \mathbb{R}Y$ and $R(D) = Y^\perp$, and that $\Delta = \rho\Gamma\mathcal{Y}$, it is easily shown that $\rho\mathcal{Y}D$ is the group inverse of Γ, so that we have $C = \rho\mathcal{Y}D$. ∎

The next proposition gives the behavior of the matrices Γ and C when some of the mass fractions vanish [Gio91] [EG94].

Proposition 7.7.5. *Keep the assumptions of Lemma 7.7.3, and introduce the sets $S^+ = \{\ k \in S,\ Y_k > 0\ \}$ and $S^0 = \{\ k \in S,\ Y_k = 0\ \}$. Let Υ be* *the permutation matrix associated with the reordering of S into (S^+, S^0). Then we have the following block decompositions for C and Γ :*

$$\Upsilon^t C \Upsilon = \begin{pmatrix} C^{++} & C^{+0} \\ 0 & C^{00} \end{pmatrix}, \qquad \Upsilon^t \Gamma \Upsilon = \begin{pmatrix} \Gamma^{++} & \Gamma^{+0} \\ 0 & \Gamma^{00} \end{pmatrix}, \qquad (7.7.14)$$

where C^{00} and Γ^{00} are diagonal with positive entries and inverse of each other, and C^{++} and Γ^{++} are exactly the matrices that would be obtained by considering the S^+ submixture of nonzero mass fractions.

Proof. The decomposition of $\Upsilon^t \Gamma \Upsilon$ directly results from the expressions (7.7.7). In addition, it is straightforward to check that $\Upsilon^t C \Upsilon$ is the generalized inverse of $\Upsilon^t \Gamma \Upsilon$ with prescribed range $\mathcal{U}^\perp$ and prescribed nullspace $\mathbb{R}\Upsilon^t Y = \mathbb{R}(Y^+, Y^0)^t$. Upon decomposing the matrix $\Upsilon^t C \Upsilon$ into the four blocks C^{++}, C^{+0}, C^{0+}, and C^{00}, associated with the reordering of S into (S^+, S^0), we obtain from the identity

$$(\Upsilon^t \Gamma \Upsilon)(\Upsilon^t C \Upsilon) = I - \Upsilon^t Y \otimes \mathcal{U} / \langle \mathcal{U}, Y \rangle, \qquad (7.7.15)$$

that $\Gamma^{00} C^{0+} = 0$, so that $C^{0+} = 0$ and $\Gamma^{00} C^{00} = I$ since $Y^0 = 0$.

Now let $z^+ \in \mathbb{R}^p$, such that $z^+ \in R(C^{++})$. Then $z^+ = C^{++} x^+$ for some $x^+ \in \mathbb{R}^p$. Letting now $z = \Upsilon(z^+, 0)^t$ and $x = \Upsilon(x^+, 0)^t$, we get $z = Cx$ and thus $z \in R(C) = \mathcal{U}^\perp$. Therefore, $\langle \mathcal{U}, z \rangle = \langle \mathcal{U}^+, z^+ \rangle = 0$ and $z^+ \in (\mathcal{U}^+)^\perp$. Conversely, if $z^+ \in (\mathcal{U}^+)^\perp$, then, for $z = \Upsilon(z^+, 0)^t$, we have $z \in \mathcal{U}^\perp = R(C)$ and thus $z = Cx$ for some $x \in \mathbb{R}^n$. Now, since $C^{0+} = 0$ and C^{00} is invertible, we get $z^0 = C^{00} x^0 = 0$, so that $x^0 = 0$ and $x = (x^+, 0)$, $z^+ = C^{++} x^+$, and $z^+ \in R(C^{++})$. We have thus shown that $R(C^{++}) = (\mathcal{U}^+)^\perp$, and, since $CY = 0$, we deduce that $C^{++} Y^+ = 0$ and thus $N(C^{++}) = \mathbb{R} Y^+$. On the other hand, we get from (7.7.15) that $C^{++} \Gamma^{++} = \Gamma^{++} C^{++} = I - Y^+ \otimes \mathcal{U}^+ / \langle \mathcal{U}, Y \rangle$. Finally, multiplying this equality by C^{++} and Γ^{++}, we obtain $C^{++} \Gamma^{++} C^{++} = C^{++}$ and $\Gamma^{++} C^{++} \Gamma^{++} = \Gamma^{++}$, and thus C^{++} is the group inverse of Γ^{++}. Therefore, C^{++} is exactly the diffusion matrix that would be obtained by considering the S^+ submixture of positive mass fractions. ∎

7.7.3. Diagonal first-order diffusion

In this section, we investigate the diagonal diffusion problem for first-order diffusion matrices. More accurate characterizations of diagonal diffusion processes are obtained since Γ and Δ are known explicitly.

Proposition 7.7.6. *The matrix C coincides with a diagonal matrix on the subspace $\mathcal{U}^\perp$ if and only if the numbers $\mathcal{D}_{kl}^{\text{bin}} m_k m_l$, $k, l \in S$, $k \neq l$, are equal. In this situation, the relations* (7.5.2) *and* (7.5.3) *hold.*

Proposition 7.7.7. *The matrix CE coincides with a diagonal matrix on the subspace $\mathcal{U}^{\perp}$ if and only if the numbers $\mathcal{D}_{kl}^{\text{bin}}$, $k,l \in S$, $k \neq l$, are equal. In this situation, the relations* (7.5.6), (7.5.7), *and* (7.5.8) *hold.*

Proof. Proposition 7.7.7 is straightforward by identifying the nondiagonal coefficients of Γ in (7.5.6) and (7.7.7), and the proof of 7.7.6 is similar. ∎

Similar results are easily rewritten in terms of the matrix D [Gio91]. More generally, we now investigate diagonal diffusion processes with respect to the first $n-1$ species. The proof of Propositions 7.7.8 and 7.7.9 is similar to the one of Propositions 7.7.6 and 7.7.7.

Proposition 7.7.8. *The matrix C coincides with a diagonal matrix on the subspace $\mathcal{U}^{\perp}$ for the first $n-1$ species if and only if the numbers $\mathcal{D}_{kl}^{\text{bin}} m_k m_l$ are equal for $(k,l),(l,k) \in S^{+\prime} \times S$, $k \neq l$, where $S^{+\prime}$ denotes the elements of $\{1,\dots,n-1\}$ such that $Y_k > 0$.*

Proposition 7.7.9. *The matrix CE coincides with a diagonal matrix on the subspace $\mathcal{U}^{\perp}$ for the first $n-1$ species if and only if the numbers $\mathcal{D}_{kl}^{\text{bin}}$ are equal for $(k,l),(l,k) \in S^{+\prime} \times (S^{+\prime} \cup \{n\})$, $k \neq l$, and the following compatibility conditions hold:*

$$\frac{\langle Y,\mathcal{U}\rangle}{\rho\mathcal{D}_{kl}^{\text{bin}}} = \frac{m_l \dfrac{1}{\alpha} + (m_n - m_l)x_n \dfrac{\langle Y,\mathcal{U}\rangle}{\rho\mathcal{D}_{nl}^{\text{bin}}}}{(1-x_n)m_l + x_n m_n}, \tag{7.7.16}$$

for $(k,l) \in S^{+\prime} \times S^{0\prime}$, $k \neq l$, where $S^{+\prime}$ denotes the elements of $\{1,\dots,n-1\}$, such that $Y_k > 0$, $S^{0\prime}$ the elements of $\{1,\dots,n-1\}$, such that $Y_k = 0$, where $x_n = X_n/\langle Y,\mathcal{U}\rangle$, and $1/\alpha$ denotes the constant value $1/\alpha = \langle Y,\mathcal{U}\rangle\rho/\mathcal{D}_{ij}^{\text{bin}}$ for $(i,j) \in S^{+\prime} \times (S^{+\prime} \cup \{n\})$, $i \neq j$.

Note, in particular, that the only way to exclude compatibility relations among the binary diffusion coefficients in Propositions 7.7.8 or 7.7.9 is to assume that $Y_1 = \dots = Y_{n-1} = 0$, since then $S^{+\prime} = \emptyset$.

7.7.4. Asymptotic expansions of C and D

In this section, we investigate asymptotic expansions of diffusion matrices in order to define approximate transport models. More specifically, we investigate the convergence of iterative methods for the Stefan–Maxwell diffusion equations (7.7.5) and (7.7.10).

Since these equations can be projected along each coordinate direction in the three-dimensional space, it is sufficient to consider the case in which velocities and diffusion driving forces are scalars, and we denote by $\mathcal{V} = (\mathcal{V}_1,\dots,\mathcal{V}_n)^t$, $\mathcal{F} = (\mathcal{F}_1,\dots,\mathcal{F}_n)^t$, and $d = (d_1,\dots,d_n)^t$ the corresponding vectors. Therefore, for a given vector $d \in \mathcal{U}^{\perp}$, we want to solve the consistent singular system $d = -\Delta \mathcal{V}$ by iteration techniques, and we want to obtain the only solution $\mathcal{V}$ which is in $Y^\perp$. Similarly, we want to solve the system $d = -\Gamma \mathcal{F}$ by iteration techniques, and we want to obtain the only solution $\mathcal{F}$ which is in $\mathcal{U}^\perp$.

Theorem 7.7.10. *Let Δ be as in (7.7.2), and keep the assumptions of Proposition 7.7.1. Let $\mathcal{M} = diag(\mathcal{M}_1, \ldots, \mathcal{M}_n)$ be such that $\mathcal{M}_k > \Delta_{kk}$, for $k \in S$, denote $P = P_{Y^\perp, \mathbb{R}\mathcal{U}}$, and $\mathcal{T} = \mathcal{M}^{-1}(\mathcal{M} - \Delta) = I - \mathcal{M}^{-1}\Delta$, let $x_0 \in \mathbb{R}^n$, $y_0 = Px_0$, and $d \in \mathcal{U}^\perp$ and define*

$$x_{i+1} = \mathcal{T}x_i + \mathcal{M}^{-1}(-d), \qquad i \geq 0, \tag{7.7.17}$$

$$y_{i+1} = P\mathcal{T}y_i + P\mathcal{M}^{-1}(-d), \qquad i \geq 0. \tag{7.7.18}$$

Then, $y_i = Px_i$, for all $i \geq 0$, $\mathrm{Srd}(T) = 1$, $\mathrm{Srd}(PT) < 1$, where $\mathrm{Srd}(A)$ is the spectral radius of a matrix A, and

$$\mathcal{V} = P\Big(\lim_{i\to\infty} x_i\Big) = \lim_{i\to\infty} y_i, \tag{7.7.19}$$

where $\mathcal{V}$ is the unique solution of $\Delta\mathcal{V} = -d$ in the subspace $Y^\perp$. Moreover, if D is the generalized inverse of Δ with prescribed range $Y^\perp$ and nullspace $\mathbb{R}Y$, then we have

$$D = \sum_{k\in\mathbb{N}} (P\mathcal{T})^k P\mathcal{M}^{-1}P^t, \tag{7.7.20}$$

and each partial sum $D_{[i]} = \sum_{k=0}^{i}(P\mathcal{T})^k P\mathcal{M}^{-1}P^t$ of this series is symmetric, satisfies $D_{[i]}Y = 0$, and is positive definite on $\mathcal{U}^\perp$.

Two different proofs have been given in [Gio91] and [EG97], and they are not repeated here. It is interesting to note that the splitting

$$\mathcal{M}_k = \frac{\Delta_{kk}}{1 - \dfrac{Y_k}{\langle Y, \mathcal{U}\rangle}} = \frac{X_k}{D_k^*}, \qquad D_k^* = \Big(1 - \frac{Y_k}{\langle Y, \mathcal{U}\rangle}\Big)\Big/ \sum_{\substack{l\in S\\ l\neq k}} \frac{X_l}{\mathcal{D}_{kl}^{\mathrm{bin}}},$$

is well defined and satisfies the hypotheses of Theorem 7.7.10 since $Y > 0$, and thus $0 < 1 - Y_k / \langle Y, \mathcal{U}\rangle < 1$ and $0 < \Delta_{kk} < \mathcal{M}_k$.

For this particular splitting, the vector $\mathcal{M}^{-1}(-d)$ corresponds to the so-called Hirshfelder–Curtiss approximate diffusion velocities $\mathcal{V}$ [HC49] and the vector $P\mathcal{M}^{-1}(-d)$ exactly corresponds to the Hirschfelder–Curtiss approximate velocities with a species independent mass correction velocity [CH81] [Gio91]. This shows that the widely used approximations $\mathcal{V} \simeq P\mathcal{M}^{-1}(-d)$, which are often considered as ad hoc approximations, have indeed a rigorous justification. Note that, for $d \in \mathcal{U}^\perp$, the latter approximation can be written in the symmetric form $P\mathcal{M}^{-1}P^t(-d)$. Remark also that a justification for choosing this particular splitting is, for instance, to substitute the approximation $D \simeq \mathcal{M}^{-1}$ into the relation $\Delta D = I - Y \otimes \mathcal{U} / \langle \mathcal{U}, Y \rangle$, and to identify the corresponding diagonals. For this particular splitting, the iterative scheme (7.7.17) has been introduced by Oran and Boris [OB81] and Jones and Boris [JB81]. The projected algorithm (7.7.18), asymptotic expansion (7.7.20), and convergence results have been obtained in [Gio91].

A similar result can also be obtained for the flux-type Stefan–Maxwell equations [Gio91] [EG94].

Theorem 7.7.11. *Let Γ be as in (7.7.7), and keep the assumptions of Proposition 7.7.3. Let $\mathcal{L} = \mathrm{diag}(\mathcal{L}_1, \ldots, \mathcal{L}_n)$ be such that $\mathcal{L}_k > \Gamma_{kk}$, if $Y_k > 0$, and $\mathcal{L}_k \geq \Gamma_{kk}$, if $Y_k = 0$, for $k \in S$. Denote $Q = P_{\mathcal{U}^\perp, \mathbb{R}Y}$ and $\mathcal{S} = \mathcal{L}^{-1}(\mathcal{L} - \Gamma) = I - \mathcal{L}^{-1}\Gamma$, let $x_0 \in \mathbb{R}^n$, $y_0 = Qx_0$, and $d \in \mathcal{U}^\perp$, and define*

$$x_{i+1} = \mathcal{S} x_i + \mathcal{L}^{-1}(-d), \qquad i \geq 0, \tag{7.7.21}$$

$$y_{i+1} = Q\mathcal{S} y_i + Q\mathcal{L}^{-1}(-d), \qquad i \geq 0. \tag{7.7.22}$$

Then, $y_i = Qx_i$, for all $i \geq 0$, $\mathrm{Srd}(\mathcal{S}) = 1$, $\mathrm{Srd}(Q\mathcal{S}) < 1$, and

$$\mathcal{F} = Q\Big(\lim_{i\to\infty} x_i\Big) = \lim_{i\to\infty} y_i, \tag{7.7.23}$$

where $\mathcal{F}$ is the unique solution of $\Gamma\mathcal{F} = -d$ in the subspace $\mathcal{U}^\perp$. Moreover, if C is the group inverse of Γ, then we have

$$C = \sum_{k\in\mathbb{N}} (Q\mathcal{S})^k Q\mathcal{L}^{-1} Q. \tag{7.7.24}$$

When there are at least two nonzero mass fractions, then the splitting

$$\mathcal{L}_k = \frac{\Gamma_{kk}}{1 - \dfrac{Y_k}{\langle Y, \mathcal{U}\rangle}} = \frac{\overline{m}}{m_k}\frac{1}{D_k^*}, \qquad D_k^* = \Big(1 - \frac{Y_k}{\langle Y, \mathcal{U}\rangle}\Big) \Big/ \sum_{\substack{l\in S\\ l\neq k}} \frac{X_l}{\mathcal{D}_{kl}^{\mathrm{bin}}},$$

is well defined since $0 < 1 - Y_k / \langle Y, \mathcal{U}\rangle$, for $k \in S$, and satisfies the hypotheses of Theorem 7.7.11 since $1 - Y_k / \langle Y, \mathcal{U}\rangle < 1$ if and only if $Y_k > 0$. The vector $\mathcal{L}^{-1}(-d)$ also corresponds to the Hirshfelder–Curtiss approximate diffusion fluxes $\mathcal{F}$, and, for $d \in \mathcal{U}^\perp$, the vector $Q\mathcal{L}^{-1}(-d)$ exactly corresponds to the Hirschfelder–Curtiss approximations with mass correction fluxes proportional to the mass fractions Y. This shows that the widely used approximations $\mathcal{F} \simeq Q\mathcal{L}^{-1}(-d)$ also have a rigorous justification, provided that there are at least two nonzero mass fractions in the mixture. Moreover, in the case of a pure species state of the mixture $Y = (1, 0, \ldots, 0)$, we have $\Gamma_{11} = 0$, and it is easy to check that any $\mathcal{L}$ in the particular form $\mathcal{L} = \mathrm{diag}(\Upsilon, \Gamma_{22}, \ldots, \Gamma_{nn})$, where $\Upsilon > 0$ is arbitrary, leads to a one-step convergence of the sequence (7.7.22). Although the coefficients D_k^*, $k \in S$, do not provide such a splitting, since then D_1^* is undefined, a typical numerical procedure for evaluating these mixture diffusion coefficients consists in evaluating perturbed coefficients $D_k^*(\epsilon)$ defined by

$$D_k^*(\epsilon) = \sum_{\substack{l \in S \\ l \neq k}} \frac{m_l(X_l + \epsilon)}{\overline{m}\,\langle Y, \mathcal{U} \rangle} \Big/ \sum_{\substack{l \in S \\ l \neq k}} \frac{(X_l + \epsilon)}{\mathcal{D}_{kl}^{\text{bin}}},$$

where ϵ is a small positive constant, typically of the order of the machine precision. Now for a pure species state $Y = (1, 0, \dots, 0)$, this formula yields $D_k^*(\epsilon) = D_k^* + O(\epsilon)$, for $k \in \{2, \dots, n\}$, whereas it gives an arbitrary but positive $D_1^*(\epsilon) > 0$ for $k = 1$. Defining now $\mathfrak{L}_k = \overline{m}/\big(m_k D_k^*(\epsilon)\big)$, for $k \in S$, we obtain a one-step convergence of the algorithm (7.7.22), with an arbitrary but positive $\mathfrak{L}_1 = \Upsilon > 0$ [Gio91].

7.8. Notes

7.1. The fundamental diffusion inequalities (7.6.1) and (7.6.3) reveal that the natural entropy production norm is not the classical W_2^1 norm, but contains mass fraction gradients at the denominator of mass fraction gradients [Gio97] [Gio98] [Gio99].

7.9. References

[BG74] A. Ben-Israel and T. N. E. Greville, *Generalized Inverses, Theory and Applications,* John Wiley & Sons, Inc., New York, (1974).

[BP79] A. Bermann and R. J. Plemmons, *Nonnegative Matrices in the Mathematical Science,* Academic Press, New York, (1979).

[CC70] S. Chapman and T. G. Cowling, *The Mathematical Theory of Non-Uniform Gases,* Cambridge University Press, Cambridge, (1970).

[CH81] T. P. Coffee and J. M. Heimerl, *Transport Algorithms for Premixed, Laminar Steady-State Flames,* Comb. Flame, **43**, (1981), pp. 273–289.

[EG94] A. Ern and V. Giovangigli, *Multicomponent Transport Algorithms,* Lecture Notes in Physics, New Series "Monographs", **m 24**, Springer-Verlag, Berlin, 1994.

[EG97] A. Ern and V. Giovangigli, *Projected Iterative Algorithms with Application to Multicomponent Transport,* Linear Algebra and its Applications, **250**, (1997), pp. 289–315.

[FK72] J. H. Ferziger and H. G. Kaper, *Mathematical Theory of Transport Processes in Gases*, North Holland Publishing Company, Amsterdam, (1972).

[Gio90] V. Giovangigli, *Mass Conservation and Singular Multicomponent Diffusion Algorithms,* IMPACT Comput. Sci. Eng., **2**, (1990), pp. 73–97.

[Gio91] V. Giovangigli, *Convergent Iterative Methods for Multicomponent Diffusion,* IMPACT Comput. Sci. Eng., **3**, (1991), pp. 244–276.

[Gio97] V. Giovangigli, *Plane Flames with Multicomponent Transport and Complex Chemistry,* Internal Report, **366**, Centre de Mathématiques Appliquées de l'Ecole Polytechnique, (1997).

[Gio98] V. Giovangigli, *Flames Planes avec Transport Multi-espece et Chimie Complexe,* C. R. Acad. Sci. Paris, **326**, Série I, (1998), p. 775–780.

[Gio99] V. Giovangigli, *Plane Flames with Multicomponent Transport and Complex Chemistry,* Math. Mod. Meth. Appl. Sci., **9**, (1999), pp. 337–378.

[GM98a] V. Giovangigli and M. Massot, *Asymptotic Stability of Equilibrium States for Multicomponent Reactive Flows,* Math. Mod. Meth. Appl. Sci., **8**, (1998), pp. 251–297.

[HC49] J. O. Hirschfelder and C. F. Curtiss, *Flame Propagation in Explosive Gas Mixtures,* in Third Symposium (International) on Combustion, Reinhold, New York, (1949), pp. 121–127.

[HCB54] J. O. Hirschfelder, C. F. Curtiss, and R. B. Bird, *Molecular Theory of Gases and Liquids,* John Wiley & Sons, Inc., New York, (1954).

[JB81] W. W. Jones and J. P. Boris, *An algorithm for multispecies diffusion fluxes,* Comput. Chem., **5**, (1981), pp. 139–146.

[OB81] E. S. Oran and J. P. Boris, *Detailed modeling of combustion systems,* Prog. Energy Combust. Sci., **7**, (1981), pp. 1–72.

[Var62] R. S. Varga, *Matrix iterative Analysis,* Prentice-Hall, Englewood Cliffs, NJ, (1962).

[Wil58] F. A. Williams, *Elementary Derivation of the Multicomponent Diffusion Equation,* Amer. J. Phys., **26**, (1958), pp. 467–469.

[Wal47] L. Waldmann, *Der Diffusionthermoeffekt II,* Zeitschr. Physik., **124**, (1947), pp. 175–195.

[Wal58] L. Waldmann, *Transporterscheinungen in Gasen von Mittlerem Druck,* Handbuch der Physik, S. Flügge, Ed., **12**, Springer-Verlag, Berlin, (1958), pp. 295–514.

[WT62] L. Waldmann und E. Trübenbacher, *Formale Kinetische Theorie von Gasgemischen aus Anregbaren Molekülen,* Zeitschr. Naturforschg., **17a**, (1962), pp. 363–376.

8

Symmetrization

8.1. Introduction

In this chapter we investigate the mathematical structure of the system of partial differential equations governing multicomponent reactive flows. We obtain, in particular, various symmetrized forms of this system, making use of entropic variables and normal variables. Symmetric forms are a fundamental step towards existence results for systems of partial differential equations [VH72] [Maj84] [SK85] [Ser96a] [Ser96b]. These symmetrized forms will be used, in particular, in Chapter 9 and Chapter 10 to obtain global existence theorems around constant equilibrium states. Symmetrized forms can also be used for finite element numerical simulations based on streamline upwind Petrov–Galerkin techniques [HFM86] [CHS90] [HH98].

Symmetrizability is first discussed for an abstract system of conservation laws. This property is closely related to the existence of an entropy function, as shown by Kawashima and Shizuta in the case of hyperbolic–parabolic systems [KS88], extending previous work on hyperbolic systems [God62] [FL71] [Moc80]. Starting from an entropy related conservative symmetric form, Kawashima and Shizuta have further investigated normal forms, that is, symmetric hyperbolic–parabolic composite forms. These authors have shown, in particular, that symmetric systems of conservation laws, such that the nullspace naturally associated with dissipation matrices is invariant, can be recast into a normal form. In the framework of their theory, all normal forms for systems satisfying the invariance property are characterized.

As an application, we exhibit an entropy function for the system of equations modeling multicomponent flows. By using the corresponding entropic variable, we derive a symmetric conservative formulation of the system. Chalot, et al. [CHS90] have carried out similar calculations in the case of flows in thermochemical nonequilibrium. However, they have used a multicomponent diffusion matrix that is not symmetric [HCB54] and prohibits complete symmetrization. These authors have thus advocated Onsager's phenomenological coefficients in order to achieve symmetrization. On the contrary, by using the symmetric form of the transport fluxes and of the diffusion coefficients presented in Chapter 7, we obtain a naturally symmetric conservative formulation [GM98a].

The symmetrized multicomponent reactive flows governing equations are then shown to satisfy the nullspace invariance property and recast into various normal forms. The first normal form has simple matrix coefficients and generalizes previous results of Kawashima and Shizuta [KS88]. The second natural normal form has dissipative terms in conservative form and leaves unchanged the structure of the source term. These forms can be used for the asymptotic stability of constant equilibrium states investigated in Chapter 9. The third normal form is intermediate between these two and will be used in Chapter 10 for chemical equilibrium flows.

8.2. Vector notation

The governing equations have been presented in Chapter 2, and the mathematical properties of thermodynamic functions and transport coefficients have been obtained in Chapter 6 and Chapter 7, respectively. In this section, we introduce a compact notation that will be used throughout Chapters 8, 9, and 10.

8.2.1. Conservative and natural variables

We introduce the conservative variable U given by

$$U = \Big(\rho_1, \ldots, \rho_n, \rho v_1, \rho v_2, \rho v_3, \rho e^{\mathrm{tot}}\Big)^t, \tag{8.2.1}$$

and the natural variable Z

$$Z = \Big(\rho_1, \ldots, \rho_n, v_1, v_2, v_3, T\Big)^t, \tag{8.2.2}$$

where ρ_k is the density of the kth species, n is the number of species, S is the set of species indices $S = \{1, \ldots, n\}$, ρ is the total density, v_i is the mass averaged flow velocity in the ith direction in such a way that the velocity vector is $\boldsymbol{v} = (v_1, v_2, v_3)^t$, e^{tot} is the total energy per unit mass of the mixture, ^{t} is the transposition symbol, and T is the absolute temperature.

The components of U naturally appear as conserved quantities in the system of partial differential equations governing multicomponent flows. On the other hand, the components of the natural variable Z are more practical to use in actual calculations of differential identities.

8.2.2. Vector equations

The equations modeling multicomponent reactive flows express the conservation of species mass, momentum, and energy. These equations have been presented in Chapter 2 and can be written in the compact form

$$\partial_t U + \sum_{i\in C} \partial_i F_i + \sum_{i\in C} \partial_i F_i^{\mathrm{dis}} = \Omega, \tag{8.2.3}$$

where ∂_t is the time derivative operator, ∂_i is the space derivative operator in the ith direction, so that $\boldsymbol{\partial_x} = (\partial_1, \partial_2, \partial_3)^t$, $C = \{1,2,3\}$ is the set of direction indices, F_i is the convective flux in the ith direction, F_i^{dis} is the dissipative flux in the ith direction, and Ω is the source term.

From Section 2.2 the convective flux F_i in the ith direction is given by

$$F_i = \Big(\rho_1 v_i, \ldots, \rho_n v_i,\ \rho v_1 v_i + \delta_{i1} p,\ \rho v_2 v_i + \delta_{i2} p,\ \rho v_3 v_i + \delta_{i3} p,\ \rho e^{\mathrm{tot}} v_i + p v_i\Big)^t, \tag{8.2.4}$$

where p is the thermodynamic pressure. For convenience, the dissipative flux F_i^{dis} in the ith direction is split between the mass and heat diffusion flux $F_i^{D\lambda}$ and the viscous flux $F_i^{\kappa\eta}$, so that

$$F_i^{\mathrm{dis}} = F_i^{D\lambda} + F_i^{\kappa\eta}. \tag{8.2.5}$$

From Section 2.2 the fluxes $F_i^{D\lambda}$ and $F_i^{\kappa\eta}$ in the ith direction are given by

$$F_i^{\kappa\eta} = \Big(0, \ldots, 0,\ \Pi_{i1},\ \Pi_{i2},\ \Pi_{i3},\ \sum_{j\in C} \Pi_{ij} v_j\Big)^t \tag{8.2.6}$$

and

$$F_i^{D\lambda} = \Big(\mathcal{F}_{1i}, \ldots, \mathcal{F}_{ni},\ 0,\ 0,\ 0,\ \mathcal{Q}_i\Big)^t, \tag{8.2.7}$$

where $\boldsymbol{\Pi} = (\Pi_{ij})_{i,j\in C}$ is the viscous stress tensor, $\boldsymbol{\mathcal{F}}_k = (\mathcal{F}_{k1}, \mathcal{F}_{k2}, \mathcal{F}_{k3})^t$ is the diffusion flux of the kth species, and $\boldsymbol{\mathcal{Q}} = (\mathcal{Q}_1, \mathcal{Q}_2, \mathcal{Q}_3)^t$ is the heat flux vector. On the other hand, the source term Ω is given by

$$\Omega = \Big(m_1\omega_1, \ldots, m_n\omega_n,\ \rho g_1,\ \rho g_2,\ \rho g_3,\ \rho \sum_{i\in C} g_i v_i\Big)^t, \tag{8.2.8}$$

where m_k is the molar mass of the kth species, ω_k is the molar production rate of the kth species, and $\boldsymbol{g} = (g_1, g_2, g_3)^t$ is the external force per unit mass acting on the species. Note that we assume in this chapter that the specific forces $\boldsymbol{g}$ acting on the species do not depend on the species.

These equations have to be completed by the relations expressing the transport fluxes $\boldsymbol{\Pi}$, $\boldsymbol{\mathcal{F}}_k$, $k \in S$, and $\boldsymbol{\mathcal{Q}}$, thermodynamic properties ρ, p and e^{tot}, chemical source terms ω_k, $k \in S$, and specific force $\boldsymbol{g}$. These relations have already been presented in Chapters 2, 6, and 7 and are not repeated here. These relations have been given in terms of the natural variables and are used in the next section to rewrite the system as a quasilinear system in terms of the variable U. The mathematical assumptions concerning thermochemistry are the assumptions (Th_1)–(Th_4) presented in Chapter 6, and those concerning transport properties are the assumptions (Tr_1)–(Tr_{4a}) presented in Chapter 7. These properties are assumed to hold throughout the chapter. Note that we do not need assumptions (Tr_{4b}) and (Tr_5), which will only be used in Chapter 11. Finally, since $(\rho_1, \dots, \rho_n)$ appear as the natural species variables in the system (8.2.3), the thermochemistry formalism presented in Section 6.2 will be used.

8.3. Quasilinear form

In this section, we rewrite (8.2.3) as a quasilinear system of second-order partial differential equations in terms of the conservative variable U.

8.3.1. The map $Z \to U$

In order to express the natural variable Z in terms of the conservative variable U, we investigate the map $Z \to U$ and its range.

Proposition 8.3.1. *The map $Z \longmapsto U$ is a C^∞ diffeomorphism from the open set $\mathcal{O}_Z = (0,\infty)^n \times \mathbb{R}^3 \times (0,\infty)$ onto an open set $\mathcal{O}_U$. The open set $\mathcal{O}_U$ is convex and given by*

$$\mathcal{O}_U = \{\ u \in \mathbb{R}^{n+4};\, u_i > 0,\ 1 \le i \le n, \quad u_{n+4} - \phi(u_1, \dots, u_{n+3}) > 0\ \}, \tag{8.3.1}$$

where

$$\phi(u_1, \dots, u_{n+3}) = \tfrac{1}{2}\frac{u_{n+1}^2 + u_{n+2}^2 + u_{n+3}^2}{\sum_{i\in S} u_i} + \sum_{i\in S} u_i e_i^0,$$

and e_i^0 is the energy of formation of the ith species at zero temperature.

Proof. From Assumption (Th_1) and (8.2.1), we first deduce that the map $Z \to U$ is C^∞ over the domain $\mathcal{O}_Z$. On the other hand, it is straightforward to show that the map $Z \to U$ is one to one and that the matrix $\partial_Z U$ is nonsingular over $\mathcal{O}_Z$, thanks to its triangular structure. From the inverse function theorem, we deduce that $Z \to U$ is a C^∞ diffeomorphism onto an open set $\mathcal{O}_U$. From

$$\rho e^{\text{tot}} = \mathcal{E} + \tfrac{1}{2}\rho \boldsymbol{v}\cdot\boldsymbol{v},$$

the expressions of $\mathcal{E}$ given by (6.2.7) and (6.2.10), and from (Th_1), it is then established that $\mathcal{O}_U$ is given by (8.3.1). The convexity of $\mathcal{O}_U$ is finally a direct consequence of the convexity of ϕ, which is established by evaluating its second derivative. ∎

8.3.2. Dissipation matrices and quasilinear form

In Section 2.5, the transport fluxes $\boldsymbol{\Pi}$, $\mathcal{F}_k$, $k \in S$, and $\boldsymbol{\mathcal{Q}}$ and, thus, the dissipative fluxes F_i^{dis}, $i \in C$, have been expressed in terms of the gradient of the natural variable Z. By using Proposition 8.3.1, these dissipation fluxes can thus be expressed as functions of the conservative variable gradients

$$F_i^{\mathrm{dis}} = -\sum_{j\in C} B_{ij}(U)\partial_j U, \tag{8.3.2}$$

where $B_{ij}(U)$, $i,j \in C$, are the dissipation matrices. The matrix $B_{ij}(U)$ is a square matrix of dimension $n+4$, which relates the dissipative flux in the ith direction to the gradient of U in the jth direction. Defining similarly the dissipation matrices $B_{ij}^{\kappa\eta}$ and $B_{ij}^{D\lambda}$, $i,j \in C$, from

$$\mathcal{F}_i^{\kappa\eta} = -\sum_{j\in C} B_{ij}^{\kappa\eta}(U)\partial_j U,$$

$$\mathcal{F}_i^{D\lambda} = -\sum_{j\in C} B_{ij}^{D\lambda}(U)\partial_j U,$$

we have the decomposition of dissipation matrices

$$B_{ij} = B_{ij}^{\kappa\eta} + B_{ij}^{D\lambda}, \qquad i,j \in C. \tag{8.3.3}$$

We may further introduce the Jacobian matrices A_i, $i \in C$, of the convective fluxes F_i, $i \in C$, that is,

$$A_i = \partial_U F_i, \qquad i \in C, \tag{8.3.4}$$

and finally rewrite the system (8.2.3) in the quasilinear form

$$\partial_t U + \sum_{i\in C} A_i(U)\partial_i U = \sum_{i,j\in C} \partial_i\big(B_{ij}(U)\partial_j U\big) + \Omega(U), \tag{8.3.5}$$

where the matrix coefficients are defined on the open convex set $\mathcal{O}_U$. The detailed form of the coefficient matrices $A_i(U)$, $i \in C$, and $B_{ij}(U)$, $i,j \in C$, will not be needed in the following, and, therefore, will not be given.

8.4. Symmetrization and entropic variables

For hyperbolic systems of conservation laws, the existence of a conservative symmetric formulation has been shown to be equivalent to the existence of an entropy function [God62] [FL71] [Moc80]. These results have been generalized to the case of second-order quasilinear systems of equations by Kawashima and Shizuta [Kaw84] [KS88]. In this section, we first restate these results about conservative symmetrizability. Following Kawashima and Shizuta [KS88], we then investigate normal forms of the system, i.e., symmetric hyperbolic–parabolic composite forms. Kawashima and Shizuta have shown that, when the nullspace naturally associated with dissipation matrices is a fixed subspace, a symmetric system of conservation equations can be put into a normal form. In the framework of their theory, we have characterized all normal forms for such symmetric systems of conservation laws [GM96a] [GM96b] [GM98a].

8.4.1. Symmetric conservative forms

We consider an abstract second-order quasilinear system in the form

$$\partial_t U^* + \sum_{i\in C^*} \partial_i F_i^*(U^*) = \sum_{i,j\in C^*} \partial_i\big(B_{ij}^*(U^*)\partial_j U^*\big) + \Omega^*(U^*), \qquad (8.4.1)$$

where $U^* \in \mathcal{O}_{U^*}$, $\mathcal{O}_{U^*}$ is an open convex set of $\mathbb{R}^{n^*}$, and $C^* = \{1,\dots,d\}$ is the set of direction indices of $\mathbb{R}^d$. Note that the superscript * is used to distinguish between the abstract second-order system (8.4.1) of size n^* in $\mathbb{R}^d$ and the particular multicomponent reactive flows system (8.3.5) of size $n+4$ in $\mathbb{R}^3$. All quantities associated with the abstract system have the corresponding superscript *, so that, for instance, the unknown vector is U^*. We assume that the following properties hold for system (8.4.1) .

(Edp$_1$) The convective fluxes F_i^*, $i \in C^*$, dissipation matrices B_{ij}^*, $i,j \in C^*$, and source term Ω^* are smooth functions of the variable $U^* \in \mathcal{O}_{U^*}$, where $\mathcal{O}_{U^*}$ is a convex open set of $\mathbb{R}^{n^*}$.

The following definition of a symmetric (conservative) form for the system (8.4.1), is due to Kawashima and Shizuta [KS88].

Definition 8.4.1. *Assume that $V^* \to U^*$ is a diffeomorphism from $\mathcal{O}_{V^*}$ onto $\mathcal{O}_{U^*}$, and consider the system in the V^* variable*

$$\widetilde{A}_0^*\partial_t V^* + \sum_{i\in C^*} \widetilde{A}_i^*\partial_i V^* = \sum_{i,j\in C^*} \partial_i\Big(\widetilde{B}_{ij}^*\partial_j V^*\Big) + \widetilde{\Omega}^*, \qquad (8.4.2)$$

where

$$\begin{cases} \widetilde{A}_0^* = \partial_{V^*}U^*, & \widetilde{A}_i^* = A_i^*\partial_{V^*}U^* = \partial_{V^*}F_i^*, \\ \widetilde{B}_{ij}^* = B_{ij}^*\partial_{V^*}U^*, & \widetilde{\Omega}^* = \Omega^*. \end{cases} \qquad (8.4.3)$$

The system (8.4.2) is said to be of the symmetric form if the matrix coefficients satisfy the following properties.

(S_1) *The matrix* $\widetilde{A}_0^*(V^*)$ *is symmetric positive definite for* $V^* \in \mathcal{O}_{V^*}$.

(S_2) *The matrices* $\widetilde{A}_i^*(V^*)$, $i \in C^*$, *are symmetric for* $V^* \in \mathcal{O}_{V^*}$.

(S_3) *We have* $\widetilde{B}_{ij}^*(V^*)^t = \widetilde{B}_{ji}^*(V^*)$ *for* $i, j \in C^*$ *and* $V^* \in \mathcal{O}_{V^*}$.

(S_4) *The matrix* $\widetilde{B}^*(V^*, w) = \sum_{i,j \in C^*} \widetilde{B}_{ij}^*(V^*) w_i w_j$ *is symmetric and positive semidefinite, for* $V^* \in \mathcal{O}_{V^*}$ *and* $w \in \Sigma^{d-1}$, *where* Σ^{d-1} *is the unit sphere in* d *dimensions.*

Note that both first-order and second-order derivative terms are in conservative form in (8.4.2) in the sense that $\widetilde{A}_0^*$ and $\widetilde{A}_i^*$, $i \in C^*$, are Jacobian matrices.

8.4.2. Entropic variables

The following generalized definition of an entropy function has been given by Kawashima [Kaw84] [KS88].

Definition 8.4.2. *A real-valued smooth function* $\sigma^*(U^*)$ *defined on a convex set* $\mathcal{O}_{U^*}$ *is said to be an entropy function for the system* (8.4.1) *if the following properties hold.*

(E_1) *The function* σ^* *is a strictly convex function on* $\mathcal{O}_{U^*}$ *in the sense that the Hessian matrix is positive definite on* $\mathcal{O}_{U^*}$.

(E_2) *There exists real-valued smooth functions* $q_i^* = q_i^*(U^*)$ *such that*

$$(\partial_{U^*}\sigma^*)\, A_i^* = \partial_{U^*} q_i^*, \quad i \in C^*, \qquad U^* \in \mathcal{O}_{U^*}. \tag{8.4.4}$$

(E_3) *We have the property that, for any* $U^* \in \mathcal{O}_{U^*}$,

$$\left(\partial_{U^*}^2 \sigma^*\right)^{-1} (B_{ij}^*)^t = B_{ji}^* \left(\partial_{U^*}^2 \sigma^*\right)^{-1}, \quad i, j \in C^*. \tag{8.4.5}$$

(E_4) *The matrix* $\widetilde{B}^* = \sum_{i,j \in C^*} B_{ij}^*(U^*) \left(\partial_{U^*}^2 \sigma^*(U^*)\right)^{-1} w_i w_j$ *is symmetric positive semidefinite for* $U^* \in \mathcal{O}_{U^*}$ *and* $w \in \Sigma^{d-1}$.

8.4.3. The equivalence theorem

Kawashima and Shizuta have established the equivalence between conservative symmetrizability and the existence of an entropy function for the system (8.4.1) [KS88].

Theorem 8.4.3. *The system* (8.4.1) *can be symmetrized on the open convex set* $\mathcal{O}_{U^*}$ *if and only if the system admits an entropy function* σ^* *on* $\mathcal{O}_{U^*}$. *In this situation, the symmetrizing variable* V^* *can be expressed in terms of the gradient of the entropy function* σ^* *as*

$$V^* = \left(\partial_{U^*}\sigma^*\right)^t. \tag{8.4.6}$$

Proof. Assume first that there exists an entropy σ^*, and let $V^* = \left(\partial_{U^*}\sigma^*\right)^t$ be the symmetrizing variable (8.4.6). The map $U^* \to V^*$ is then a diffeomorphism since $\mathcal{O}_{U^*}$ is convex and $\partial_{U^*} V^* = \partial^2_{U^*}\sigma^*$ is positive definite. We can thus define the smooth functions

$$\mathfrak{s}^*(V^*) = \langle U^*, V^*\rangle - \sigma^*(U^*) \tag{8.4.7}$$

and

$$\mathfrak{q}_i^*(V^*) = \langle F_i^*, V^*\rangle - q_i^*(U^*), \qquad i \in C^*. \tag{8.4.8}$$

Differentiating (8.4.7) and (8.4.8) then yields the relations $\left(\partial_{V^*}\mathfrak{s}^*\right)^t = U^*$ and $\left(\partial_{V^*}\mathfrak{q}_i^*\right)^t = F_i^*$, making use of Properties (E_1)–(E_4). This implies that $\widetilde{A}_0^* = \partial_{V^*}U^* = \partial^2_{V^*}\mathfrak{s}^*$ and $\widetilde{A}_i^* = \partial_{V^*}F_i^* = \partial^2_{V^*}\mathfrak{q}_i^*$, $i \in C^*$, so that these matrices are symmetric. Moreover, we directly get from Properties (E_1)–(E_4) that the matrices

$$\widetilde{B}_{ji}^* = B_{ji}^*\left(\partial^2_{U^*}\sigma^*\right)^{-1}, \qquad i,j \in C^*,$$

satisfy the reciprocity relations $(\widetilde{B}_{ji}^*)^t = \widetilde{B}_{ij}^*$, $i,j \in C^*$, and are such that

$$\widetilde{B}^*(V^*, w) = \sum_{i,j\in C^*} \widetilde{B}_{ij}^*(V^*)\, w_i w_j$$

is symmetric positive semidefinite for $U^* \in \mathcal{O}_{U^*}$ and $w \in \Sigma^{d-1}$, making use of

$$\widetilde{A}_0^* = \partial_{V^*}U^* = (\partial_{U^*}V^*)^{-1} = (\partial^2_{U^*}\sigma^*)^{-1}.$$

Conversely, assume that the system is symmetrizable in the sense of Definition 8.4.1. Since $\partial_{V^*}U^*$ and $\partial_{V^*}F_i^*$ are symmetric and $\mathcal{O}_{V^*}$ is simply connected, there exists $\mathfrak{s}^*$ and $\mathfrak{q}_i^*$, $i \in C^*$, defined over $\mathcal{O}_{V^*}$, such that $\left(\partial_{V^*}\mathfrak{s}^*\right)^t = U^*$ and $\left(\partial_{V^*}\mathfrak{q}_i^*\right)^t = F_i^*$, $i \in C^*$. We can thus define the functions

$$\sigma^*(U^*) = \langle U^*, V^*\rangle - \mathfrak{s}^*(V^*)$$

and

$$q_i^*(U^*) = \langle F_i^*, V^*\rangle - \mathfrak{q}_i^*(V^*), \qquad i \in C^*.$$

Differentiating these identities, and using Properties (S_1)–(S_4), it is then straightforward to establish that σ^* is an entropy with fluxes q_i^*, $i \in C^*$, such that (8.4.6) holds. ∎

The mathematical entropy σ^* is generally taken to be the opposite of the physical entropy per unit volume. The variable V^* is usually termed the entropic variable associated with the conservative variable U^*.

Remark 8.4.4. Note that, for convenience, we have considered source terms in the previous definitions, which are minor modifications of [KS88]. Properties of entropy functions associated with source terms are discussed in Section 9.3 of Chapter 9. ∎

8.5. Normal forms

In this section we investigate normal forms for the abstract system (8.4.1) of second-order partial differential equations.

8.5.1. Definition of normal forms

We assume that the abstract second-order quasilinear system (8.4.1) is symmetrizable in the sense of Definition 8.4.1. In other words, using Theorem 8.4.3, we assume that the following property holds.

(Edp_2) The system (8.4.1) admits an entropy function σ^* over the convex domain $\mathcal{O}_{U^*}$.

Introducing the symmetrizing variable $V^* = (\partial_{U^*}\sigma^*)^t$, the corresponding symmetric system (8.4.2) then satisfies Properties (S_1)–(S_4). However, depending on the range of the dissipation matrices $\widetilde{B}_{ij}^*$, this system lies between the two limit cases of a hyperbolic system and a strongly parabolic system. In this section, we use a sufficient condition on the matrices $\widetilde{B}_{ij}^*$, $i, j \in C^*$, essentially the condition N introduced by Kawashima and Shizuta [KS88], under which the system can be recast in the form of a symmetric hyperbolic–parabolic composite system, defined as a normal form of the system. We then characterize all normal forms for symmetric systems of conservation laws satisfying condition N.

Introducing a new variable W^*, associated with a diffeomorphism from $\mathcal{O}_{W^*}$ onto $\mathcal{O}_{V^*}$, and multiplying the conservative symmetric form (8.4.2) on the left side by the transpose of the matrix $\partial_{W^*}V^*$, we then get a new system in the variable W^* and have the following definition of a normal form [KS88].

Definition 8.5.1. *Consider a system in symmetric form, as in Definition 8.4.1, and a diffeomorphism $W^* \to V^*$ from $\mathcal{O}_{W^*}$ to $\mathcal{O}_{V^*}$. The system in the new variable W^**

$$\overline{A}_0^* \partial_t W^* + \sum_{i\in C^*} \overline{A}_i^* \partial_i W^* = \sum_{i,j\in C^*} \partial_i \Big(\overline{B}_{ij}^* \partial_j W^* \Big) + \overline{\mathcal{T}}^* + \overline{\Omega}^*, \tag{8.5.1}$$

where

$$\begin{cases} \overline{A}_0^* = (\partial_{W^*} V^*)^t \, \widetilde{A}_0^* \, (\partial_{W^*} V^*), & \overline{B}_{ij}^* = (\partial_{W^*} V^*)^t \, \widetilde{B}_{ij}^* \, (\partial_{W^*} V^*), \\ \overline{A}_i^* = (\partial_{W^*} V^*)^t \, \widetilde{A}_i^* \, (\partial_{W^*} V^*), & \overline{\Omega}^* = (\partial_{W^*} V^*)^t \widetilde{\Omega}^*, \\ \overline{\mathcal{T}}^* = - \sum_{i,j\in C^*} \partial_i (\partial_{W^*} V^*)^t \, \widetilde{B}_{ij}^* \, (\partial_{W^*} V^*) \partial_j W^*, & \end{cases} \tag{8.5.2}$$

satisfies

($\bar{\mathrm{S}}_1$) *The matrix $\overline{A}_0^*(W^*)$ is symmetric positive definite for $W^* \in \mathcal{O}_{W^*}$.*

($\bar{\mathrm{S}}_2$) *The matrices $\overline{A}_i^*(W^*)$, $i \in C^*$, are symmetric for $W^* \in \mathcal{O}_{W^*}$.*

($\bar{\mathrm{S}}_3$) *We have $\overline{B}_{ij}^*(W^*)^t = \overline{B}_{ji}^*(W^*)$ for $i,j \in C^*$ and $W^* \in \mathcal{O}_{W^*}$.*

($\bar{\mathrm{S}}_4$) *The matrix $\overline{B}^*(W^*, w) = \sum_{i,j\in C^*} \overline{B}_{ij}^*(W^*) w_i w_j$ is symmetric positive semidefinite, for $W^* \in \mathcal{O}_{W^*}$ and $w \in \Sigma^{d-1}$, where Σ^{d-1} is the unit sphere in d dimensions.*

This system is then said to be of the normal form if there exists a partition of $\{1, \dots, n^*\}$ into $I = \{1, \dots, n_0^*\}$ and $II = \{n_0^* + 1, \dots, n^*\}$, such that the following properties hold.

(Nor_1) *The matrices $\overline{A}_0^*$ and $\overline{B}_{ij}^*$ have the block structure*

$$\overline{A}_0^* = \begin{pmatrix} \overline{A}_0^{*I,I} & 0 \\ 0 & \overline{A}_0^{*II,II} \end{pmatrix}, \qquad \overline{B}_{ij}^* = \begin{pmatrix} 0 & 0 \\ 0 & \overline{B}_{ij}^{*II,II} \end{pmatrix}.$$

(Nor_2) *The matrix $\overline{B}^{*II,II}(W^*, w) = \sum_{i,j\in C^*} \overline{B}_{ij}^{*II,II}(W^*) w_i w_j$ is positive definite, for $W^* \in \mathcal{O}_{W^*}$ and $w \in \Sigma^{d-1}$.*

(Nor_3) *Denoting $\partial_x = (\partial_1, \dots, \partial_d)^t$, we have*

$$\overline{\mathcal{T}}^* (W^*, \partial_x W^*) = \Big(\overline{\mathcal{T}}_I^* (W^*, \partial_x W_{II}^*), \overline{\mathcal{T}}_{II}^* (W^*, \partial_x W^*) \Big)^t,$$

where we have used the vector and matrix block structure induced by the partitioning of $\{1, \dots, n^*\}$ into $I = \{1, \dots, n_0^*\}$ and $II = \{n_0^* + 1, \dots, n^*\}$, so that we have $W^* = (W_I^*, W_{II}^*)^t$, for instance.

Remark 8.5.2. Note that, for convenience, we have kept the dissipative terms $\overline{B}^*_{ij}$ in conservation form and have considered a source term $\overline{\Omega}^*$ in Definition 8.4.5, which are minor modifications of [KS88]. ∎

8.5.2. Nullspace invariance property

A sufficient condition for system (8.4.2) to be recast into a normal form is that the nullspace naturally associated with dissipation matrices is a fixed subspace of $\mathbb{R}^{n^*}$ common to all dissipation matrices. This is essentially condition N introduced by Kawashima and Shizuta, which is now assumed to hold.

(Edp_3) The nullspace of the matrix

$$\widetilde{B}^*(V^*, w) = \sum_{i,j \in C^*} \widetilde{B}^*_{ij}(V^*) w_i w_j,$$

does not depend on $V^* \in \mathcal{O}_{V^*}$ and $w \in \Sigma^{d-1}$, and we denote by n_0^* its dimension $n_0^* = \dim\big(N(\widetilde{B}^*)\big)$. Furthermore, we have $\widetilde{B}^*_{ij}(V^*) N(\widetilde{B}^*) = 0$, $i, j \in C^*$.

In order to characterize more easily normal forms for symmetric systems of conservation laws satisfying (Edp_1)–(Edp_3), we introduce the auxiliary variables $U^{*\prime}$ and $V^{*\prime}$, depending linearly on U^* and V^*, respectively. The dissipation matrices corresponding to these auxiliary variables have nonzero coefficients only in the lower right block of size $n^* - n_0^*$, where $n_0^* = \dim\big(N(\widetilde{B}^*)\big)$. Normal symmetric forms are then equivalently—and more easily—obtained from the $V^{*\prime}$ symmetric equation.

Lemma 8.5.3. *Consider a system of conservation laws* (8.4.2) *that is symmetric in the sense of Definition* 8.4.1 . *Denote by* σ^* *the associated entropy function and by* $V^* = (\partial_{U^*}\sigma^*)^t$ *the symmetrizing variable, and assume that the nullspace invariance property* (Edp_3) *is satisfied over* $\mathcal{O}_{V^*}$.

Further consider any constant nonsingular matrix P *of dimension* n^*, *such that its first* n_0^* *columns span the nullspace* $N(\widetilde{B}^*)$. *More specifically, assume that* P *is such that*

$$span\Big\{ (P_{1j}, \ldots, P_{nj})^t;\ 1 \le j \le n_0^* \Big\} = N(\widetilde{B}^*).$$

Then the auxiliary variable

$$U^{*\prime} = P^t U^*$$

satisfies the equation

$$\partial_t U^{*\prime} + \sum_{i \in C^*} A_i^{*\prime} \partial_i U^{*\prime} = \sum_{i,j \in C^*} \partial_i \Big(B_{ij}^{*\prime} \partial_j U^{*\prime} \Big) + \Omega^{*\prime}, \tag{8.5.3}$$

where $A_i^{*\prime} = P^t A_i^* (P^t)^{-1}$, $B_{ij}^{*\prime} = P^t B_{ij}^* (P^t)^{-1}$, *and* $\Omega^{*\prime} = P^t \Omega^*$. *The corresponding entropy is then the functional* $U^{*\prime} \to \sigma^*\big((P^t)^{-1}U^{*\prime}\big)$, *and the associated entropic variable* $V^{*\prime} = (\partial_{U^{*\prime}}\sigma^*)^t$ *is given by*

$$V^{*\prime} = P^{-1}V^*$$

and satisfies the equation

$$\widetilde{A}_0^{*\prime}\partial_t V^{*\prime} + \sum_{i\in C^*} \widetilde{A}_i^{*\prime}\partial_i V^{*\prime} = \sum_{i,j\in C^*} \partial_i\Big(\widetilde{B}_{ij}^{*\prime}\partial_j V^{*\prime}\Big) + \widetilde{\Omega}^{*\prime}, \tag{8.5.4}$$

where $\widetilde{A}_0^{*\prime} = P^t\widetilde{A}_0^*P$, $\widetilde{A}_i^{*\prime} = P^t\widetilde{A}_i^*P$, $\widetilde{B}_{ij}^{*\prime} = P^t\widetilde{B}_{ij}^*P$, *and* $\widetilde{\Omega}^{*\prime} = P^t\widetilde{\Omega}^*$. *In particular,* $\widetilde{B}_{ij}^{*\prime}$ *is in the form*

$$\widetilde{B}_{ij}^{*\prime} = \begin{pmatrix} 0_{n_0^*\times n_0^*} & 0_{n_0^*\times(n^*-n_0^*)} \\ 0_{(n^*-n_0^*)\times n_0^*} & \widetilde{B}_{ij}^{*\prime II,II} \end{pmatrix}, \tag{8.5.5}$$

and the matrix $\widetilde{B}^{*\prime II,II}(V^{*\prime},w) = \sum_{i,j\in S} \widetilde{B}_{ij}^{*\prime II,II} w_i w_j$ *is positive definite over* $\mathcal{O}_{V^{*\prime}} \times \Sigma^{d-1}$.

Finally, the normal form (8.5.1) *is equivalently obtained by multiplying the* V^* *equation* (8.4.2) *by* $(\partial_{W^*}V^*)^t$ *or the* $V^{*\prime}$ *equation* (8.5.4) *by* $(\partial_{W^*}V^{*\prime})^t$.

Proof. Equation (8.5.3) is easily established by multiplying (8.4.1) on the left by P^t. This also yields the matrix relations $A_i^{*\prime} = P^tA_i^*(P^t)^{-1}$ and $B_{ij}^{*\prime} = P^tB_{ij}^*(P^t)^{-1}$, and $\Omega^{*\prime} = P^t\Omega^*$. It is also easily checked that the functional $U^{*\prime} \to \sigma^*\big((P^t)^{-1}U^{*\prime}\big)$ is the corresponding entropy. From the definition $V^{*\prime} = (\partial_{U^{*\prime}}\sigma^*)^t$ and the chain rule, we then get that $V^{*\prime} = P^{-1}V^*$ and (8.5.4) is obtained as in (8.4.1)–(8.4.2). Since $\widetilde{B}^{*\prime} = P^t\widetilde{B}^*P$ and the first n_0^* columns of P span $N(\widetilde{B}^*)$, we next deduce that $\widetilde{B}^{*\prime}$ is in the form

$$\widetilde{B}^{*\prime} = \begin{pmatrix} 0_{n_0^*\times n_0^*} & 0_{n_0^*\times(n^*-n_0^*)} \\ 0_{(n^*-n_0^*)\times n_0^*} & \widetilde{B}^{*\prime II,II} \end{pmatrix}, \tag{8.5.6}$$

and, similarly, all matrices $\widetilde{B}_{ij}^{*\prime}$, $i,j \in C^*$, are also in the form (8.5.5). Moreover, the matrix $\widetilde{B}^{*\prime II,II}(V^{*\prime},w)$ is positive definite since the $n^* - n_0^*$ last columns of P span a subspace complementary to $N(\widetilde{B}^*)$. ■

Remark 8.5.4. Condition N introduced by Kawashima has been slightly strengthened in (Edp$_3$). The relations $\widetilde{B}_{ij}^*(V^*)N(\widetilde{B}^*) = 0$, $i,j \in C^*$, are needed in order to obtain the block structure (8.5.5), as was overlooked in [GM98a]. ■

8.5.3. Description of normal variables

Normal forms for symmetrizable systems of conservation laws satisfying the nullspace invariance property (Edp_3) are now completely characterized in the following theorem, in terms of the auxiliary variables $U^{*\prime}$ and $V^{*\prime}$ [GM98a].

Theorem 8.5.5. *Keeping the assumptions and notation of Lemma 8.5.3, any normal form of the system (8.4.2) is given by a change of variable in the form*

$$W^* = \left(\psi_I(U_I^{*\prime}), \phi_{II}(V_{II}^{*\prime})\right)^t, \tag{8.5.7}$$

where ψ_I and ϕ_{II} are two diffeomorphisms of $\mathbb{R}^{n_0^*}$ and $\mathbb{R}^{n^*-n_0^*}$, respectively. The n_0^* components of $U_I^{*\prime}$ can thus be termed hyperbolic components, and the $n^*-n_0^*$ components of $V_{II}^{*\prime}$ can be termed parabolic components. Furthermore, we have

$$\overline{\mathcal{T}}^*(W^*, \partial_x W^*) = \left(0, \overline{\mathcal{T}}^*_{II}(W^*, \partial_x W^*_{II})\right)^t,$$

and, when ϕ_{II} is a constant linear mapping, we also have $\overline{\mathcal{T}}^*_{II} = 0$ and the dissipative terms are in conservative form.

Proof. The mapping $U^{*\prime} \longmapsto W^*$ is smooth by assumption and a straightforward calculation yields

$$\partial_{U^{*\prime}} W^* = \begin{pmatrix} \partial_{U_I^{*\prime}} \psi_I & 0_{(n^*-n_0^*)\times n^*} \\ \partial_{V_{II}^{*\prime}} \phi_{II} \left(\widetilde{A}_0^{*\prime -1}\right)^{II,I} & \partial_{V_{II}^{*\prime}} \phi_{II} \left(\widetilde{A}_0^{*\prime -1}\right)^{II,II} \end{pmatrix},$$

so that $\partial_{U^{*\prime}} W^*$ is invertible. On the other hand, the derivative of the mapping $U_{II}^{*\prime} \longmapsto V_{II}^{*\prime}(U_I^{*\prime}, U_{II}^{*\prime})$, for a fixed value of $U_I^{*\prime}$, is given by $\partial_{U_{II}^{*\prime}} V_{II}^{*\prime} = \left(\widetilde{A}_0^{*\prime -1}\right)^{II,II}$ and is positive definite. Since $\mathcal{O}_{U^{*\prime}}$ is an open convex set, we deduce that $U^{*\prime} \longmapsto W^*$ is one to one. As a consequence, the mapping $U^{*\prime} \longmapsto W^*$ is a diffeomorphism from $\mathcal{O}_{U^{*\prime}} = P^t \mathcal{O}_{U^*}$ onto an open set denoted by $\mathcal{O}_{W^*}$ and, similarly, the mapping $V^{*\prime} \longmapsto W^*$ is a diffeomorphism from $\mathcal{O}_{V^{*\prime}} = P^{-1}\mathcal{O}_{V^*}$ onto $\mathcal{O}_{W^*}$.

Evaluating and inverting the matrix $\partial_{V^{*\prime}} W^*$, we next obtain the following expression for $\partial_{W^*} V^{*\prime}$:

$$\partial_{W^*} V^{*\prime} = \begin{pmatrix} \left(\widetilde{A}_0^{*\prime\, I,I}\right)^{-1} \left(\partial_{U_I^{*\prime}} \psi_I\right)^{-1} & -\left(\widetilde{A}_0^{*\prime\, I,I}\right)^{-1} \widetilde{A}_0^{*\prime\, I,II} \left(\partial_{V_{II}^{*\prime}} \phi_{II}\right)^{-1} \\ 0_{(n^*-n_0^*)\times n^*} & \left(\partial_{V_{II}^{*\prime}} \phi_{II}\right)^{-1} \end{pmatrix}, \tag{8.5.8}$$

and a direct calculation using (8.5.8) then shows that Properties (Nor_1)–(Nor_3) are satisfied, keeping in mind that normal forms are equivalently obtained from the $V^{*\prime}$ equation (8.5.4) or the V^* equation (8.4.2).

Consider then a diffeomorphism

$$V^{*\prime} \longmapsto \mathcal{Z}^* = \big(\phi_I(V_I^{*\prime}, V_{II}^{*\prime}), \phi_{II}(V_I^{*\prime}, V_{II}^{*\prime})\big)^t,$$

and assume that the system (8.5.1) in the variable $\mathcal{Z}^*$ is of the normal form. By using the definition of $\overline{B}^*$ and the auxiliary variable $V^{*\prime}$, it is easily established that

$$\widetilde{B}^{*\prime} = \big(\partial_{V^{*\prime}}\mathcal{Z}^*\big)^t\,\overline{B}^*\,\big(\partial_{V^{*\prime}}\mathcal{Z}^*\big).$$

Using the block structure of $\overline{B}^*$ and $\widetilde{B}^{*\prime}$ then yields

$$\begin{pmatrix} \big(\partial_{V_I^{*\prime}}\mathcal{Z}_{II}^*\big)^t\,\overline{B}^{*II,II}\big(\partial_{V_I^{*\prime}}\mathcal{Z}_{II}^*\big) & \big(\partial_{V_I^{*\prime}}\mathcal{Z}_{II}^*\big)^t\,\overline{B}^{*II,II}\big(\partial_{V_{II}^{*\prime}}\mathcal{Z}_{II}^*\big) \\ \big(\partial_{V_{II}^{*\prime}}\mathcal{Z}_{II}^*\big)^t\,\overline{B}^{*II,II}\big(\partial_{V_I^{*\prime}}\mathcal{Z}_{II}^*\big) & \big(\partial_{V_{II}^{*\prime}}\mathcal{Z}_{II}^*\big)^t\,\overline{B}^{*II,II}\big(\partial_{V_{II}^{*\prime}}\mathcal{Z}_{II}^*\big) \end{pmatrix} = \begin{pmatrix} 0 & 0 \\ 0 & \widetilde{B}^{*\prime II,II} \end{pmatrix}, \tag{8.5.9}$$

which implies that $\partial_{V_I^{*\prime}}\mathcal{Z}_{II}^* = 0$ since $\widetilde{B}^{*II,II}$ is positive definite, and thus that

$$\mathcal{Z}_{II}^* = \phi_{II}(V_{II}^{*\prime}). \tag{8.5.10}$$

On the other hand, by using the definition of $\overline{A}_0^*$, it is easily established that

$$\overline{A}_0^*\big(\partial_{U^{*\prime}}\mathcal{Z}^*\big) = \big(\partial_{\mathcal{Z}^*}V^{*\prime}\big)^t.$$

Using the block structure of $\overline{A}_0^*$ and the relation $\partial_{\mathcal{Z}_I^*}V_{II}^{\prime *} = 0$, derived from $\partial_{V_I^{\prime *}}\mathcal{Z}_{II}^* = 0$, we then obtain that $\partial_{U_{II}^{*\prime}}\mathcal{Z}_I^* = 0$ and thus that

$$\mathcal{Z}_I^* = \psi_I(U_I^{*\prime}), \tag{8.5.11}$$

where we have defined $\phi_I(V_I^{*\prime}, V_{II}^{*\prime}) = \psi_I(U_I^{*\prime}, U_{II}^{*\prime})$. This shows that any diffeomorphism associated with a normal form is of type (8.5.7).

Moreover, a direct calculation yields

$$\overline{\mathcal{T}}^* = \Big(0, -\sum_{i,j\in C^*} \partial_i\big((\partial_{V_{II}^{*\prime}}\phi_{II})^{-1}\big)^t\,\widetilde{B}_{ij}^{*\prime II,II}\big(\partial_{V_{II}^{*\prime}}\phi_{II}\big)^{-1}\partial_j W_{II}^*\Big)^t. \tag{8.5.12}$$

This shows that $\overline{\mathcal{T}}_I^* = 0$ and $\overline{\mathcal{T}}_{II}^* = 0$ when ϕ_{II} is constant linear mapping and the proof is complete. ∎

Remark 8.5.6. Theorem 8.5.5 shows, in particular, that the general form (8.5.7) is independent of the choice of P. It is also possible, however, to check it directly. Consider indeed another matrix Q, as in Lemma 8.5.3, and define $U^{*\prime\prime} = Q^t U^*$ and $V^{*\prime\prime} = Q^{-1}V^*$. Denoting by $\mathfrak{P}$ the matrix $\mathfrak{P} = P^{-1}Q$, we thus get $U^{*\prime\prime} = \mathfrak{P}^t U^{*\prime}$ and $V^{*\prime\prime} = \mathfrak{P}^{-1}V^{*\prime}$. Since the nullspace $N(\widetilde{B})$ is spanned by the n_0^* first columns of P and Q, it is easily checked that $\mathfrak{P}^{II,I} = 0$. This implies that $\mathfrak{P}^{I,I}$ and $\mathfrak{P}^{II,II}$ are invertible, so that $U_I^{*\prime\prime} = (\mathfrak{P}^{I,I})^t U_I^{*\prime}$ and $V_{II}^{*\prime\prime} = (\mathfrak{P}^{II,II})^{-1}V_{II}^{*\prime}$, and we recover that the general form (8.5.7) is independent of P. ∎

8.6. Symmetrization for multicomponent flows

We now apply the general results of Section 8.4 to the system of equations governing multicomponent reactive flows (8.3.5). As stated in Section 8.3, the assumptions (Th_1)–(Th_4) of Chapter 6 concerning thermochemistry and the assumptions (Tr_1)–(Tr_{4a}) of Chapter 7 concerning transport properties are assumed to hold. We exhibit an entropy function and derive the corresponding conservative symmetric form. This symmetric form is used in the next section to derive normal forms.

8.6.1. Entropy and symmetric conservative form

We first note that Property (Edp_1) is a direct consequence of Assumptions (Th_1)–(Th_4). We next define the mathematical entropy function σ as the opposite of the physical mixture entropy per unit volume

$$\sigma = -\mathcal{S} = -\sum_{k\in S} \rho_k \mathcal{S}_k. \tag{8.6.1}$$

The corresponding entropic variable

$$V = \left(\partial_U \sigma\right)^t \tag{8.6.2}$$

is then easily obtained

$$V = \frac{1}{T}\Big(\mathcal{G}_1 - \tfrac{1}{2}\boldsymbol{v}\cdot\boldsymbol{v}, \ldots, \mathcal{G}_n - \tfrac{1}{2}\boldsymbol{v}\cdot\boldsymbol{v},\, v_1,\, v_2,\, v_3,\, -1\Big)^t, \tag{8.6.3}$$

where $\mathcal{G}_k$ is the Gibbs function of the kth species.

Proposition 8.6.1. *The change of variable $U \longmapsto V$ from the open convex set $\mathcal{O}_U$ onto the open set $\mathcal{O}_V = \mathbb{R}^{n+3} \times (-\infty, 0)$ is a C^∞ diffeomorphism.*

Proof. From Proposition 8.3.1, $Z \to U$ is a C^∞ diffeomorphism from $\mathcal{O}_Z$ onto $\mathcal{O}_U$, so that we only have to show that $Z \to V$ is a C^∞ diffeomorphism from $\mathcal{O}_Z$ onto the open set $\mathcal{O}_V$. From Assumption (Th_1) and the expressions presented in Section 6.2, we first deduce that the map $Z \to V$ is C^∞ over the domain $\mathcal{O}_Z$. It is then straightforward to show that the map $Z \to V$ is one to one—since $-\mathcal{S}_k$ is an increasing function of ρ_k at fixed T—and that its range is $\mathcal{O}_V$. In addition, the matrix $\partial_Z V$ is easily shown to be nonsingular over $\mathcal{O}_Z$, thanks to its triangular structure, and the proof is complete, thanks to the inverse function theorem. ∎

The conservative symmetric form is now investigated in the following theorem.

Theorem 8.6.2. *The function σ is an entropy for system (8.3.5), that is, σ satisfies Properties* (E_1)–(E_4) *of Definition* 8.4.2. *The system associated with the entropic variable $V \in \mathcal{O}_V$ can then be written as*

$$\widetilde{A}_0 \partial_t V + \sum_{i \in C} \widetilde{A}_i \partial_i V = \sum_{i,j \in C} \partial_i \Big(\widetilde{B}_{ij} \partial_j V \Big) + \widetilde{\Omega}, \tag{8.6.4}$$

and satisfies Properties (S_1)–(S_4) *of Definition* 8.4.1. *The matrix $\widetilde{A}_0$ is given by*

$$\widetilde{A}_0 = \begin{pmatrix} \left(\dfrac{\rho_k}{r_k} \delta_{kl} \right)_{k,l \in S} & & Sym \\ \left(\dfrac{\rho_l}{r_l} v_i \right)_{i \in C, l \in S} & (\Sigma_\rho v_i v_j + \rho T \delta_{ij})_{i,j \in C} & \\ \left(\dfrac{\rho_l}{r_l} e_l^{\mathrm{tot}} \right)_{l \in S} & (\Sigma_e v_j + \rho T v_j)_{j \in C} & \Upsilon_e \end{pmatrix}, \tag{8.6.5}$$

where

$$\Sigma_\rho = \sum_{k \in S} \frac{\rho_k}{r_k}, \qquad \Sigma_e = \sum_{k \in S} \frac{\rho_k}{r_k} e_k^{\mathrm{tot}}, \qquad \Upsilon_e = \sum_{k \in S} \frac{\rho_k}{r_k} e_k^{\mathrm{tot}\,2} + \rho T \left(\boldsymbol{v} \cdot \boldsymbol{v} + c_v T \right).$$

Since this matrix is symmetric, we only give its block lower triangular part and write "Sym" in the upper triangular part. On the other hand, denoting by $\boldsymbol{\xi} = (\xi_1, \xi_2, \xi_3)^t$ an arbitrary vector of $\mathbb{R}^3$, the matrices $\widetilde{A}_i$, $i \in C$, are given by

$$\sum_{i \in C} \widetilde{A}_i \xi_i = \begin{pmatrix} \left(\delta_{kl} \dfrac{\rho_l}{r_l} \boldsymbol{v} \cdot \boldsymbol{\xi} \right)_{k,l \in S} & & Sym \\ \left(\rho_l T \xi_i + \dfrac{\rho_l}{r_l} v_i \, \boldsymbol{v} \cdot \boldsymbol{\xi} \right)_{i \in C, l \in S} & \Sigma_{\rho, \boldsymbol{v}} & \\ \left(h_l^{\mathrm{tot}} \dfrac{\rho_l}{r_l} \boldsymbol{v} \cdot \boldsymbol{\xi} \right)_{l \in S} & \Sigma_{h, \boldsymbol{v}} & \Upsilon_h \, \boldsymbol{v} \cdot \boldsymbol{\xi} \end{pmatrix}, \tag{8.6.6}$$

where

$$\Sigma_{\rho, \boldsymbol{v}} = \Sigma_\rho \boldsymbol{v} \cdot \boldsymbol{\xi} \, \boldsymbol{v} \otimes \boldsymbol{v} + \rho T (\boldsymbol{v} \otimes \boldsymbol{\xi} + \boldsymbol{\xi} \otimes \boldsymbol{v} + \boldsymbol{v} \cdot \boldsymbol{\xi} \, \boldsymbol{I}),$$

$$\Sigma_{h, \boldsymbol{v}} = (\Sigma_h + \rho T) \boldsymbol{v} \cdot \boldsymbol{\xi} \, \boldsymbol{v}^t + \rho T h^{\mathrm{tot}} \boldsymbol{\xi}^t,$$

$$\Sigma_h = \sum_{k \in S} \frac{\rho_k}{r_k} h_k^{\mathrm{tot}}, \qquad \Upsilon_h = \sum_{k \in S} \frac{\rho_k}{r_k} h_k^{\mathrm{tot}\,2} + \rho T \left(\boldsymbol{v} \cdot \boldsymbol{v} + (c_v + r) T \right).$$

and $r = R / \overline{m}$. Furthermore, concerning the dissipation matrices, we have the usual decomposition

$$\widetilde{B}_{ij} = \widetilde{B}_{ij}^{\kappa \eta} + \widetilde{B}_{ij}^{D \lambda}. \tag{8.6.7}$$

The viscous matrices $\widetilde{B}_{ij}^{\kappa\eta}$, $i,j \in C$, have the following structure:

$$\widetilde{B}_{ij}^{\kappa\eta} = \begin{pmatrix} 0_{n\times n} & 0_{n\times 4} \\ 0_{4\times n} & \widetilde{\mathcal{K}}_{ij} \end{pmatrix}, \tag{8.6.8}$$

so that we only need the expressions of $\widetilde{\mathcal{K}}_{ij}$, $i,j \in C$. For sake of brevity, we only express $\widetilde{\mathcal{K}}_{11}$ and $\widetilde{\mathcal{K}}_{12}$

$$\widetilde{\mathcal{K}}_{11} = T \begin{pmatrix} (\kappa + \frac{4}{3}\eta) & 0 & 0 & (\kappa + \frac{4}{3}\eta)v_1 \\ 0 & \eta & 0 & \eta v_2 \\ 0 & 0 & \eta & \eta v_3 \\ (\kappa + \frac{4}{3}\eta)v_1 & \eta v_2 & \eta v_3 & (\kappa + \frac{1}{3}\eta)v_1^2 + \eta \boldsymbol{v}\cdot\boldsymbol{v} \end{pmatrix},$$

$$\widetilde{\mathcal{K}}_{12} = T \begin{pmatrix} 0 & (\kappa - \frac{2}{3}\eta) & 0 & (\kappa - \frac{2}{3}\eta)v_2 \\ \eta & 0 & 0 & \eta v_1 \\ 0 & 0 & 0 & 0 \\ \eta v_2 & (\kappa - \frac{2}{3}\eta)v_1 & 0 & (\kappa + \frac{1}{3}\eta)v_1 v_2 \end{pmatrix},$$

the other matrices being obtained by circular permutation and using the relations

$$\widetilde{\mathcal{K}}_{12} = \widetilde{\mathcal{K}}_{21}^t, \qquad \widetilde{\mathcal{K}}_{13} = \widetilde{\mathcal{K}}_{31}^t, \qquad \widetilde{\mathcal{K}}_{23} = \widetilde{\mathcal{K}}_{32}^t.$$

On the other hand, the heat and mass diffusion matrices $\widetilde{B}_{ij}^{D\lambda}$, $i,j \in C$, satisfy

$$\widetilde{B}_{11}^{D\lambda} = \widetilde{B}_{22}^{D\lambda} = \widetilde{B}_{33}^{D\lambda} = \widetilde{B}^{D\lambda}, \qquad \widetilde{B}_{ij}^{D\lambda} = 0, \quad i \neq j,$$

where $\widetilde{B}^{D\lambda}$ is given by

$$\begin{pmatrix} (\mathcal{D}_{kl})_{k,l\in S} & & Sym \\ 0_{3\times n} & 0_{3\times 3} & \\ \Big(\sum_{k\in S} \mathcal{D}_{kl} h_k + \rho_l \theta_l T\Big)_{l\in S} & 0_{1\times 3} & \Upsilon_{\widetilde{B}} \end{pmatrix}, \tag{8.6.9}$$

where

$$\Upsilon_{\widetilde{B}} = \widehat{\lambda} T^2 + 2\sum_{k\in S} \rho_k \theta_k h_k T + \sum_{k,l\in S} \mathcal{D}_{kl} h_k h_l,$$

and the symmetric matrix $\mathcal{D}$ has been defined from the multicomponent diffusion matrix D by

$$\mathcal{D}_{kl} = \rho_k \rho_l D_{kl} / r\rho. \tag{8.6.10}$$

Finally, the source term $\widetilde{\Omega}$ is given by

$$\widetilde{\Omega} = \Omega. \tag{8.6.11}$$

Proof. The calculation of the matrices $\widetilde{A}_0$, $\widetilde{A}_i$, $i \in C$, and $\widetilde{B}_{ij}$, $i,j \in C$, is lengthy but straightforward and, therefore, is omitted [Mas96]. This calculation is easily conducted by using the natural variable Z as an intermediate variable. The symmetry properties of $\widetilde{A}_0$, $\widetilde{A}_i$, $i \in C$, and $\widetilde{B}_{ij}$, $i,j \in C$, required in (S_1)–(S_4), are then obtained. We also have the identity $\widetilde{\Omega} = \Omega$, since source terms are invariants during a change of variable.

Consider then a vector $x \in \mathbb{R}^{n+4}$, with components $(x_1, \ldots, x_{n+4})$. After a little algebra, we obtain

$$x^t \widetilde{A}_0 x = \rho T \left((x_{n+1} + v_1 x_{n+4})^2 + (x_{n+2} + v_2 x_{n+4})^2 + (x_{n+3} + v_3 x_{n+4})^2 \right) +$$

$$\sum_{k \in S} \frac{\rho_k}{r_k} \left(x_k + v_1 x_{n+1} + v_2 x_{n+2} + v_3 x_{n+3} + e_k^{\text{tot}} x_{n+4} \right)^2 + \rho c_v T^2 x_{n+4}^2,$$

so that from (Th_1) and the positivity of ρ_k, $k \in S$, and T, we deduce that $\widetilde{A}_0$ is positive definite.

On the other hand, by using the properties of transport coefficients, one can establish that

$$x^t \widetilde{B}^{D\lambda} x = \sum_{k,l \in S} \mathcal{D}_{kl} \left(x_k + (h_k + \frac{p\,\chi_k}{\rho_k}) x_{n+4} \right) \left(x_l + (h_l + \frac{p\,\chi_l}{\rho_l}) x_{n+4} \right) + \lambda T^2 x_{n+4}^2, \tag{8.6.12}$$

which shows that $\widetilde{B}^{D\lambda}$ is positive semidefinite, thanks to the properties of the modified diffusion matrix $\mathcal{D}$ (8.6.10).

Furthermore, a straightforward calculation leads to the following expression for the quadratic form associated with $\widetilde{B}(V,w)$:

$$x^t \widetilde{B}(V,w) x = T(\kappa + \tfrac{1}{3}\eta)(o_1 w_1 + o_2 w_2 + o_3 w_3)^2 + T\eta(o_1^2 + o_2^2 + o_3^2) + x^t \widetilde{B}^{D\lambda} x, \tag{8.6.13}$$

where $o_i = x_{n+i} + v_i x_{n+4}$, $i = 1,2,3$, and $w_1^2 + w_2^2 + w_3^2 = 1$. We thus obtain that the matrix $\widetilde{B}$ is symmetric, since it is the sum of symmetric matrices. Furthermore, $\widetilde{B}$ is positive semidefinite from the positivity properties of transport coefficients. Finally, σ also satisfies (E_1)–(E_4) as is easily checked and is strictly convex since $\widetilde{A}_0$ is positive definite over the open convex set $\mathcal{O}_U$. ∎

8.7. Normal forms for multicomponent flows

We now apply the general results of Section 8.5 to the system of equations governing multicomponent reactive flows (8.2.3) which admits the symmetric form (8.6.4) from Theorem 8.6.2 and thus satisfies (Edp_1) and (Edp_2). We establish that (Edp_3) is satisfied and recast the symmetrized system

(8.6.4) into several normal forms. The first normal form has simpler matrix coefficients and generalizes the normal form of the Navier–Stokes equations previously obtained by Kawashima and Shizuta [KS88]. This form also perturbs the structure of the source term. The second normal form is more natural, but also more complex, and has dissipative terms in conservative form. We also obtain a third intermediate normal form, which will be used in the context of chemical equilibrium flows investigated in Chapter 10. Finally, any of these forms can be used for the asymptotic stability of constant equilibrium states investigated in Chapter 9.

8.7.1. Nullspace of dissipation matrices

In the following lemma, we establish that the nullspace invariance property holds for multicomponent reactive flow models.

Proposition 8.7.1. *The nullspace of the matrix $\widetilde{B}$ associated with system (8.6.4) is one-dimensional and given by*

$$N(\widetilde{B}) = span(1,\ldots,1,0,0,0,0)^t, \tag{8.7.1}$$

and we have $\widetilde{B}_{ij}N(\widetilde{B}) = 0$, for $i,j \in C$.

Proof. According to (8.6.12) and (8.6.13), the matrix $\widetilde{B}$ is positive semidefinite, so that its nullspace is constituted by the vectors x of $\mathbb{R}^{n+4}$ such that $x^t\widetilde{B}x = 0$. However, we have

$$x^t\widetilde{B}(V,w)x = \left(\kappa + \tfrac{1}{3}\eta\right) T(o_1w_1 + o_2w_2 + o_3w_3)^2 + \eta T(o_1^2 + o_2^2 + o_3^2)$$

$$+ \sum_{k,l\in S} \mathcal{D}_{kl}\left(x_k + (h_k + \frac{p\chi_k}{\rho_k})x_{n+4}\right)\left(x_l + (h_l + \frac{p\chi_l}{\rho_l})x_{n+4}\right) + \lambda T^2 x_{n+4}^2,$$

where $o_i = x_{n+i} + v_i x_{n+4}$, $i = 1,2,3$. As a consequence, $x^t\widetilde{B}x = 0$ implies that $x_{n+4} = 0$ and $x_{n+i} = 0$, $i = 1,2,3$, so that we obtain

$$\sum_{k,l\in S} \mathcal{D}_{kl}x_kx_l = 0. \tag{8.7.2}$$

Using now $N(D) = \mathbb{R}(\rho_1,\ldots,\rho_n)^t$ and (8.6.10), we obtain that the nullspace of $\widetilde{B}(V,w)$ is one-dimensional and spanned by $(1,\ldots,1,0,0,0,0)^t$, and it is thus independent of $V \in \mathcal{O}_V$ and $w \in \Sigma^2$. Finally, one can easily check that $\widetilde{B}_{ij}(1,\ldots,1,0,0,0,0)^t = 0$, for $i,j \in C$. ∎

Since the system of equations governing multicomponent reacting flows satisfies (Edp_1)–(Edp_3), we can now introduce the auxiliary variables U'

and V' of Lemma 8.5.3. From Proposition 8.7.1 and Lemma 8.5.3, the matrix P can be taken to be

$$P = \begin{pmatrix} 1 & 0 & \cdots & \cdots & 0 & 0 & 0 & 0 & 0 \\ 1 & 1 & \ddots & & \vdots & \vdots & \vdots & \vdots & \vdots \\ \vdots & 0 & \ddots & \ddots & \vdots & \vdots & \vdots & \vdots & \vdots \\ \vdots & \vdots & \ddots & \ddots & 0 & \vdots & \vdots & \vdots & \vdots \\ 1 & 0 & \cdots & 0 & 1 & 0 & 0 & 0 & 0 \\ 0 & \cdots & \cdots & \cdots & 0 & 1 & 0 & 0 & 0 \\ 0 & \cdots & \cdots & \cdots & 0 & 0 & 1 & 0 & 0 \\ 0 & \cdots & \cdots & \cdots & 0 & 0 & 0 & 1 & 0 \\ 0 & \cdots & \cdots & \cdots & 0 & 0 & 0 & 0 & 1 \end{pmatrix}, \qquad (8.7.3)$$

in such a way that

$$U' = P^t U = \Big(\rho,\, \rho_2, \ldots,\, \rho_n,\, \rho v_1,\, \rho v_2,\, \rho v_3,\, \rho e^{\mathrm{tot}}\Big)^t. \qquad (8.7.4)$$

The associated entropic variable is then $V' = P^{-1}V$, where V is given by (8.6.3), and the corresponding symmetric system is easily obtained from the symmetrized system (8.6.4).

Proposition 8.7.2. *The system in the new dependent variable V',*

$$V' = \frac{1}{T}\Big(\mathcal{G}_1 - \tfrac{1}{2}\boldsymbol{v}\cdot\boldsymbol{v},\, \mathcal{G}_2 - \mathcal{G}_1, \ldots,\, \mathcal{G}_n - \mathcal{G}_1,\, v_1,\, v_2,\, v_3,\, -1\Big)^t, \qquad (8.7.5)$$

can be written as

$$\widetilde{A}'_0 \partial_t V' + \sum_{i\in C} \widetilde{A}'_i \partial_i V' = \sum_{i,j\in C} \partial_i\Big(\widetilde{B}'_{ij}\partial_j V'\Big) + \widetilde{\Omega}', \qquad (8.7.6)$$

where $V' \in \mathcal{O}_{V'} = \mathbb{R}^{n+3} \times (-\infty, 0)$, $\widetilde{A}'_0 = P^t \widetilde{A}_0 P$, $\widetilde{A}'_i = P^t \widetilde{A}_i P$, $i = 1,2,3$, $\widetilde{B}'_{ij} = P^t \widetilde{B}_{ij} P$, $i,j = 1,2,3$, and $\widetilde{\Omega}' = P^t \Omega = (0, \Omega_{II})^t$.

In particular, Properties (S$_1$)–(S$_4$) *of Definition* 8.4.1 *are satisfied and the dissipation matrices are given by*

$$\widetilde{B}'_{ij} = \begin{pmatrix} 0 & 0_{1\times(n+3)} \\ 0_{(n+3)\times 1} & \widetilde{B}^{II,II}_{ij} \end{pmatrix}, \qquad (8.7.7)$$

where $\widetilde{B}^{II,II}_{ij}$ is the lower right block of size $n+3$ of $\widetilde{B}$.

8.7.2. First normal form

We now investigate normal forms for the system (8.6.4), or, equivalently, for the system (8.7.6). We first use the possibility of mixing parabolic components—the V'_{II} components—established in Theorem 8.5.5, in order to simplify the analytic expression of the normal variable and, consequently, of the matrix coefficients appearing in the normal form. More specifically, we consider the variable

$$W = \Big(\rho,\, \log(\rho_2^{r_2}/\rho_1^{r_1}), \ldots,\, \log(\rho_n^{r_n}/\rho_1^{r_1}),\, v_1,\, v_2,\, v_3,\, T\Big)^t, \qquad (8.7.8)$$

easily obtained by combining the V'_{II} components, and derive the corresponding normal form of the governing equations. When there is only one gas, this normal form is identical with the one previously obtained by Kawashima and Shizuta [KS88].

Theorem 8.7.3. *The system in the variable $W = (W_I, W_{II})^t$, on the open convex set $\mathcal{O}_W = (0,\infty) \times \mathbb{R}^{n-1} \times \mathbb{R}^3 \times (0,\infty)$, with hyperbolic variable*

$$W_I = \rho \qquad (8.7.9)$$

and parabolic variable

$$W_{II} = \big(\log(\rho_2^{r_2}/\rho_1^{r_1}), \ldots,\, \log(\rho_n^{r_n}/\rho_1^{r_1}),\, v_1,\, v_2,\, v_3,\, T\big)^t, \qquad (8.7.10)$$

can be written in the form

$$\overline{A}_0^{I,I}\,\partial_t W_I \;+ \sum_{i\in C} \overline{A}_i^{I,I}\,\partial_i W_I \;+ \sum_{i\in C} \overline{A}_i^{I,II}\,\partial_i W_{II} \;= 0, \qquad (8.7.11)$$

$$\overline{A}_0^{II,II}\,\partial_t W_{II} + \sum_{i\in C} \overline{A}_i^{II,I}\,\partial_i W_I + \sum_{i\in C} \overline{A}_i^{II,II}\,\partial_i W_{II} =$$

$$\sum_{i,j\in C} \partial_i\Big(\overline{B}_{ij}^{II,II}\,\partial_j W_{II}\Big) + \overline{\mathcal{T}}_{II} + \overline{\Omega}_{II}, \qquad (8.7.12)$$

and is of the normal form. The matrix $\overline{A}_0$ is given by

$$\overline{A}_0 = \begin{pmatrix} \dfrac{1}{\Sigma_\rho} & & & 0 \\ & \mathfrak{R} & & \\ & & \dfrac{\rho}{T}\boldsymbol{I} & \\ 0 & & & \dfrac{\rho c_v}{T^2} \end{pmatrix}, \qquad (8.7.13)$$

where $\mathfrak{R}$ is a square matrix of dimension $n-1$ given by

$$\mathfrak{R}_{kl} = \delta_{kl}\frac{\rho_k}{r_k} - \frac{\rho_k\rho_l}{r_k r_l}\frac{1}{\Sigma_\rho}, \qquad k,l \in \{2,\ldots,n\}. \tag{8.7.14}$$

Denoting by $\boldsymbol{\xi} = (\xi_1,\xi_2,\xi_3)^t$ an arbitrary vector of $\mathbb{R}^3$, the matrices $\overline{A}_i$, $i=1,2,3$, are given by

$$\sum_{i\in C}\overline{A}_i\xi_i = \begin{pmatrix} \dfrac{\boldsymbol{v}\cdot\boldsymbol{\xi}}{\Sigma_\rho} & & & Sym \\ 0_{(n-1)\times 1} & \mathfrak{R}\,\boldsymbol{v}\cdot\boldsymbol{\xi} & & \\ \dfrac{\rho}{\Sigma_\rho}\boldsymbol{\xi} & \boldsymbol{\xi}\otimes\mathfrak{p} & \dfrac{\rho}{T}\boldsymbol{v}\cdot\boldsymbol{\xi}\,\boldsymbol{I} & \\ 0 & 0_{1\times(n-1)} & \dfrac{\rho r}{T}\boldsymbol{\xi}^t & \dfrac{\rho c_v}{T^2}\boldsymbol{v}\cdot\boldsymbol{\xi} \end{pmatrix}, \tag{8.7.15}$$

where $r = R/\overline{m}$ and $\mathfrak{p}$ is a vector of dimension $n-1$ given by

$$\mathfrak{p}_l = \rho_l - \frac{\rho_l\,\rho}{r_l\,\Sigma_\rho}, \qquad l \in \{2,\ldots,n\}.$$

For the heat and mass diffusion matrices, $\overline{B}_{ij}^{D\lambda}$, $i,j=1,2,3$, we have

$$\overline{B}_{11}^{D\lambda} = \overline{B}_{22}^{D\lambda} = \overline{B}_{33}^{D\lambda} = \overline{B}^{D\lambda}, \tag{8.7.16}$$

where

$$\overline{B}^{D\lambda} = \begin{pmatrix} 0 & & & Sym \\ 0_{(n-1)\times 1} & (\mathcal{D}_{kl})_{k,l\geq 2} & & \\ 0_{3\times 1} & 0_{3\times(n-1)} & 0_{3\times 3} & \\ 0 & \frac{1}{T}\Big(\sum_{k\in S}\mathcal{D}_{kl}r_k + \rho_l\theta_l\Big)_{l\geq 2} & 0_{1\times 3} & \Upsilon_{\overline{B}} \end{pmatrix}$$

and

$$\Upsilon_{\overline{B}} = \frac{\widehat{\lambda}}{T^2} + 2\sum_{k\in S}\frac{\rho_k\theta_k r_k}{T^2} + \sum_{k,l\in S}\frac{\mathcal{D}_{kl}r_k r_l}{T^2},$$

whereas the nondiagonal terms vanish $\overline{B}_{ij}^{D\lambda} = 0$, $i \neq j$. The dissipation matrices due to the viscous effects $\overline{B}_{ij}^{\kappa\eta}$, $i,j=1,2,3$, still have the structure (8.6.8), and the corresponding matrices $\overline{\mathcal{K}}_{ij}$, $i,j=1,2,3$, are given by

$$\overline{\mathcal{K}}_{11} = \frac{1}{T}\begin{pmatrix} \kappa+\frac{4}{3}\eta & 0 & 0 & 0 \\ 0 & \eta & 0 & 0 \\ 0 & 0 & \eta & 0 \\ 0 & 0 & 0 & 0 \end{pmatrix}, \qquad \overline{\mathcal{K}}_{12} = \frac{1}{T}\begin{pmatrix} 0 & \kappa-\frac{2}{3}\eta & 0 & 0 \\ \eta & 0 & 0 & 0 \\ 0 & 0 & 0 & 0 \\ 0 & 0 & 0 & 0 \end{pmatrix}, \tag{8.7.17}$$

with the other ones deduced by circular permutation and from the relations

$$\overline{\mathcal{K}}_{12} = \overline{\mathcal{K}}_{21}^t, \qquad \overline{\mathcal{K}}_{13} = \overline{\mathcal{K}}_{31}^t, \qquad \overline{\mathcal{K}}_{23} = \overline{\mathcal{K}}_{32}^t. \tag{8.7.18}$$

Finally, the term $\overline{\mathcal{T}}_{II}$ *is easily computed from* (8.5.2), *whereas the corresponding source term* $\overline{\Omega} = (\partial_{\mathcal{W}^*} V^*)^t \, \Omega$ *is given by*

$$\overline{\Omega} = \Big(0,\, m_2\omega_2, \ldots, m_n\omega_n,\, \frac{\rho g_1}{T}, \frac{\rho g_2}{T}, \frac{\rho g_3}{T}, -\frac{1}{T^2}\sum_{k\in S} e_k m_k \omega_k\Big)^t. \tag{8.7.19}$$

Proof. The calculations are lengthy but straightforward and make use of Theorem 8.6.2, Proposition 8.7.2, and Assumptions (Th_1)–(Th_4) and (Tr_1)–(Tr_{4a}). ■

Remark 8.7.4. Note that, even if $g = 0$ and the source term Ω remains in a fixed subspace of $\mathbb{R}^n \times 0_{\mathbb{R}^4}$, the source term $\overline{\Omega}$ is no longer in a *fixed* subspace of $\mathbb{R}^{n+4}$ of the same dimension because of the coefficients e_k/T^2 in the term $\sum_{k\in S} e_k m_k \omega_k / T^2$, which introduce an explicit dependence on the state variables. ■

8.7.3. Natural normal form

We consider here the normal form of the multicomponent reactive flow equations obtained with the "natural" normal variable $\mathcal{W} = (U_I', V_{II}')^t$ given by

$$\mathcal{W} = \Big(\rho, \frac{\mathcal{G}_2 - \mathcal{G}_1}{T}, \ldots, \frac{\mathcal{G}_n - \mathcal{G}_1}{T}, \frac{v_1}{T}, \frac{v_2}{T}, \frac{v_3}{T}, \frac{-1}{T}\Big)^t, \tag{8.7.20}$$

and suggested by Theorem 8.5.5. This normal form has dissipative terms in conservative form and leaves unchanged the last components of the source term Ω_{II}. This form, however, has a more complex expression than the one obtained in Theorem 8.7.3.

Theorem 8.7.5. *The system in the variable* $\mathcal{W} = (\mathcal{W}_I, \mathcal{W}_{II})^t$, *on the open convex set* $\mathcal{O}_{\mathcal{W}} = (0, \infty) \times \mathbb{R}^{n-1} \times \mathbb{R}^3 \times (-\infty, 0)$, *with hyperbolic variable*

$$\mathcal{W}_I = \rho \tag{8.7.21}$$

and parabolic variable

$$\mathcal{W}_{II} = \frac{1}{T}\Big(\mathcal{G}_2 - \mathcal{G}_1, \ldots, \mathcal{G}_n - \mathcal{G}_1, v_1, v_2, v_3, -1\Big)^t, \tag{8.7.22}$$

can be written in the form

$$\overline{\mathcal{A}}_0^{I,I}\partial_t\mathcal{W}_I + \sum_{i\in C}\overline{\mathcal{A}}_i^{I,I}\partial_i\mathcal{W}_I + \sum_{i\in C}\overline{\mathcal{A}}_i^{I,II}\partial_i\mathcal{W}_{II} = 0, \tag{8.7.23}$$

$$\overline{\mathcal{A}}_0^{II,II}\partial_t\mathcal{W}_{II} + \sum_{i\in C}\overline{\mathcal{A}}_i^{II,I}\partial_i\mathcal{W}_I + \sum_{i\in C}\overline{\mathcal{A}}_i^{II,II}\partial_i\mathcal{W}_{II} = \sum_{i,j\in C}\partial_i\Big(\overline{\mathcal{B}}_{ij}^{II,II}\partial_j\mathcal{W}_{II}\Big) + \Omega_{II}, \tag{8.7.24}$$

and is of the normal form. The matrix $\overline{\mathcal{A}}_0$ *is given by*

$$\overline{\mathcal{A}}_0 = \begin{pmatrix} 1/\Sigma_\rho & & & Sym \\ 0 & \mathfrak{R} & & \\ 0 & 0 & \rho T \boldsymbol{I} & \\ 0 & \Xi_e^t & \rho T \boldsymbol{v}^t & \Upsilon_e' \end{pmatrix}, \tag{8.7.25}$$

where Ξ_e *is a vector given by*

$$\Xi_{el} = \frac{\rho_l}{r_l}\left(e_l^{\rm tot} - \frac{\Sigma_e}{\Sigma_\rho}\right), \qquad l\in\{2,\ldots,n\}, \tag{8.7.26}$$

and

$$\Upsilon_e' = \sum_{k\in S}\frac{\rho_k}{r_k}e_k^{{\rm tot}\,2} + \rho T\left(\boldsymbol{v}\cdot\boldsymbol{v} + c_v T\right) - \frac{\Sigma_e^2}{\Sigma_\rho}. \tag{8.7.27}$$

The matrices $\overline{\mathcal{A}}_i$, $i=1,2,3$, *are also given by*

$$\sum_{i\in C}\overline{\mathcal{A}}_i\xi_i = \begin{pmatrix} \dfrac{\boldsymbol{v}\cdot\boldsymbol{\xi}}{\Sigma_\rho} & & & Sym \\ 0_{(n-1)\times 1} & \mathfrak{R}\boldsymbol{v}\cdot\boldsymbol{\xi} & & \\ \dfrac{\rho T}{\Sigma_\rho}\boldsymbol{\xi} & T\boldsymbol{\xi}\otimes\mathfrak{p} & \rho T\boldsymbol{v}\cdot\boldsymbol{\xi}\,\boldsymbol{I} & \\ \dfrac{\rho T}{\Sigma_\rho}\boldsymbol{v}\cdot\boldsymbol{\xi} & (\boldsymbol{v}\cdot\boldsymbol{\xi})\Xi_h^t & \rho T(\boldsymbol{v}\cdot\boldsymbol{\xi})\boldsymbol{v}^t + \rho T\Xi_{h0}\boldsymbol{\xi}^t & \Upsilon_h'\boldsymbol{v}\cdot\boldsymbol{\xi} \end{pmatrix}, \tag{8.7.28}$$

where Ξ_{h0} *and* Ξ_{hl} *are defined by*

$$\Xi_{h0} = h^{\rm tot} - \frac{\Sigma_e}{\Sigma_\rho}, \qquad \Xi_{hl} = \frac{\rho_l}{r_l}\left(h_l^{\rm tot} - \frac{\Sigma_h}{\Sigma_\rho}\right), \quad l\in\{2,\ldots,n\},$$

and

$$\Upsilon_h' = \sum_{k\in S}\frac{\rho_k}{r_k}h_k^{{\rm tot}\,2} + \rho T\left(\boldsymbol{v}\cdot\boldsymbol{v} + (c_v + r)T\right) - \frac{\Sigma_h^2}{\Sigma_\rho} + \frac{(\rho T)^2}{\Sigma_\rho}.$$

For the dissipative part, we also have

$$\overline{\mathcal{B}}_{ij}^{II,II} = \widetilde{B}_{ij}^{II,II}, \tag{8.7.29}$$

and an explicit expression for $\widetilde{B}_{ij}^{II,II}$ *is given in Theorem* 8.7.3. *Finally, concerning the source term, we have no contribution in the hyperbolic equations, whereas the parabolic component of the source term is left unchanged and coincides with* $\widetilde{\Omega}'_{II} = \Omega'_{II} = \Omega_{II}$.

8.7.4. Intermediate normal form

We consider here an intermediate normal form of the multicomponent reactive flow equations obtained with the "intermediate" normal variable $\mathcal{W}$—not to be confused with the natural variable $\mathcal{W}$—and given by

$$\mathcal{W} = \Big(\rho, \frac{\mathcal{G}_2 - \mathcal{G}_1}{T}, \dots, \frac{\mathcal{G}_n - \mathcal{G}_1}{T}, v_1, v_2, v_3, T\Big)^t. \tag{8.7.30}$$

This variable introduces some decoupling between the last components, but still contains the Gibbs potentials. The corresponding normal form will be especially practical for investigating chemical equilibrium flows in Chapter 10.

Theorem 8.7.6. *The system in the variable* $\mathcal{W} = (\mathcal{W}_I, \mathcal{W}_{II})^t$, *on the open convex set* $\mathcal{O}_{\mathcal{W}} = (0,\infty) \times \mathbb{R}^{n-1} \times \mathbb{R}^3 \times (0,\infty)$, *with hyperbolic variable*

$$\mathcal{W}_I = \rho \tag{8.7.31}$$

and parabolic variable

$$\mathcal{W}_{II} = \Big(\frac{\mathcal{G}_2 - \mathcal{G}_1}{T}, \dots, \frac{\mathcal{G}_n - \mathcal{G}_1}{T}, v_1, v_2, v_3, T\Big)^t, \tag{8.7.32}$$

can be written in the form

$$\overline{\mathcal{A}}_0^{I,I}\,\partial_t \mathcal{W}_I + \sum_{i\in C} \overline{\mathcal{A}}_i^{I,I}\,\partial_i \mathcal{W}_I + \sum_{i\in C} \overline{\mathcal{A}}_i^{I,II}\,\partial_i \mathcal{W}_{II} = 0, \tag{8.7.33}$$

$$\overline{\mathcal{A}}_0^{II,II}\,\partial_t \mathcal{W}_{II} + \sum_{i\in C} \overline{\mathcal{A}}_i^{II,I}\,\partial_i \mathcal{W}_I + \sum_{i\in C} \overline{\mathcal{A}}_i^{II,II}\,\partial_i \mathcal{W}_{II} = \sum_{i,j\in C} \partial_i\Big(\overline{\mathcal{B}}_{ij}^{II,II}\,\partial_j \mathcal{W}_{II}\Big) + \overline{\mathcal{T}}_{II} + \overline{\Omega}_{II}, \tag{8.7.34}$$

and is of the normal form. The matrix $\overline{\mathcal{A}}_0$ *is given by*

$$\overline{\mathcal{A}}_0 = \begin{pmatrix} \dfrac{1}{\Sigma_\rho} & & & Sym \\ 0 & \mathfrak{R} & & \\ 0 & 0 & \dfrac{\rho}{T}\boldsymbol{I} & \\ 0 & \dfrac{\Xi_e^t}{T^2} & 0 & \dfrac{\Upsilon'_e}{T^4} \end{pmatrix}. \tag{8.7.35}$$

The matrices $\overline{\mathcal{A}}_i$*,* $i = 1, 2, 3$*, are also given by*

$$\sum_{i\in C} \overline{\mathcal{A}}_i \xi_i = \begin{pmatrix} \dfrac{\boldsymbol{v}\cdot\boldsymbol{\xi}}{\Sigma_\rho} & & & Sym \\ 0_{(n-1)\times 1} & \mathfrak{R}\boldsymbol{v}\cdot\boldsymbol{\xi} & & \\ \dfrac{\rho}{\Sigma_\rho}\boldsymbol{\xi} & \boldsymbol{\xi}\otimes\mathfrak{p} & \dfrac{\rho}{T}\boldsymbol{v}\cdot\boldsymbol{\xi}\,\boldsymbol{I} & \\ 0 & (\boldsymbol{v}\cdot\boldsymbol{\xi})\dfrac{\Xi_e^t}{T^2} & \dfrac{\rho\Xi_{h0}}{T^2}\boldsymbol{\xi}^t & (\Upsilon'_e - \rho T\boldsymbol{v}\cdot\boldsymbol{v})\dfrac{\boldsymbol{v}\cdot\boldsymbol{\xi}}{T^4} \end{pmatrix}. \tag{8.7.36}$$

On the other hand, the viscous dissipation matrices coincide with those of the simple normal form

$$\overline{\mathcal{B}}_{ij}^{II,II\kappa\eta} = \overline{B}_{ij}^{II,II\kappa\eta}, \tag{8.7.37}$$

whereas the heat and mass diffusion matrices are very similar to that of the natural normal form

$$\overline{\mathcal{B}}_{ij}^{II,II D\lambda} = \begin{pmatrix} (\mathcal{D}_{kl})_{k,l\geq 2} & & Sym \\ 0_{3\times(n-1)} & 0_{3\times 3} & \\ \dfrac{1}{T^2}\Big(\sum_{k\in S}\mathcal{D}_{kl}h_k + \rho_l\theta_l T\Big)_{l\geq 2} & 0_{1\times 3} & \dfrac{\Upsilon_{\widetilde{B}}}{T^4} \end{pmatrix}. \tag{8.7.38}$$

Finally, the term $\overline{\mathcal{T}}_{II}$ *is easily computed from* (8.5.2)*, whereas the corresponding source term* $\overline{\Omega}$ *is given by*

$$\overline{\Omega} = \Big(0,\, m_2\omega_2, \ldots, m_n\omega_n,\, \frac{\rho g_1}{T},\, \frac{\rho g_2}{T},\, \frac{\rho g_3}{T},\, 0\Big)^t. \tag{8.7.39}$$

8.8. Notes

8.1. Up to trivial modifications, the mathematical entropy for the compressible Navier–Stokes equations is unique, as can be seen by combining the results of [Vul97] and [HFM86]. More specifically, entropies satifying Property (E_2) are necessarily functions of S [Vul97], and the reciprocity relations (E_3) imply that these functions are affine whenever heat conduction is present [HFM86]. This is a fortiori the case for multicomponent reactive flow equations, which include even more dissipative effects.

8.2. In order to perform actual calculations of various differential relations, the most useful matrices are the derivatives with respect to the natural variable Z. This variables can indeed be used as an intermediate variable in all calculations.

8.3. Symmetrization is also feasible with the thermodynamic formalism of Section 6.3, but requires the use of symmetric generalized inverses to handle the case of singular Hessians.

8.9. References

[CHS90] F. Chalot, T.J.R. Hughes, and F. Shakib, *Symmetrization of Conservation Laws with Entropy for High-Temperature Hypersonic Computations,* Comp. Syst. Eng., **1**, (1990), pp. 495–521.

[FL71] K. O. Friedrichs and P. D. Lax, *Systems of Conservation Laws with a Convex Extension,* Proc. Nat. Acad. Sci. USA, **68**, (1971), pp. 1686–1688.

[GM96a] V. Giovangigli and M. Massot, *Les Mélanges Gazeux Réactifs, (I) Symétrisation et Existence Locale,* C. R. Acad. Sci. Paris, **323**, Série I, (1996), pp. 1153–1158.

[GM96b] V. Giovangigli and M. Massot, *Les Mélanges Gazeux Réactifs, (II) Stabilité Asymptotique des États d'Équilibres,* C. R. Acad. Sci. Paris, **323**, Série I, (1996), pp. 1207–1212.

[GM98a] V. Giovangigli and M. Massot, *Asymptotic Stability of Equilibrium States for Multicomponent Reactive Flows,* Math. Mod. Meth. Appl. Sci., **8**, (1998), pp. 251–297.

[God62] S.K. Godunov, *The Problem of a Generalized Solution in the Theory of Quasilinear Equations and in Gas Dynamics,* Russian Math. Surveys, **17**, (1962), pp. 145–156.

[HCB54] J. O. Hirschfelder, C. F. Curtiss, and R. B. Bird, *Molecular Theory of Gases and Liquids,* John Wiley & Sons, Inc., New York, (1954).

[HFM86] T. J. R. Hughes, L. P. Franca, and M. Mallet, *A New Finite Element Formulation for Computational Fluid Dynamics: I. Symmetric Forms of the Compressible Euler and Navier-Stokes Equations and the Second Law of Thermodynamics,* Comp. Meth. Appl. Mech. Eng., **54**, (1986), pp. 223–234.

[HH98] G. Hauke and T. J. R. Hughes, *A Comparative Study of Different Sets of Variables for Solving Compressible and Incompressible Flows,* Comp. Meth. Appl. Mech. Eng., **153**, (1998), pp. 1–44.

[Kaw84] S. Kawashima, *Systems of Hyperbolic-Parabolic Composite Type, with Application to the Equations of Magnetohydrodynamics,* Doctoral Thesis, Kyoto University, (1984).

[KS88] S. Kawashima and Y. Shizuta, *On the Normal Form of the Symmetric Hyperbolic-Parabolic Systems Associated with the Conservation Laws,* Tôhoku Math. J., **40**, (1988), pp. 449–464.

[Mas96] M. Massot, *Modélisation Mathématique et Numérique de la Combustion des Mélanges Gazeux,* Doctoral Thesis, Ecole Polytechnique, (1996).

[Maj84] A. Majda, *Compressible Fluid Flow and Systems of Conservation Laws in Several Space Variables,* Appl. Math. Sci., **53**, Springer-Verlag, New York, (1984).

[Moc80] M. Mock, *Systems of Conservation Laws of Mixed Type,* J. Diff. Eq., **37**, (1980), pp. 70–88.

[SK85] Y. Shizuta and S. Kawashima, *Systems of Equations of Hyperbolic-Parabolic Type with Applications to the Discrete Boltzmann Equation,* Hokkaido Math. J., **14**, (1985), pp. 249–275.

[Ser96a] D. Serre, *Systèmes de Loi de Conservation I,* Diderot, Ed., Arts et Sciences, Paris, (1996).

[Ser96b] D. Serre, *Systèmes de Loi de Conservation II,* Diderot, Ed., Arts et Sciences, Paris, (1996).

[VH72] A.I. Vol'pert and S.I. Hudjaev, *On the Cauchy Problem for Composite Systems of Nonlinear Differential Equations,* Math USSR Sbornik, **16**, (1972), pp. 517–544.

[Vul97] L. G. Vulkov, *On the Conservation Laws of the Compressible Euler Equations,* Applicable Analysis, **64**, (1997), pp. 255–271.

9

Asymptotic Stability

9.1. Introduction

In this chapter we investigate global existence results and asymptotic stability of equilibrium states for the system of partial differential equations governing multicomponent reactive flows [GM96a] [GM96b] [GM98a].

We first consider an abstract quasilinear system of second-order which admits an entropy function and satisfies the nullspace invariance property, so that it can be recast into a normal form. Under stability conditions on the source term, and assuming conditions that guarantee the local dissipative structure of the linearized normal system around the constant equilibrium state, we obtain global existence and asymptotic stability of the stationary state. As stability conditions, we assume that the chemical entropy production is nonnegative and that the source term lies in the range of its derivative at equilibrium. Our method of proof relies on Kawashima's theory [Kaw84], a priori estimates provided by the entropy conservation law, and stability properties of the source term. Decay estimates towards the constant stationary state are also obtained in all space dimensions.

We then apply these results to the system modeling multicomponent reactive flows. The existence of equilibrium points is deduced from Chapter 6. From the structure properties of chemical production rates—using, in particular, thermodynamic stability inequalities—and the structure of multicomponent transport fluxes presented in Chapter 7, we then establish the local dissipativity of linearized normal forms around equilibrium states. Therefore, we obtain global existence and asymptotic stability of the constant equilibrium states. Decay estimates are also obtained for multicomponent flows in all space dimensions.

9.2. Governing equations

9.2.1. Abstract system

In this section, we further consider the abstract quasilinear system of second order in general form (8.4.1), and we assume that Properties (Edp_1)–(Edp_3) of Chapter 8 hold. In particular, this system can be written in symmetric form (8.4.2) and in normal form (8.5.1).

9.2.2. Equilibrium point

We further assume in the following that system (8.4.1) possesses a constant equilibrium state.

(Edp_4) There exists a constant equilibrium state $U^{*\mathrm{e}} \in \mathcal{O}_{U^*}$ such that

$$\Omega^*(U^{*\mathrm{e}}) = 0.$$

The equilibrium states corresponding to the various variables are also denoted with the superscript ^{e}, so that the equilibrium states in the variables V^* and W^*, for instance, are denoted by $V^{*\mathrm{e}}$ and $W^{*\mathrm{e}}$, respectively.

9.2.3. Entropy equation

In order to establish existence theorems, we will need a priori estimates for solutions of (8.4.1). For this purpose, we establish a balance equation for the generalized entropy function σ^*. This equation is easily obtained by taking the scalar product of (8.4.2) with the vector V^*, and it reads as

$$\begin{aligned}\partial_t \sigma^* + \sum_{i \in C^*} \partial_i q_i^* &= \sum_{i,j \in C^*} \partial_i \langle V^*, \widetilde{B}_{ij}^* \partial_j V^* \rangle \\ &- \sum_{i,j \in C^*} \langle \partial_i V^*, \widetilde{B}_{ij}^* \partial_j V^* \rangle + \langle V^*, \widetilde{\Omega}^*(V^*) \rangle,\end{aligned} \tag{9.2.1}$$

where q_i^* is the convective entropy flux in the ith direction. Note that the quantity $-\sum_{j \in C^*} \langle V^*, \widetilde{B}_{ij}^* \partial_j V^* \rangle$ represents the entropy diffusive flux in the ith direction, whereas $-\sum_{i,j \in C^*} \langle \partial_i V^*, \widetilde{B}_{ij}^* \partial_j V^* \rangle$ represents the entropy production due to macroscopic variable gradients and $\langle V^*, \widetilde{\Omega}^*(V^*) \rangle$ the entropy production due to source terms.

Since we are interested in the behavior of solutions near the state $U^{*\mathrm{e}}$, we introduce the associated function $\mathfrak{h}(U^*, U^{*\mathrm{e}})$ defined by

$$\mathfrak{h}(U^*, U^{*\mathrm{e}}) = \sigma^*(U^*) - \sigma^*(U^{*\mathrm{e}}) - \langle V^{*\mathrm{e}}, U^* - U^{*\mathrm{e}} \rangle, \qquad U^* \in \mathcal{O}_{U^*}. \tag{9.2.2}$$

Thanks to the strict convexity of σ^*, this function plays the role of a distance between U^* and the stationary state U^{*e}, and the following proposition is easily established.

Proposition 9.2.1. *The function $\mathfrak{h}$ is a positive, smooth, strictly convex function of $U^* \in \mathcal{O}_{U^*}$, satisfying $\mathfrak{h}(U^*, U^{*e}) = 0$ if and only if $U^* = U^{*e}$. There exists also a neighborhood $\mathfrak{O} = \{ z \in \mathbb{R}^n;\ |z - W^{*e}| < r \}$ of W^{*e} with r small enough such that, for any W^* in $\mathfrak{O}$ and $U^* = U^*(W^*)$,*

$$c|W^* - W^{*e}|^2 \le \mathfrak{h}(U^*, U^{*e}) \le C|W^* - W^{*e}|^2, \tag{9.2.3}$$

where c and C are positive constants. In addition, the function $\mathfrak{h}$ satisfies the conservation equation

$$\begin{aligned}\partial_t \mathfrak{h} + \sum_{i \in C^*} \partial_i \Big(q_i^*(U^*) - q_i^*(U^{*e}) - \langle V^{*e}, F_i^*(U^*) - F_i^*(U^{*e}) \rangle \Big) = \\ \sum_{i,j \in C^*} \partial_i \langle V^* - V^{*e}, \widetilde{B}_{ij}^* \partial_j V^* \rangle \\ - \sum_{i,j \in C^*} \langle \partial_i V^*, \widetilde{B}_{ij}^* \partial_j V^* \rangle + \langle V^* - V^{*e}, \widetilde{\Omega}^*(V^*) \rangle.\end{aligned} \tag{9.2.4}$$

Proof. The proof is straightforward from the positivity of $\partial^2_{U^*}\sigma^*$. ∎

9.2.4. Functional spaces

In this chapter, we will use the classical functional spaces $L^p(\mathbb{R}^d)$ with norm

$$\|\phi\|_{0,p} = |\phi|_{0,p} = \left(\int_{\mathbb{R}^d} |\phi(x)|^p dx \right)^{\frac{1}{p}}, \quad p \ge 1,$$

and the Sobolev spaces $W_p^l(\mathbb{R}^d)$, $1 \le p \le \infty$, with norm

$$\|\phi\|_{l,p} = \sum_{k \in [0,l]} |\phi|_{k,p}, \quad |\phi|_{k,p} = \sum_{|\beta| = k} \|\partial^\beta \phi\|_{0,p}.$$

We also extend these definitions to vector functions by using the Euclidean norm of $\mathbb{R}^d$, and, for more details about Sobolev spaces, we refer the reader to [Ada75].

9.3. Local dissipative structure

In order to establish global existence in time and asymptotic stability of constant stationary states, decay estimates for linearized equations are needed. A condition that guarantees decay properties for the linearized system is the local dissipative structure introduced by Kawashima [Kaw84]. This dissipative structure is completed here by stability properties of the source term [GM98a].

9.3.1. Linearized equations

If we linearize system (8.5.1) around the constant stationary state W^{*e}, we obtain the following linear system in terms of the variable $z = W^* - W^{*e}$:

$$\overline{A}_0^*(W^{*e})\partial_t z + \sum_{i\in C^*} \overline{A}_i^*(W^{*e})\partial_i z = \sum_{i,j\in C^*} \overline{B}_{ij}^*(W^{*e})\partial_i\partial_j z - \overline{L}^*(W^{*e})z, \tag{9.3.1}$$

where $\overline{L}^*$ is defined by $\overline{L}^* = -\partial_{W^*}\overline{\Omega}^*$. We assume that this linearized system has a dissipative structure in the sense of Kawashima.

(Dis$_1$) The matrix $\overline{A}_0^*(W^{*e})$ is symmetric positive definite, the matrices $\overline{A}_i^*(W^{*e})$, $i \in C^*$ are symmetric, we have the reciprocity relations $\overline{B}_{ij}^*(W^{*e})^t = \overline{B}_{ji}^*(W^{*e})$ for $i, j \in C^*$, and the matrix $\overline{L}^*(W^{*e})$ is symmetric positive semidefinite,

(Dis$_2$) There exist compensating matrices K^j, $j \in C^*$, such that the products $K^j\overline{A}_0^*(W^{*e})$, $j \in C^*$, are skew-symmetric and the matrix

$$\sum_{i,j\in C^*} K^j\overline{A}_i^*(W^{*e})w_iw_j + \overline{B}^*(W^{*e}, w) + \overline{L}^*(W^{*e})$$

is positive definite for $w \in \Sigma^{d-1}$, where Σ^{d-1} is the unit sphere in d dimensions and $\overline{B}^*(W^*, w)$ denotes the matrix

$$\overline{B}^*(W, w) = \sum_{i,j\in C^*} \overline{B}_{ij}^*(W^*)w_iw_j.$$

Remark 9.3.1. The existence of the compensating matrices K^j, $j \in C^*$, implies that the linearized normal form is strictly dissipative in the sense that the eigenvalues $\lambda^*(\zeta, w)$ of the problem

$$\lambda^*\overline{A}_0^*(W^{*e})\phi + [\zeta \sum_{i\in C^*} \overline{A}_i^*(W^{*e})w_i - \zeta^2\overline{B}^*(W^{*e}, w) + \overline{L}^*(W^{*e})]\phi = 0,$$

for $\zeta \in \mathrm{i}\,\mathbb{R}$ and $w \in \Sigma^{d-1}$, have a negative real part [SK85]. However, the converse is not known to be true [SK85]. The latter property only implies the existence of a combined compensating matrix $K(w)$, for $w \in \Sigma^{d-1}$, such that $w \to K(w)$ is C^∞ on Σ^{d-1}, $K(w)$ is real, the product $K(w)\overline{A}_0^*(W^{*\mathrm{e}})$ is skew-symmetric, $K(-w) = -K(w)$, for $w \in \Sigma^{d-1}$, and the matrix

$$\sum_{i\in C^*} K(w)\overline{A}_i^*(W^{*\mathrm{e}})w_i + \overline{B}^*(W^{*\mathrm{e}}, w) + \overline{L}^*(W^{*\mathrm{e}})$$

is positive definite for $w \in \Sigma^{d-1}$. It is not known, however, if the matrix $K(w)$ is of the form $\sum_{j\in C^*} K^j w_j$ [SK85]. Nevertheless, all results obtained in this chapter can be proved without the existence of matrices K^j, $j \in C$, by only using the combined compensating matrix $K(w)$, $w \in \Sigma^{d-1}$, that is, by only using the strict dissipativity of the system. Nevertheless, in practical applications, it is generally possible to obtain compensating matrices K^j, $j \in C^*$, and to set $K(w) = \sum_{j\in C} K^j w_j$. ■

9.3.2. Locally stable source terms

We have already assumed that the matrix $\overline{L}^*(W^{*\mathrm{e}})$ is symmetric in (Dis_1). We now further introduce local stability assumptions concerning the source term Ω^*.

(Dis_3) The smallest linear subspace containing the source term vector $\widetilde{\Omega}^*(V^*) = \Omega^*\big(U^*(V^*)\big)$, for all $V^* \in \mathcal{O}_{V^*}$, is included in the range of $\widetilde{L}^*(V^{*\mathrm{e}}) = -(\partial_{V^*}\widetilde{\Omega}^*)(V^{*\mathrm{e}})$.

(Dis_4) There exists a neighborhood of $V^{*\mathrm{e}}$ in $\mathcal{O}_{V^*}$ and a positive constant δ such that, for any V^* in this neighborhood, we have

$$\delta\,\big|\,\widetilde{\Omega}^*(V^*)\big|^2 \;\le\; -\big\langle V^* - V^{\mathrm{e}*},\, \widetilde{\Omega}^*(V^*)\big\rangle.$$

We can then choose ϵ small enough such that this inequality holds with $V^* = V^*(W^*)$ and W^* in $\mathfrak{O} = \{\, z \in \mathbb{R}^n;\ |z - W^{*\mathrm{e}}| < \epsilon \,\}$.

Note that Properties (Dis_3) and (Dis_4) only concern the source term $\widetilde{\Omega}^*$. Property (Dis_3) will be used for decay estimates, whereas Property (Dis_4) will be needed for the existence theorem.

9.3.3. Global dissipative structure

The physical meaning of the entropy conservation equation (9.2.1) is that when

$$\sum_{i,j\in C^*} \langle \partial_i V^*, \widetilde{B}_{ij}^* \partial_j V^* \rangle \ge 0 \tag{9.3.2}$$

and

$$\langle V^*, \widetilde{\Omega}(V^*)\rangle \leq 0, \tag{9.3.3}$$

then the integral $\int_{\mathbb{R}^d} \sigma^* \, dx$ is nonincreasing in time, which corresponds to the second principle of thermodynamics. This reveals, in particular, the close link between the second principle of thermodynamics and the parabolic nature of systems of conservation laws.

Property (S_4), however, does not imply the stronger condition (9.3.2). Similarly, we have only assumed, with Property (Dis_4), that the source term is locally stable, and it does not imply (9.3.3) globally. For multicomponent reactive flows, however, Properties (9.3.2) and (9.3.3) are globally satisfied, as they should for any physically reasonable model. This suggests the following definition of a strong entropy, which could be used to obtain global estimates, not necessarily in the neighborhood of a constant state.

Definition 9.3.2. *A function σ^* is said to be a strong entropy for system* (8.4.1) *if it is an entropy according to Definition* 8.4.2 *of Chapter* 8 *and if the inequalities* (9.3.2) *and* (9.3.3) *hold over $\mathcal{O}_{V^*}$.*

Property (9.3.2) simply corresponds to the positive semidefiniteness of the matrix $[\widetilde{B}^*]$ of size $d \times n^*$ whose blocks are the dissipation matrices $\widetilde{B}^*_{ij}$, $i, j \in C^*$. Note that when this property holds, condition N directly implies that $\widetilde{B}^*_{ij} N(\widetilde{B}^*) = 0$, $i, j \in C^*$, as easily checked.

9.4. Global existence theorem

Now that we have stated Assumptions (Edp_1)–(Edp_4) and Dissipative conditions (Dis_1)–(Dis_4), concerning the various forms of the governing equations, we investigate global existence of solutions around the stationary state W^{*e} and its asymptotic stability. Of course, these Assumptions (Edp_1)–(Edp_4) imply Properties (S_1)–(S_4), ($\bar{S}_1$)–($\bar{S}_4$), and (Nor_1)–(Nor_3) of Chapter 8. Asymptotic decay towards the equilibrium state W^{*e} will be investigated in Section 9.5.

9.4.1. Main result

Theorem 9.4.1. *Consider the quasilinear system in normal form* (8.5.1) *and assume that Conditions* (Edp_1)–(Edp_4) *and* (Dis_1)–(Dis_4) *hold. Let $d \geq 1$, $l \geq [d/2] + 2$, and $W^{*0}(x)$, such that*

$$W^{*0} - W^{*e} \in W_2^l(\mathbb{R}^d). \tag{9.4.1}$$

Then if $\|W^{*0} - W^{*e}\|_{l,2}$ *is small enough, there exists a unique global solution to the Cauchy problem*

$$\overline{A}_0^* \partial_t W^* + \sum_{i \in C^*} \overline{A}_i^* \partial_i W^* = \sum_{i,j \in C^*} \partial_i \left(\overline{B}_{ij}^* \partial_j W^* \right) + \overline{\mathcal{T}}^* + \overline{\Omega}^*, \tag{9.4.2}$$

with initial conditions

$$W^*(0, x) = W^{*0}(x), \tag{9.4.3}$$

such that

$$\begin{cases} W_{I}^* - W_{I}^{*e} \in C^0\left([0, \infty); W_2^l(\mathbb{R}^d)\right) \cap C^1\left([0, \infty); W_2^{l-1}(\mathbb{R}^d)\right), \\ W_{II}^* - W_{II}^{*e} \in C^0\left([0, \infty); W_2^l(\mathbb{R}^d)\right) \cap C^1\left([0, \infty); W_2^{l-2}(\mathbb{R}^d)\right), \end{cases} \tag{9.4.4}$$

and

$$\begin{cases} W_{I}^* - W_{I}^{*e} \in L^2(0, \infty; W_2^l(\mathbb{R}^d)), \\ W_{II}^* - W_{II}^{*e} \in L^2(0, \infty; W_2^{l+1}(\mathbb{R}^d)). \end{cases} \tag{9.4.5}$$

This solution W^* *satisfies the estimate*

$$\|W^*(t) - W^{*e}\|_{l,2}^2 + \int_0^t \left(\|\partial_x W_{I}^*(\tau)\|_{l-1,2}^2 + \|\partial_x W_{II}^*(\tau)\|_{l,2}^2 \right) d\tau$$
$$\leq C \|W^{*0} - W^{*e}\|_{l,2}^2, \tag{9.4.6}$$

and $\sup_{\mathbb{R}^d} |W^*(t) - W^{*e}|$ *goes to zero as* $t \to \infty$.

In order to establish this result, following Kawashima [Kaw84], we restate a local existence theorem, derive a priori estimates for the local solution of the Cauchy problem, and show that this local solution can be extended indefinitely.

9.4.2. Local existence

In this section we restate a local existence theorem due to Kawashima [Kaw84]. Local existence is proved for initial data near the stationary state, with a control on the distance between the solution and the constant state.

Theorem 9.4.2. *Assume that Properties* (Edp_1)–(Edp_4) *are satisfied. Let* $d \geq 1$ *and* $l \geq [d/2] + 2$ *be integers. Let* $\mathcal{O}_0$ *be a bounded convex open set, such that* $\overline{\mathcal{O}}_0 \subset \mathcal{O}_{W^*}$*, let* $d_1 < d(\mathcal{O}_0, \partial\mathcal{O}_{W^*})$*, let* b_0 *be positive, and let* $\mathcal{O}_1 = \{ z \in \mathbb{R}^n;\, d(z, \mathcal{O}_0) < d_1 \}$*. Then there exists a positive constant*

T_1, depending only on $\mathcal{O}_0$, d_1, and b_0, such that, for any $W^{*0}(x)$ with $W^{*0} - W^{*e} \in W_2^l(\mathbb{R}^d)$ and

$$\|W^{*0} - W^{*e}\|_{l,2} \leq b_0, \qquad W^{*0}(x) \in \mathcal{O}_0, \quad x \in \mathbb{R}^d, \tag{9.4.7}$$

the system (8.5.1) *or* (9.4.2) *with initial condition W^{*0} has a unique solution W^* satisfying*

$$\begin{cases} W_I^* - W_I^{*e} \in C^0\big([0,T_1]; W_2^l(\mathbb{R}^d)\big) \cap C^1\big([0,T_1]; W_2^{l-1}(\mathbb{R}^d)\big), \\ W_{II}^* - W_{II}^{*e} \in C^0\big([0,T_1]; W_2^l(\mathbb{R}^d)\big) \cap C^1\big([0,T_1]; W_2^{l-2}(\mathbb{R}^d)\big), \end{cases} \tag{9.4.8}$$

and

$$\begin{cases} W_I^* - W_I^{*e} \in L^2\big(0,T_1; W_2^l(\mathbb{R}^d)\big), \\ W_{II}^* - W_{II}^{*e} \in L^2\big(0,T_1; W_2^{l+1}(\mathbb{R}^d)\big), \end{cases} \tag{9.4.9}$$

and $W^*(t,x) \in \mathcal{O}_1$, for $(t,x) \in [0,T_1] \times \mathbb{R}^d$. This solution also satisfies the estimate, for $t \in [0,T_1]$

$$N_l(t) \leq C_0 \|W^{*0} - W^{*e}\|_{l,2}, \tag{9.4.10}$$

where $C_0 > 1$ is a constant, depending only on $\mathcal{O}_0$, d_1, and b_0, where we have defined $N_l(t) = N_l(0,t)$ and

$$N_l(t_1,t_2)^2 = \sup_{t_1 \leq \tau \leq t_2} \|W^*(\tau) - W^{*e}\|_{l,2}^2 + \int_{t_1}^{t_2} \Big(\|\partial_x W_I^*(\tau)\|_{l-1,2}^2 + \|\partial_x W_{II}^*(\tau)\|_{l,2}^2 \Big)\, d\tau. \tag{9.4.11}$$

This theorem is essentially obtained by first considering linear equations and establishing a priori estimates. Successive iterations are then shown to be convergent, and we refer the reader to [Kaw84] for more details and to [Maj84] [Ser96a] [Ser96b] for the hyperbolic case.

9.4.3. A priori estimates

We first remark that the norm $N_l(T)$ can be used to control the solution in a given neighborhood of W^{*e} in $\mathcal{O}_{W^*}$.

Lemma 9.4.3. *Let $l \geq [d/2] + 1$, and let $\mathcal{B}$ be a bounded neighborhood of W^{*e} in $\mathcal{O}_{W^*}$. Then there exists a constant $\beta_0(\mathcal{B})$ such that, if the solution W^* satisfies $N_l(T) \leq \beta_0(\mathcal{B})$, then $W^*(t,x) \in \mathcal{B}$ for any $(t,x) \in [0,T] \times \mathbb{R}^d$.*

In particular, when $N_l(T) \le \beta_0(\mathfrak{O})$, *then inequality* (9.2.3) *of Proposition* 9.2.1 *and the estimate of* (Dis_4) *hold.*

We now want to estimate the quantity $N_l(T)$ when it satisfies a smallness assumption. We first restate a result of Kawashima concerning the norms of higher derivatives of W_I^* and W_{II}^*. In the next lemmas, we complete this estimate by considering lower derivatives. These estimates involve the orthogonal projector $\mathcal{P}$ onto the range of $\overline{L}^*(W^{*e})$. The following result has been established by Kawashima [Kaw84] and is given here without proof.

Lemma 9.4.4. *Assume that Properties* (Edp_1)–(Edp_4) *and* (Dis_1)–(Dis_2) *hold. Let* $d \ge 1$ *and* $l \ge [d/2]+2$ *be integers, and let the initial data satisfy* $W^{*0} - W^{*e} \in W_2^l(\mathbb{R}^d)$. *Let* b *be positive, and let* W^* *be a solution of* (9.4.2) (9.4.3) *satisfying*

$$\begin{cases} W_I^* - W_I^{*e} \in C^0\big([0,T]; W_2^l(\mathbb{R}^d)\big) \cap C^1\big([0,T]; W_2^{l-1}(\mathbb{R}^d)\big), \\ W_{II}^* - W_{II}^{*e} \in C^0\big([0,T]; W_2^l(\mathbb{R}^d)\big) \cap C^1\big([0,T]; W_2^{l-2}(\mathbb{R}^d)\big), \end{cases}$$

and

$$\begin{cases} W_I^* - W_I^{*e} \in L^2\big(0,T; W_2^l(\mathbb{R}^d)\big), \\ W_{II}^* - W_{II}^{*e} \in L^2\big(0,T; W_2^{l+1}(\mathbb{R}^d)\big), \end{cases}$$

and $N_l(T) \le b$. *Then there exists a constant* $C_1 = C_1(b)$ *such that the following estimate holds for* $t \in [0,T]$:

$$\begin{aligned} &\|\partial_x W^*(t)\|_{l-1,2}^2 \\ &\quad + \int_0^t \Big(\|\mathcal{P}\partial_x W^*(\tau)\|_{l-1,2}^2 + \|\partial_x^2 W_{II}^*(\tau)\|_{l-1,2}^2 + \|\partial_x^2 W_I^*(\tau)\|_{l-2,2}^2 \Big)\, d\tau \\ &\qquad\qquad \le C_1 \Big(\|\partial_x W^{*0}\|_{l-1,2}^2 + N_l(T)^3 \Big). \end{aligned} \tag{9.4.12}$$

We next estimate $\|W^* - W^{*e}\|_{0,2}^2 + \int_0^t (\|\partial_x W_{II}^*(\tau)\|_{0,2}^2 + \|\overline{\Omega}^*(\tau)\|_{0,2}^2)\, d\tau$ by means of the modified entropy function $\mathfrak{h}$.

Proposition 9.4.5. *Keep the hypotheses of Lemma* 9.4.4, *and assume that* (Dis_4) *holds and that* $b \le \beta_0(\mathfrak{O})$. *Then there exists a positive constant* $C_2 = C_2(b)$ *such that the following estimate holds for* $t \in [0,T]$:

$$\begin{aligned} &\|W^*(t) - W^{*e}\|_{0,2}^2 + \int_0^t \Big(\|\partial_x W_{II}^*(\tau)\|_{0,2}^2 + \|\overline{\Omega}^*(\tau)\|_{0,2}^2 \Big)\, d\tau \\ &\qquad\qquad \le C_2 \Big(\|W^{*0} - W^{*e}\|_{0,2}^2 + N_l(T)^3 \Big). \end{aligned} \tag{9.4.13}$$

Proof. Integrating the conservation equation (9.2.4) for $\mathfrak{h}$ over $[0,T]\times\mathbb{R}^d$ using inequality (9.2.3) and (Dis_4), we easily obtain

$$\|W^*(t) - W^{*\mathrm{e}}\|_{0,2}^2 + \sum_{i,j\in C^*} \int_0^t\!\!\int_{\mathbb{R}^d} \langle \partial_i V^*, \widetilde{B}_{ij}^* \partial_j V^*\rangle \, dx d\tau$$

$$+ \int_0^t\!\!\int_{\mathbb{R}^d} \big| \widetilde{\Omega}^*(V^*)\big|^2 \, dx d\tau \;\le\; C\|W^{*0} - W^{*\mathrm{e}}\|_{0,2}^2. \tag{9.4.14}$$

From Property (Nor_1) we also have $\langle \partial_i V^*, \widetilde{B}_{ij}^* \partial_j V^*\rangle = \langle \partial_i V_{II}^{*\prime}, \widetilde{B}_{ij}^{*\prime II,II} \partial_j V_{II}^{*\prime}\rangle$ since $V^{*\prime} = P^{-1}V^*$ and $\widetilde{B}_{ij}^{*\prime} = P^t \widetilde{B}_{ij}^* P$, and there exists constants c and C such that

$$\sum_{i,j\in C^*} \int_0^t\!\!\int_{\mathbb{R}^d} \langle \partial_i V_{II}^{*\prime}, \widetilde{B}_{ij}^{*\prime II,II} \partial_j V_{II}^{*\prime}\rangle \, dx d\tau \;\ge\; c \int_0^t\!\!\int_{\mathbb{R}^d} |\partial_x V_{II}^{*\prime}|^2 \, dx d\tau \;-\; C\, N_l(t)^3,$$

since $\sum_{i,j\in C^*} \widetilde{B}_{ij}^{*\prime II,II}(W^{*\mathrm{e}}) w_i w_j$ is positive definite. The proof is then complete, since $W_{II}^* = \phi_{II}(V_{II}^{*\prime})$, where ϕ_{II} is a diffeomorphism of $\mathbb{R}^{n-n_0}$ and $\overline{\Omega}^* = (\partial_{W^*} V^*)^t\, \widetilde{\Omega}^*$. ■

9.4.4. More a priori estimates

We now focus on a priori estimates concerning $\int_0^t \|\partial_x W_I^*(\tau)\|_{0,2}^2 d\tau$.

Lemma 9.4.6. *Keep the assumptions of Lemma* 9.4.4. *Then there exists a positive constant* $C_3 = C_3(b)$ *such that the following estimate holds for* $t \in [0,T]$:

$$\int_0^t \|\partial_x W_I^*(\tau)\|_{0,2}^2 d\tau - C_3 \bigg(\|W^*(t) - W^{*\mathrm{e}}\|_{1,2}^2 + \int_0^t \|\partial_x \mathcal{P} W^*(\tau)\|_{0,2}^2 \, d\tau$$

$$+ \int_0^t \Big(\|\partial_x W_{II}^*(\tau)\|_{1,2}^2 + \|\overline{\Omega}^*(\tau)\|_{0,2}^2 \Big) \, d\tau \bigg)$$

$$\le C_3 \Big(\|W^{*0} - W^{*\mathrm{e}}\|_{1,2}^2 + N_l(T)^3 \Big). \tag{9.4.15}$$

Proof. We rewrite the system (9.4.2) in the form

$$\overline{A}_0(W^{*\mathrm{e}}) \partial_t W^* + \sum_{i\in C} \overline{A}_i^*(W^{*\mathrm{e}}) \partial_i W^* - \sum_{i,j\in C} \overline{B}_{ij}^*(W^{*\mathrm{e}}) \partial_i \partial_j W^*$$

$$- \overline{A}_0^*(W^{*\mathrm{e}}) \big(\overline{A}_0^*(W^*)\big)^{-1} \overline{\Omega}^* = \mathrm{h}, \tag{9.4.16}$$

where

$$\mathrm{h} = \sum_{i\in C^*} \overline{A}_0^*(W^{*e})\Big[(\overline{A}_0^*(W^{*e}))^{-1}\overline{A}_i(W^{*e}) - (\overline{A}_0^*(W^*))^{-1}\overline{A}_i^*(W^*)\Big]\partial_i W^*$$

$$-\sum_{i,j\in C^*} \overline{A}_0^*(W^{*e})\Big[(\overline{A}_0^*(W^{*e}))^{-1}\overline{B}_{ij}^*(W^{*e}) - (\overline{A}_0^*(W^*))^{-1}\overline{B}_{ij}(W^*)\Big]\partial_i\partial_j W^*$$

$$+ \overline{A}_0^*(W^{*e})(\overline{A}_0^*(W^*))^{-1}\Big(\sum_{i,j\in C} \partial_{W^*}\overline{B}_{ij}^*(W^*)\partial_i W^*\partial_j W^* + \overline{\mathcal{T}}^*\Big).$$

Multiplying this equation by K^j, taking the scalar product with $\partial_j W^*$, summing with respect to j, and integrating over $[0,t]\times\mathbb{R}^d$ yields the estimate (9.4.15) up to the last term, which reads $\int_0^t \mathcal{R}(\tau)\,d\tau$ with $\mathcal{R}$ given by

$$\mathcal{R}(t) = \sum_{j\in C^*}\int_{\mathbb{R}^d} \langle \partial_j W^*, K^j \mathrm{h}\rangle\, dx. \tag{9.4.17}$$

This results from the skew-symmetry of $K^j\overline{A}_0^*(W^{*e})$, which implies that

$$\Big\langle \partial_j W^*,\, K^j\overline{A}_0^*(W^{*e})\partial_t W^*\Big\rangle = \tfrac{1}{2}\partial_t\Big\langle \partial_j W^*,\, K^j\overline{A}_0^*(W^{*e})(W^* - W^{*e})\Big\rangle$$

$$+\ \tfrac{1}{2}\partial_j\Big\langle (W^* - W^{*e}),\, K^j\overline{A}_0^*(W^{*e})\partial_t W^*\Big\rangle,$$

from Property (Dis$_2$) and from the structure of $\overline{B}^*(W^{*e})_{ij}$, $i,j\in C^*$, and $\overline{L}^*(W^{*e})$, which yield

$$\sum_{i,j\in C^*}\int_{\mathbb{R}^d} \langle \partial_j W^*, K^j\overline{A}_i^*(W^{*e})\partial_i W^*\rangle\, dx \geq c\int_{\mathbb{R}^d} |\partial_x W^*|^2\, dx$$

$$-\ C\int_{\mathbb{R}^d} \big(|\partial_x W_{II}^*|^2 + |\partial_x \mathcal{P} W^*|^2\big)\, dx.$$

We further deduce from expression (9.4.17) that

$$\mathcal{R}(t) = O\Big(\int_{\mathbb{R}^d} \big(|\partial_x W^*|^3 + |W^* - W^{*e}|\,|\partial_x W^*|^2$$

$$+ |W^* - W^{*e}|\,|\partial_x W^*|\,|\partial_x^2 W_{II}^*|\big)\, dx\Big),$$

and, therefore, we have $\int_0^t \mathcal{R}(t)\,dt \leq C N_l(t)^3$, and the proof is complete. ∎

We can now combine the preceding lemmas in order to obtain a priori estimates for $N_l(T)$.

Proposition 9.4.7. *Keep the hypotheses of Lemma* 9.4.4, *and assume that* (Dis$_4$) *holds. Then there exist positive constants* $b_4 \le \beta_0(\mathfrak{O})$ *and* C_4, *depending only on* $\beta_0(\mathfrak{O})$, *such that, whenever* $N_l(T) \le b_4$, *then the following a priori estimate holds for* $t \in [0,T]$:

$$N_l(t) \le C_4 \|W^{*0} - W^{*e}\|_{l,2}. \tag{9.4.18}$$

Proof. Letting $b = \beta_0(\mathfrak{O})$ in Lemma 9.4.4, Proposition 9.4.5 and Lemma 9.4.6 yield the inequality $N_l(T)^2 \le C_4\big(\|W^{*0} - W^{*e}\|_{l,2}^2 + N_l(T)^3\big)$. As a consequence, for $N_l(T) < 1/(2C_4)$, we obtain the estimate (9.4.18) . ∎

9.4.5. Global existence proof

We now use repeatedly the local existence theorem and the a priori estimates [Kaw84]. We apply Theorem 9.4.2 with $\mathcal{O}_0 = \mathfrak{O}$, $d_1 < d(\mathfrak{O}, \partial\mathcal{O}_{W^*})$ and $b_0 = \beta_0(\mathfrak{O})$, and we assume that W^{*0} satisfies

$$\|W^{*0} - W^{*e}\|_{l,2} \le \bar{b} = \min\left(\frac{b_4}{C_0}, \frac{b_4}{C_4(1+C_0^2)^{\frac{1}{2}}}\right). \tag{9.4.19}$$

The solution is then defined on $[0,T_1]$, and, from (9.4.10), we obtain $N_l(t) \le C_0\bar{b} \le b_4$ over $[0,T_1]$. As a consequence, the estimates obtained in Proposition 9.4.7 hold and $N_l(T_1) \le C_4\bar{b}$. We then apply Theorem 9.4.2 again on $[T_1, 2T_1]$, with T_1 as a new initial time, still with $\mathcal{O}_0 = \mathfrak{O}$, and with the same d_1 and b_0. The solution on $[T_1, 2T_1]$ then satisfies the estimate $N_l(T_1, 2T_1) \le C_0 N_l(T_1)$. As a consequence, we obtain $N_l(2T_1) \le \big(N_l(T_1)^2 + N_l(T_1, 2T_1)^2\big)^{\frac{1}{2}} \le (1 + C_0^2)^{\frac{1}{2}} N_l(T_1)$, so that

$$N_l(2T_1) \le (1 + C_0^2)^{\frac{1}{2}} C_4 \bar{b} \le b_4 \le \beta_0(\mathfrak{O}).$$

The estimates of Proposition 9.4.7 are therefore again valid on $[0, 2T_1]$, and we can use Theorem 9.4.2 again on $[2T_1, 3T_1]$. An easy induction then yields global existence and $N_l(t) \le C_4\|W^{*0} - W^{*e}\|_{l,2}$ for $t \ge 0$.

Defining then $\Phi(t) = \|\partial_x W^*(t)\|_{l-2,2}^2$, it follows from the a priori estimates and the relations (9.4.2) that

$$\int_0^\infty |\Phi(t)|\, dt + \int_0^\infty |\partial_t \Phi(t)|\, dt \;\le\; C\|W^{*0} - W^{*e}\|_{l,2}^2,$$

which shows that $\Phi(t) \to 0$ as $t \to \infty$. From the interpolation inequality

$$\sup_{\mathbb{R}^d} |W^*(t) - W^{*e}| \;\le\; C\,\|\partial_x^{l-1} W^*(t)\|_{0,2}^{a}\, \|W^*(t)\|_{0,2}^{1-a},$$

with $a = d/\big(2(l-1)\big) > 0$, we then deduce that $\sup_{\mathbb{R}^d} |W^*(t) - W^{*\mathrm{e}}| \to 0$ as $t \to \infty$ as was to be shown. ∎

9.5. Decay estimates

In this section we establish decay estimates towards the equilibrium state.

Theorem 9.5.1. *Assume that Properties* (Edp$_1$)–(Edp$_4$) *and* (Dis$_1$)–(Dis$_4$) *hold. Let $d \geq 1$ and $l \geq [d/2] + 3$ be integers, and let $W^{*0}(x)$ be such that*

$$W^{*0} - W^{*\mathrm{e}} \in W_2^l(\mathbb{R}^d). \tag{9.5.1}$$

Assume that $\|W^{*0} - W^{*\mathrm{e}}\|_{l,2}$ is small enough so that there exists a unique global solution to the Cauchy problem from Theorem 9.4.1. Further assume that $W^{*0} - W^{*\mathrm{e}} \in W_2^l(\mathbb{R}^d) \cap L^p(\mathbb{R}^d)$, where $p = 1$, if $d = 1$, and $p \in [1,2)$, if $d \geq 2$. Defining for $\tau \leq l$

$$\|W^{*0} - W^{*\mathrm{e}}\|_{\tau,2|p} = \|W^{*0} - W^{*\mathrm{e}}\|_{\tau,2} + \|W^{*0} - W^{*\mathrm{e}}\|_{0,p}, \tag{9.5.2}$$

then, if $\|W^{*0} - W^{*\mathrm{e}}\|_{l,2|p}$ is small enough, the unique global solution to the Cauchy problem satisfies the decay estimate

$$\|W^*(t) - W^{*\mathrm{e}}\|_{l-2,2} \leq C(1+t)^{-\gamma} \|W^{*0} - W^{*\mathrm{e}}\|_{l-2,2|p}, \tag{9.5.3}$$

for $t \in [0, \infty)$, where C a positive constant and $\gamma = d \times (1/2p - 1/4)$.

Proof. Introducing the new variable $z(U^*) = \big(\partial_{U^*} W^*(U^{*\mathrm{e}})\big)\,(U^* - U^{*\mathrm{e}})$, we obtain from (9.4.2)

$$\overline{A}_0^*(W^{*\mathrm{e}})\partial_t z + \sum_{i \in C^*} \overline{A}_i^*(W^{*\mathrm{e}})\partial_i z - \sum_{i,j \in C^*} \overline{B}_{ij}^*(W^{*\mathrm{e}})\partial_i \partial_j z$$
$$+ \ \overline{L}^*(W^{*\mathrm{e}}) z \ = \ \mathrm{h}^1 + \mathrm{h}^2, \tag{9.5.4}$$

and $z(0,x) = z(U^{*0})$. The first nonlinear term h^1 is in conservative form and reads as

$$\mathrm{h}^1 = \sum_{i \in C^*} \partial_i \left(\mathfrak{g}_i + \mathfrak{g}_i'\right),$$

where

$$\begin{cases} \mathfrak{g}_i = \big(\partial_{W^*} V^*(W^{*\mathrm{e}})\big)^t \big(F_i^*(W^{*\mathrm{e}}) - F_i^*(W^*)\big) + \overline{A}_i^*(W^{*\mathrm{e}})\, z, \\ \mathfrak{g}_i' = \displaystyle\sum_{j \in C^*} \Big(\big(\partial_{W^*} V^*(W^{*\mathrm{e}})\big)^t B_{ij}^*(U^*)\, \partial_{W^*} U^*(W^{*\mathrm{e}}) - \overline{B}_{ij}^*(W^{*\mathrm{e}})\Big)\, \partial_j z. \end{cases}$$

On the other hand, the second term $\mathfrak{h}^2$ reads as

$$\mathfrak{h}^2 = \big(\partial_{W^*} V^*(W^{*e})\big)^t\, \widetilde{\Omega}^* \;+\; \overline{L}^*(W^{*e})\, z$$

and lies in the subspace $R\big(\overline{L}^*(W^{*e})\big)$, thanks to (Dis$_3$), so that we have $(I-\mathcal{P})\,\mathfrak{h}^2 = 0$.

We now introduce the symbol associated with the linear part of (9.5.4)

$$\begin{aligned}\overline{S}(\xi) = \overline{A}_0^*(W^{*e})^{-1/2}\Big\{\mathrm{i}\sum_{i\in C^*} \overline{A}_i^*(W^{*e})\xi_i + \sum_{i,j\in C^*} \overline{B}_{ij}^*(W^{*e})\xi_i\xi_j \\ + \overline{L}^*(W^{*e})\Big\}\overline{A}_0^*(W^{*e})^{-1/2},\end{aligned}$$

and we define for $\phi \in L^2(\mathbb{R}^d)$

$$\exp(-t\overline{S})\phi(x) \;=\; \frac{1}{(2\pi)^{d/2}}\int_{\mathbb{R}^d} \exp(\mathrm{i}x\cdot\xi)\, \exp\big(-t\overline{S}(\xi)\big)\, \hat{\phi}(\xi)\, d\xi,$$

where $\hat{\phi}$ is the Fourier transform of ϕ. We can then express the solution of system (9.5.4) by the relation [Kaw84]

$$\begin{aligned} z(t) = \exp(-t\overline{S})z(0) + \int_0^t \exp(-(t-\tau)\overline{S})\mathfrak{h}^1(\tau)\, d\tau \\ + \int_0^t \exp(-(t-\tau)\overline{S})\mathfrak{h}^2(\tau)\, d\tau, \end{aligned} \tag{9.5.5}$$

and we estimate $\|z(t)\|_{l-2,2}$ in two steps. We first deduce from Proposition 3.12 of [Kaw84] the estimate

$$\begin{aligned} \|\exp(-t\overline{S})z(0) + \int_0^t \exp(-(t-\tau)\overline{S})\mathfrak{h}^2(\tau)d\tau\|_{l-2,2} \\ \leq\; \beta(1+t)^{-\gamma}\|z(0)\|_{l-2,2|p}, \end{aligned} \tag{9.5.6}$$

where $\gamma = d\times(1/2p - 1/4)$ and β denotes a positive constant, since the nonlinear term $\mathfrak{h}^2$ satisfies $(I-\mathcal{P})\,\mathfrak{h}^2 = 0$.

On the other hand, by using the estimate (3.A.14) of [Kaw84] or Theorem 1.2 of [SK85], we also obtain

$$\begin{aligned} \|\int_0^t \exp(-(t-\tau)\overline{S})\mathfrak{h}^1(\tau)\, d\tau\|_{l-2,2} \;\leq\; \beta\int_0^t \exp(-\delta(t-\tau))\|\mathfrak{h}^1(\tau)\|_{l-2,2}\, d\tau \\ + \beta\int_0^t (1+t-\tau)^{-d/4-1/2}\Big(\sum_{i\in C^*}\|\,(\mathfrak{g}_i+\mathfrak{g}_i')\,(\tau)\|_{0,1}\Big)\, d\tau, \end{aligned} \tag{9.5.7}$$

where δ and β are positive constants. Since the initial data (9.4.3) is supposed to be sufficiently near W^{*e} in $W_2^l(\mathbb{R}^d) \cap L^p(X)$, the a priori estimates show that $\|z\|_{l,2}$ remains small for any time. In particular, we can write that $\mathsf{g}_i = O(|z|^2)$ and $\mathsf{g}'_i = O(|\partial_x z|)\, O(|z|)$, so that, for $l \geq [d/2] + 3$, and $\|z\|_{l,2}$ sufficiently small, we obtain

$$\sum_{i \in C^*} \Big(\|\mathsf{g}_i\|_{0,1} + \|\mathsf{g}'_i\|_{0,1} \Big) \leq \beta \|z\|_{1,2}^2, \tag{9.5.8}$$

and, similarly, we can show that

$$\sum_{i \in C^*} \Big(\|\partial_i \mathsf{g}_i\|_{l-2,2} + \|\partial_i \mathsf{g}'_i\|_{l-2,2} \Big) \leq \beta \|z\|_{l-2,2} \|z\|_{l,2}. \tag{9.5.9}$$

Combining (9.5.5) with inequalities (9.5.6), (9.5.7), (9.5.8), and (9.5.9) finally yields

$$\begin{aligned} \|z(t)\|_{l-2,2} \leq\ & \beta(1+t)^{-\gamma} \|z(0)\|_{l-2,2|p} \\ & + \beta \int_0^t \exp\big(-\delta(t-\tau)\big) \|z(\tau)\|_{l,2} \|z(\tau)\|_{l-2,2}\, d\tau \\ & + \beta \int_0^t (1+t-\tau)^{-d/4-1/2} \|z(\tau)\|_{1,2}^2\, d\tau. \end{aligned} \tag{9.5.10}$$

Defining then $|||z(t)|||_{l-2,\gamma} = \sup_{0 \leq \tau \leq t} (1+\tau)^{\gamma} \|z(\tau)\|_{l-2,2}$, we obtain

$$\begin{aligned} |||z(t)|||_{l-2,\gamma} \leq \beta \|z(0)\|_{l-2,2|p} & + \beta \mu_3(t) \|z(0)\|_{l,2} |||z(t)|||_{l-2,\gamma} \\ & + \beta \mu_4(t) |||z(t)|||_{l-2,\gamma}^2, \end{aligned}$$

where

$$\begin{cases} \mu_3(t) = \displaystyle\sup_{0 \leq \tau \leq t} (1+\tau)^{\gamma} \int_0^{\tau} \exp\big(-\delta(\tau-\tau_1)\big)(1+\tau_1)^{-\gamma}\, d\tau_1, \\[2ex] \mu_4(t) = \displaystyle\sup_{0 \leq \tau \leq t} (1+\tau)^{\gamma} \int_0^{\tau} (1+\tau-\tau_1)^{-d/4-1/2}(1+\tau_1)^{-2\gamma}\, d\tau_1. \end{cases}$$

Since $\mu_3(t)$ and $\mu_4(t)$ are uniformly bounded with respect to t, we obtain the desired estimate for $|||z(0)|||_{l-2,\gamma}$ small enough, and this completes the proof. ∎

9.6. Local dissipativity for multicomponent reactive flows

In this section, we establish that Theorems 9.4.1 and 9.5.1 apply to the multicomponent reactive flow governing equations (9.4.1). To this end, we assume that Properties (Th_1)–(Th_4) and (Tr_1)–(Tr_{4a}) hold. Note that Properties (Tr_{4b}) and (Tr_5) are not needed in this chapter since we only investigate neighborhoods of equilibrium states. We have established in Chapter 8 that the system 9.2.3 is symmetrizable into (8.6.4) and have obtained various normal forms in Theorems 8.7.3, 8.7.5, and 8.7.7. In order to apply Theorems 9.4.1 and 9.5.1, it is now necessary to establish the existence of equilibrium states, investigate the corresponding linearized source terms, and study the local dissipative structure.

9.6.1. Chemical sources

In order to investigate asymptotic stability of constant equilibrium states for multicomponent reactive flows, the source term is taken to be

$$\Omega = (m_1\omega_1, \ldots, m_n\omega_n, 0, 0, 0, 0)^t, \tag{9.6.1}$$

where m_k is the molar mass of the kth species and ω_k is the molar production rate of the kth species. More specifically, we no longer consider force terms as gravity in Ω and only chemistry effects are included in the model. The molar production rates that we consider are the Maxwellian production rates presented in Chapters 2 and 6. Keeping the notation introduced in these chapters, we have, in particular,

$$\Omega = (m_1\omega_1, \ldots, m_n\omega_n, 0, 0, 0, 0)^t \in M\mathcal{R} \times 0_{\mathbb{R}^4}, \tag{9.6.2}$$

where $M = \mathrm{diag}(m_1, \ldots, m_n)$ is the mass weight matrix (6.4.11) and $\mathcal{R}$ is the linear space spanned by the stoichiometic vectors ν_i, $i \in R$.

Various symmetrized forms of the governing equations have been investigated in Chapter 8. We present here some additional properties of the corresponding source terms. We first observe that the source term $\widetilde{\Omega}$ of the conservative symmetric form (8.6.4) is identical to that of the original formulation, that is, $\widetilde{\Omega} = \Omega$. As a consequence, we have

$$\Omega = \widetilde{\Omega} = (m_1\omega_1, \ldots, m_n\omega_n, 0, 0, 0, 0)^t \in M\mathcal{R} \times 0_{\mathbb{R}^4}. \tag{9.6.3}$$

On the other hand, the source terms for the auxiliary variables $U' = P^tU$ and $V' = P^{-1}V$, where the constant matrix P is given by (8.7.3), are given by $\Omega' = \widetilde{\Omega}' = P^t\Omega$, that is to say,

$$\Omega' = \widetilde{\Omega}' = (0, m_2\omega_2, \ldots, m_n\omega_n, 0, 0, 0, 0)^t. \tag{9.6.4}$$

In particular, Ω' is also in a fixed subspace of $\mathbb{R}^{n+4}$ of dimension $\dim(\mathcal{R})$ since

$$\Omega' \in P^t(M\mathcal{R} \times 0_{\mathbb{R}^4}). \tag{9.6.5}$$

Finally, the source terms $\overline{\Omega}$ corresponding to the variable W is given by

$$\overline{\Omega} = (0,\, m_2\omega_2, \ldots,\, m_n\omega_n,\, 0,\, 0,\, 0,\, -\frac{1}{T^2}\sum_{k\in S} e_k m_k \omega_k)^t, \tag{9.6.6}$$

but is not in a fixed subspace of dimension $\dim(\mathcal{R})$.

9.6.2. Local dissipative structure

We have established the existence of equilibrium points in Proposition 6.5.3 of Chapter 6. As a direct consequence of this Proposition 6.5.3, we obtain the following result, which implies, in particular, Property (Edp_4).

Proposition 9.6.1. *Let a temperature $T^{\mathrm{e}} > 0$, a velocity $\boldsymbol{v}^{\mathrm{e}} \in \mathbb{R}^3$, a mass density vector $\varrho^{\mathrm{f}} > 0$ be given, and assume that Properties* (Th_1)–(Th_4) *hold. Then there exists a unique constant equilibrium state U^{e}*

$$\Omega(U^{\mathrm{e}}) = 0, \tag{9.6.7}$$

in the form $U^{\mathrm{e}} = \big(\rho_1^{\mathrm{e}}, \ldots, \rho_n^{\mathrm{e}}, \rho^{\mathrm{e}} v_1^{\mathrm{e}}, \rho^{\mathrm{e}} v_2^{\mathrm{e}}, \rho^{\mathrm{e}} v_3^{\mathrm{e}}, \mathcal{E}(T^{\mathrm{e}}, \varrho^{\mathrm{e}}) + \frac{1}{2}\rho^{\mathrm{e}} \boldsymbol{v}^{\mathrm{e}}\cdot\boldsymbol{v}^{\mathrm{e}}\big)^t$ and such that $\varrho^{\mathrm{e}} \in (\varrho^{\mathrm{f}} + M\mathcal{R}) \cap (0,\infty)^n$. For this constant stationary state, we have $\mu^{\mathrm{e}} = (\mu_1^{\mathrm{e}}, \ldots, \mu_n^{\mathrm{e}})^t \in (M\mathcal{R})^\perp$ and $Z^{\mathrm{e}} = (\rho_1^{\mathrm{e}}, \ldots, \rho_n^{\mathrm{e}}, v_1^{\mathrm{e}}, v_2^{\mathrm{e}}, v_3^{\mathrm{e}}, T^{\mathrm{e}})^t$.

We will denote by V^{e} and W^{e} the equilibrium states in the variables V and W, respectively. In order to establish a global existence theorem, we further need to investigate the local dissipative structure and establish that Property (Dis_4) holds.

Proposition 9.6.2. *Let $U^{\mathrm{e}} = U(Z^{\mathrm{e}})$ with $Z^{\mathrm{e}} = (\rho_1^{\mathrm{e}}, \ldots, \rho_n^{\mathrm{e}}, v_1^{\mathrm{e}}, v_2^{\mathrm{e}}, v_3^{\mathrm{e}}, T^{\mathrm{e}})^t$ be a constant equilibrium state in $\mathcal{O}_U$ constructed as in Proposition* 9.6.1. *Then we have $V^{\mathrm{e}} \in (M\mathcal{R} \times 0_{\mathbb{R}^4})^\perp$ and there exists a neighborhood $\mathfrak{V}$ of V^{e} and a positive constant δ such that*

$$\delta \big|\, \widetilde{\Omega}(V)\big|^2 \le -\langle V - V^{\mathrm{e}}, \widetilde{\Omega}(V)\rangle, \qquad V \in \mathfrak{V}. \tag{9.6.8}$$

Proof. From the expression (8.6.3) of V, we obtain

$$(V_1, \ldots, V_n)^t = R\mu - \tfrac{1}{2}(\boldsymbol{v}\cdot\boldsymbol{v}/T)\,\mathcal{U}.$$

As a consequence, we have $V^{\mathrm{e}} \in (M\mathcal{R} \times 0_{\mathbb{R}^4})^\perp$ since $\mu^{\mathrm{e}} \in (M\mathcal{R})^\perp$ and $\mathcal{U} \in (M\mathcal{R})^\perp$, and, therefore, $\langle V^{\mathrm{e}}, \widetilde{\Omega}(V)\rangle = 0$ for any $V \in \mathcal{O}_V$. Similarly, we obtain that

$$\langle V, \widetilde{\Omega}(V)\rangle = R\langle \mu, M\omega\rangle.$$

As a consequence, we have

$$\langle V - V^{\mathrm{e}}, \widetilde{\Omega}(V)\rangle = R\langle \mu, M\omega\rangle, \tag{9.6.9}$$

and using Proposition 6.7.1 or Corollary 6.7.2 completes the proof. ∎

9.6.3. Linearized source term

In this section we investigate the linearized chemical source term around equilibrium states and establish Property (Dis_3).

Proposition 9.6.3. *The linearized source term* $\widetilde{L}(V^{\mathrm{e}}) = -(\partial_V \widetilde{\Omega})(V^{\mathrm{e}})$ *around the stationary state* V^{e} *constructed in Proposition* 9.6.1 *is given by*

$$\widetilde{L}(V^{\mathrm{e}}) = \begin{pmatrix} (\mathcal{L}_{kl})_{k,l\in S} & 0_{n\times 4} \\ & \\ 0_{4\times n} & 0_{4\times 4} \end{pmatrix}, \tag{9.6.10}$$

where

$$\mathcal{L}_{kl} = \sum_{i\in R} \Lambda_i \, m_k \nu_{ki} \, m_l \nu_{li}, \qquad k,l \in S, \tag{9.6.11}$$

and

$$\Lambda_i = \frac{K_i^{\mathrm{f}}(T^{\mathrm{e}})}{R} \prod_{k\in S} (\gamma_k^{\mathrm{e}})^{\nu_{ki}^{\mathrm{f}}} = \frac{K_i^{\mathrm{b}}(T^{\mathrm{e}})}{R} \prod_{k\in S} (\gamma_k^{\mathrm{e}})^{\nu_{ki}^{\mathrm{b}}}. \tag{9.6.12}$$

The matrix $\widetilde{L}(V^{\mathrm{e}})$ *can also we written in the form*

$$\widetilde{L}(V^{\mathrm{e}}) = \sum_{i\in R} \Lambda_i \, (M^\star \nu_i^\star) \otimes (M^\star \nu_i^\star), \tag{9.6.13}$$

where $M^\star \nu_i^\star$ *denotes the vector* $M^\star \nu_i^\star = (M\nu_i, 0, 0, 0, 0)^t$. *This matrix* $\widetilde{L}(V^{\mathrm{e}})$ *is symmetric positive semidefinite and satisfies*

$$R\big(\widetilde{L}(V^{\mathrm{e}})\big) = M\mathcal{R} \times 0_{\mathbb{R}^4}, \tag{9.6.14}$$

in such a way that we have $\Omega\big(U(V)\big) = \widetilde{\Omega}(V) \in R\big(\widetilde{L}(V^{\mathrm{e}})\big)$, *for all* $V \in \mathcal{O}_V$.

Proof. Evaluating the matrix $\widetilde{L}(V^{\mathrm{e}})$ is straightforward from the expression (6.4.26) of the Maxwellian source terms. The expression (9.6.13) then shows that $\widetilde{L}(V^{\mathrm{e}})$ is symmetric and yields

$$\big\langle x, \widetilde{L}(V^{\mathrm{e}})x\big\rangle = \sum_{i\in R} \Lambda_i \big\langle M^\star \nu_i^\star, x\big\rangle^2 = \sum_{i\in R} \Lambda_i \Big(\sum_{k\in S} x_k m_k \nu_{ki}\Big)^2, \tag{9.6.15}$$

where $x = (x_1, \dots, x_{n+4})^t$, so that $\widetilde{L}(V^e)$ is positive semidefinite. Furthermore, the nullspace of $\widetilde{L}(V^e)$ is also constituted by the vectors orthogonal to $M^\star \nu_i^\star$, $i \in R$, so that $R(\widetilde{L}(V^e)) = N(\widetilde{L}(V^e))^\perp = M\mathcal{R} \times 0_{\mathbb{R}^4}$. ∎

9.7. Global existence for multicomponent flows

In this section we apply theorems 9.4.1 and 9.5.1 to the multicomponent reactive flow governing equations (8.2.3). To this purpose, we have to establish that Properties (Edp_1)–(Edp_4) and (Dis_1)–(Dis_4) are satisfied. In the previous sections, we have already established that the Properties (Edp_1)–(Edp_4) and (Dis_3)–(Dis_4) hold. Therefore, we only have to investigate the dissipative structure of the linearized normal form and establish that (Dis_1) and (Dis_2) hold. We will use the normal variable W introduced in Theorem 9.7.3, but other normal variables could be used as well.

9.7.1. Linearized normal form

If we linearize the symmetric hyperbolic–parabolic system (8.7.11) (8.7.12) around the constant stationary state W^e, we obtain the linear symmetric system

$$\overline{A}_0(W^e)\partial_t z + \sum_{i\in C} \overline{A}_i(W^e)\partial_i z = \sum_{i,j\in C} \overline{B}_{ij}(W^e)\partial_i\partial_j z - \overline{L}(W^e)z. \tag{9.7.1}$$

The relation $\overline{\Omega} = (\partial_W V)^t \widetilde{\Omega}$ then yields by linearization

$$\overline{L}(W^e) = \big(\partial_W V\,(W^e)\big)^t\ \widetilde{L}(V^e)\ \partial_W V(W^e), \tag{9.7.2}$$

taking into account that $\widetilde{\Omega}(V^e) = 0$. As a consequence, $\overline{L}(W^e)$ is also symmetric positive semidefinite. Taking into account that (8.7.11) (8.7.12) is a normal form, we thus obtain that Property (Dis_1) is satisfied. We now investigate the existence of compensating matrices K^j, $j \in C$.

Proposition 9.7.1. *For a sufficiently small positive α, the matrices K^j, $j \in C$, defined by*

$$\sum_{j\in C} \xi_j K^j = \alpha \begin{pmatrix} 0 & 0_{1\times(n-1)} & \boldsymbol{\xi}^t & 0 \\ 0_{(n-1)\times 1} & 0_{(n-1)\times(n-1)} & 0_{(n-1)\times 3} & 0_{(n-1)\times 1} \\ -\boldsymbol{\xi} & 0_{3\times(n-1)} & 0_{3\times 3} & 0_{3\times 1} \\ 0 & 0_{1\times(n-1)} & 0_{1\times 3} & 0 \end{pmatrix} \overline{A}_0(W^e)^{-1}, \tag{9.7.3}$$

where $\boldsymbol{\xi} = (\xi_1, \xi_2, \xi_3)^t$, *are compensating matrices. In particular, the products* $K^j \overline{A}_0(W^{\mathrm{e}})$ *are skew-symmetric and the matrix*

$$\sum_{i,j\in C} K^j \overline{A}_i(W^{\mathrm{e}}) w_i w_j + \overline{B}(W^{\mathrm{e}}, w) + \overline{L}(W^{\mathrm{e}})$$

is positive definite for $w \in \Sigma^2$.

Proof. It is obvious by construction that the products $K^j \overline{A}_0(W^{\mathrm{e}})$, $j \in C$, are skew-symmetric. On the other hand, a direct calculation yields

$$\sum_{i,j\in C} \xi_j K^j \overline{A}_i(W^{\mathrm{e}}) \xi_i = \begin{pmatrix} \dfrac{\alpha T^{\mathrm{e}}}{\Sigma_\rho^{\mathrm{e}}} |\boldsymbol{\xi}|^2 & \dfrac{\alpha T^{\mathrm{e}}}{\rho^{\mathrm{e}}} \mathfrak{p}^{\mathrm{e}t} |\boldsymbol{\xi}|^2 & \alpha(\boldsymbol{v}^{\mathrm{e}} \cdot \boldsymbol{\xi}) \boldsymbol{\xi}^t & \alpha r^{\mathrm{e}} |\boldsymbol{\xi}|^2 \\ 0_{(n-1)\times 1} & 0_{(n-1)\times(n-1)} & 0_{(n-1)\times 3} & 0_{(n-1)\times 1} \\ -\alpha(\boldsymbol{v}^{\mathrm{e}} \cdot \boldsymbol{\xi}) \boldsymbol{\xi} & 0_{3\times(n-1)} & -\alpha \rho^{\mathrm{e}} \boldsymbol{\xi} \otimes \boldsymbol{\xi} & 0_{3\times 1} \\ 0 & 0_{1\times(n-1)} & 0_{1\times 3} & 0 \end{pmatrix}, \tag{9.7.4}$$

where $\mathfrak{p}^{\mathrm{e}}$ is the vector of $\mathbb{R}^{n-1}$ given by

$$\mathfrak{p}_l = \rho_l^{\mathrm{e}} - \frac{\rho_l^{\mathrm{e}} \rho^{\mathrm{e}}}{r_l \, \Sigma_\rho^{\mathrm{e}}}, \qquad l \in \{2, \ldots, n\},$$

and the superscript ^{e} indicates that the corresponding quantity is evaluated at W^{e}. As a consequence, for $\boldsymbol{\xi} \in \Sigma^2$, we have $|\boldsymbol{\xi}| = 1$, and there exists $\beta > 0$ such that

$$\langle x^t, \sum_{ij\in C} \xi_j K^j \overline{A}_i(W^{\mathrm{e}}) \xi_i x \rangle \geq \frac{\alpha T^{\mathrm{e}}}{2\Sigma_\rho^{\mathrm{e}}} \Big(x_1^2 - \beta \sum_{l\in\{2,\ldots,n+4\}} x_l^2 \Big).$$

Using now Property (Nor$_2$), the matrix

$$\sum_{ij\in C} K^j \overline{A}_i(W^{\mathrm{e}}) \xi_i \xi_j + \overline{B}(W^{\mathrm{e}}, \xi)$$

is positive definite for $\boldsymbol{\xi} \in \Sigma^2$ and α sufficiently small. ∎

Remark 9.7.2. The particular normal form in the W variable has additional properties. More specifically, we have the relations

$$K^j \overline{A}_0(W^{\mathrm{e}}) \left(\overline{A}_0(W)\right)^{-1} \overline{\Omega} = 0, \qquad j \in C, \tag{9.7.5}$$

which can be established by a direct calculation. Indeed, from (9.6.6) and the expression of $\overline{A}_0$ in Theorem 8.7.3, we easily get

$$\left(\overline{A}_0(W)\right)^{-1} \overline{\Omega} = R\Big(0, (\frac{\omega_2}{\rho_2} - \frac{\omega_1}{\rho_1}), \ldots, (\frac{\omega_n}{\rho_n} - \frac{\omega_1}{\rho_1}), 0, 0, 0, -\sum_{k\in S} \frac{e_k m_k \omega_k}{\rho c_v}\Big)^t,$$

and, from the sparse structure of $\sum_{j\in C}\xi_j K^j \overline{A}_0(W^{\mathrm{e}})$, we immediately obtain (9.7.5). As a consequence, for this particular formulation, the term involving $\|\overline{\Omega}\|_{0,2}$ is not needed in the estimate (9.4.15). ∎

Remark 9.7.3. Similarly, with the "natural" $\mathcal{W}$ formulation of Theorem 8.7.5, one can establish the particular estimate

$$c|\mathcal{P}(\mathcal{W}-\mathcal{W}^{\mathrm{e}})| \le |(0,\Omega_{II})^t| \le C|\mathcal{P}(\mathcal{W}-\mathcal{W}^{\mathrm{e}})|,$$

where c and C are positive constants. As a consequence, the terms involving the norm $\|(0,\Omega_{II})^t\|_{0,2}$ in a priori estimates can be replaced by the corresponding terms involving the norm $\|\mathcal{P}(\mathcal{W}-\mathcal{W}^{\mathrm{e}})\|_{0,2}$. ∎

9.7.2. Global existence and asymptotic stability

In the previous sections, we have established that Properties (Edp_1)–(Edp_4) and (Dis_1)–(Dis_4) are satisfied. Therefore, Theorem 9.4.1 and Theorem 9.5.1 can be applied to the system (8.7.11) (8.7.12) governing multicomponent reactive flows, written in the $W=(W_I,W_{II})^t$ variable, with the hyperbolic variable

$$W_I = \rho \tag{9.7.6}$$

and the parabolic variable

$$W_{II} = \big(\log(\rho_2^{r_2}/\rho_1^{r_1}),\dots,\log(\rho_n^{r_n}/\rho_1^{r_1}),\, v_1,\, v_2,\, v_3,\, T\big)^t. \tag{9.7.7}$$

Theorem 9.7.4. *Consider the system* (8.7.11) (8.7.12), *let* $d\in\{1,2,3\}$ *and* $l\ge[d/2]+2$, *and let* $W^0(x)$, *such that*

$$W^0 - W^{\mathrm{e}} \in W_2^l(\mathbb{R}^d).$$

Then, if $\|W^0-W^{\mathrm{e}}\|_{l,2}$ *is small enough, there exists a unique global solution to the the Cauchy problem* (8.7.11) (8.7.12)

$$\overline{A}_0\partial_t W + \sum_{i\in C}\overline{A}_i\partial_i W = \sum_{i,j\in C}\partial_i\Big(\overline{B}_{ij}\partial_j W\Big) + \overline{\mathcal{T}} + \overline{\Omega},$$

with initial condition

$$W(0,x) = W^0(x),$$

such that

$$\begin{cases} W_I - W_I^{\mathrm{e}} \in C^0\big([0,\infty); W_2^l(\mathbb{R}^d)\big)\cap C^1\big([0,\infty); W_2^{l-1}(\mathbb{R}^d)\big),\\ W_{II} - W_{II}^{\mathrm{e}} \in C^0\big([0,\infty); W_2^l(\mathbb{R}^d)\big)\cap C^1\big([0,\infty); W_2^{l-2}(\mathbb{R}^d)\big),\end{cases} \tag{9.7.8}$$

and

$$\begin{cases} W_{I} - W_{I}^{\mathrm{e}} \in L^2(0,\infty; W_2^l(\mathbb{R}^d)), \\ W_{II} - W_{II}^{\mathrm{e}} \in L^2(0,\infty; W_2^{l+1}(\mathbb{R}^d)). \end{cases} \tag{9.7.9}$$

Furthermore, W satisfies the estimate

$$\|W(t) - W^{\mathrm{e}}\|_{l,2}^2 + \int_0^t \Big(\|\partial_{\boldsymbol{x}} \rho(\tau)\|_{l-1,2}^2 + \sum_{k\in\{2,\dots,n\}} \|\partial_{\boldsymbol{x}} \log(\rho_k^{r_k}/\rho_1^{r_1})(\tau)\|_{l,2}^2$$
$$+ \|\partial_{\boldsymbol{x}} v(\tau)\|_{l,2}^2 + \|\partial_{\boldsymbol{x}} T(\tau)\|_{l,2}^2 \Big)\, d\tau \ \le\ \beta \|W^0 - W^{\mathrm{e}}\|_{l,2}^2,$$

where β is a positive constant and $\sup_{x\in\mathbb{R}^d} |W(t) - W^{\mathrm{e}}|$ goes to zero as $t \to \infty$.

Theorem 9.7.5. *Keeping the assumptions of the preceding theorem, assume that $l \ge [d/2] + 3$ and $W^0 - W^{\mathrm{e}} \in W_2^l(\mathbb{R}^d) \cap L^p(\mathbb{R}^d)$, with $p = 1$, if $d = 1$, and $p \in [1,2)$, if $d \ge 2$. Then, if $\|W^0 - W^{\mathrm{e}}\|_{l,2|p}$ is small enough, the unique global solution to the Cauchy problem satisfies the decay estimate*

$$\|W(t) - W^{\mathrm{e}}\|_{l-2,2} \le \beta(1+t)^{-\gamma} \|W^0 - W^{\mathrm{e}}\|_{l-2,2|p}, \qquad t \in [0,\infty),$$

where β is a positive constant and $\gamma = d \times (1/2p - 1/4)$.

9.8. Notes

9.1. The results presented in this chapter could be generalized in a number of ways. A first idea would be to treat the half space case, say, with Dirichlet/Neumann boundary conditions at the boundary, and then the case of smooth bounded domains. A typical application could be that of tanks filled with gas mixtures. A second extension could include catalytic effects on the tank boundary, requiring the introduction of surfacic adsorbed species unknowns.

9.2. Various models of mathematical physics could also be investigated by using the abstract theorem. We mention, in particular, gas mixtures in full vibrational nonequilibrium [GM98b].

9.3. Another type of existence theorem for hyperbolic–parabolic symmetric composite systems is due to Vol'pert and Hudjaev [VH72]. These authors have used the spaces $W_2^l(\mathbb{R}^d)$ and $V^l(\mathbb{R}^d)$, a stronger assumption than the Legendre–Hadamard condition (Nor2), and obtained local existence theorems. Such a theorem has been applied to multicomponent flows in full vibrational desequilibrium in [GM98b].

9.9. References

[Ada75] R. A. Adams, *Sobolev Spaces,* Academic Press, Orlando, (1975).

[GM96a] V. Giovangigli and M. Massot, *Les Mélanges Gazeux Réactifs, (I) Symétrisation et Existence Locale,* C. R. Acad. Sci. Paris, **323**, Série I, (1996), pp. 1153–1158.

[GM96b] V. Giovangigli and M. Massot, *Les Mélanges Gazeux Réactifs, (II) Stabilité Asymptotique des États d'Équilibres,* C. R. Acad. Sci. Paris, **323**, Série I, (1996), pp. 1207–1212.

[GM98a] V. Giovangigli and M. Massot, *Asymptotic Stability of Equilibrium States for Multicomponent Reactive Flows,* Math. Mod. Meth. Appl. Sci., **8**, (1998), pp. 251–297.

[GM98b] V. Giovangigli and M. Massot, *The Local Cauchy Problem for Multicomponent Reactive Flows in Full Vibrational Nonequilibrium,* Math. Meth. Appl. Sci., **21**, (1998), pp. 1415–1439.

[Kaw84] S. Kawashima, *Systems of Hyperbolic-Parabolic Composite Type, with Application to the Equations of Magnetohydrodynamics,* Doctoral Thesis, Kyoto University, (1984).

[Maj84] A. Majda, *Compressible Fluid Flow and Systems of Conservation Laws in Several Space Variables,* Appl. Math. Sci., **53**, Springer-Verlag, New York, (1984).

[Ser96a] D. Serre, *Systèmes de Loi de Conservation I,* Diderot, Ed., Arts et Sciences, Paris, (1996).

[Ser96b] D. Serre, *Systèmes de Loi de Conservation II,* Diderot, Ed., Arts et Sciences, Paris, (1996).

[SK85] Y. Shizuta and S. Kawashima, *Systems of Equations of Hyperbolic-Parabolic Type with Applications to the Discrete Boltzmann Equation,* Hokkaido Math. J., **14**, (1985), pp. 249–275.

[VH72] A. I. Vol'pert and S. I. Hudjaev, *On the Cauchy Problem for Composite Systems of Nonlinear Differential Equations,* Math USSR Sbornik, **16**, (1972), pp. 517–544.

10

Chemical Equilibrium Flows

10.1. Introduction

We investigate in this chapter the governing equations for gas mixtures which are locally at chemical equilibrium. Chemical equilibrium flows are a limiting model which is of interest for various applications, such as chemical vapor deposition reactors [Gok88], flows around space vehicles [And89] [MVZ97], or diverging nozzle rocket flows [Wil85]. These simplified models provide reasonable predictions when the characteristic chemical times are small in comparison with the flow time, and the associated computational costs are significantly reduced in comparison with chemical nonequilibrium models.

The equations governing chemical equilibrium flows may be derived by using two different methods. A first possibility is to take into account chemical equilibrium directly at the molecular level, that is, at the Boltzmann level. This model is the kinetic equilibrium regime introduced in [LH60] and further investigated in [EG98a]. In this regime, by using an Enskog expansion, the corresponding macroscopic equations at chemical equilibrium are obtained. A second possibility is to start directly from the macroscopic equations at chemical nonequilibrium presented in the previous chapters and to superimpose chemical equilibrium in these equations. It turns out, however, that both methods lead to the same conservations equations, transport fluxes, thermodynamic functions, and qualitative properties of transport coefficients [EG98a]. The resulting model is investigated in this chapter. Still note that these two methods yield different quantitative values for the transport coefficients [EG98a].

For the sake of clarity, we present the governing equations by using the second method, that is, we superimpose chemical equilibrium in the conservation equations investigated in the previous chapters. To this end, we project the nonequilibrium equations onto the zero source term linear space and write the algebraic equilibrium constraint. The equilibrium constraint is then used to eliminate the reactive part of the conservative variable and reduce the equilibrium equations to a set of partial differential equations. The thermochemistry and transport properties used in this chapter are those presented in Chapters 6 and 7.

We then investigate the mathematical structure of the resulting set of partial differential equations. This structure is closely related to that of chemical nonequilibrium flows. We show that the system is symmetrizable and admits an entropy function and relate these quantities to those of chemical nonequilibrium flows. Similarly, we investigate normal forms and linearized systems around constant states. Finally, we obtain global existence results around constant states.

10.2. Governing equations

We first introduce some convenient notation and present the equations at chemical equilibrium. We then investigate the corresponding convective and diffusive fluxes. Throughout this chapter, we assume that Properties (Th_1)–(Th_4) of Chapter 6 and (Tr_1)–(Tr_{4a}) of Chapter 7 hold. Note that Properties (Tr_{4b}) and (Tr_5) are not needed since we only investigate chemical equilibrium states.

10.2.1. Notation associated with equilibrium

In order to obtain the governing equations for chemical equilibrium flows, we have to complete the notation of Chapter 6. We have denoted by $\mathfrak{a}_{kl}$ the number of lth atom in the kth species and by $\mathfrak{A} = \{1, \dots, n^{\mathrm{a}}\}$ the set of atom indices, where $n^{\mathrm{a}} \geq 1$ is the number of atoms—or elements—in the mixture. The atom vectors are $\mathfrak{a}_l = (\mathfrak{a}_{1l}, \dots, \mathfrak{a}_{nl})^t$, $l \in \mathfrak{A}$, and we introduce as usual the linear space $\mathcal{A} = \mathrm{span}\{\ \mathfrak{a}_l,\ l \in \mathfrak{A}\ \}$. In this chapter, we set $\mathcal{R} = \mathcal{A}^{\perp}$, that is, we consider full chemical equilibrium associated with any reaction scheme, such that the reaction vectors are spanning $\mathcal{A}^{\perp}$.

Denoting by $M = \mathrm{diag}(m_1, \dots, m_n)$ the mass weight matrix, the vectors of element mass per unit species mass are $\widetilde{\mathfrak{a}}_l = \widetilde{m}_l M^{-1} \mathfrak{a}_l$, $l \in \mathfrak{A}$. These vectors $\widetilde{\mathfrak{a}}_l$ are of special interest, since they are orthogonal to chemistry source terms and associated with atom mass concentrations

$$\widetilde{\rho}_l = \langle \widetilde{\mathfrak{a}}_l, \varrho \rangle, \qquad l \in \mathfrak{A}, \tag{10.2.1}$$

which naturally appear as the species variables for chemical equilibrium flows. We assume that the element vectors are independent, so that $\widetilde{\mathfrak{a}}_l$, $l \in \mathfrak{A}$, form a basis of $M^{-1}\mathcal{A}$. Similarly, we complete this vector set by any basis $\widetilde{\nu}_{n^{\mathrm{a}}+1}, \dots, \widetilde{\nu}_n$ of $M\mathcal{R} = (M^{-1}\mathcal{A})^{\perp}$.

In order to manipulate the chemistry and flow equations, we now have to add velocity and energy components to these vectors. To this purpose, we define the basis vectors a_l, $l \in \{1, \ldots, n+4\}$, of $\mathbb{R}^{n+4}$, from

$$a_l = (\widetilde{\mathfrak{a}}_l, 0, 0, 0, 0)^t, \qquad l \in \mathfrak{A}, \tag{10.2.2}$$

$$a_l = (\widetilde{\nu}_l, 0, 0, 0, 0)^t, \qquad l \in S \backslash \mathfrak{A}, \tag{10.2.3}$$

and

$$a_{n+1} = e^{n+1}, \quad a_{n+2} = e^{n+2}, \quad a_{n+3} = e^{n+3}, \quad a_{n+4} = e^{n+4}. \tag{10.2.4}$$

We also define for convenience

$$\mathfrak{A}^\star = \mathfrak{A} \cup \{n+1, n+2, n+3, n+4\}, \tag{10.2.5}$$

$$R = \{n^{\mathrm{a}} + 1, \ldots, n\}, \tag{10.2.6}$$

and we define the augmented mass matrix

$$M^\star = \mathrm{diag}(m_1, \ldots, m_n, 1, 1, 1, 1), \tag{10.2.7}$$

which operates on vectors of $\mathbb{R}^{n+4}$.

We can then define

$$\mathcal{A}^\star = \mathcal{A} \times \mathbb{R}^4, \tag{10.2.8}$$

$$\mathcal{R}^\star = \mathcal{R} \times 0_{\mathbb{R}^4}, \tag{10.2.9}$$

so that

$$(M^\star)^{-1}\mathcal{A}^\star = M^{-1}\mathcal{A} \times \mathbb{R}^4 = \mathrm{span}\{\, a_l,\ l \in \mathfrak{A}^\star \,\}, \tag{10.2.10}$$

$$M^\star\mathcal{R}^\star = M\mathcal{R} \times 0_{\mathbb{R}^4} = \mathrm{span}\{\, a_l,\ l \in R \,\}. \tag{10.2.11}$$

As a consequence, we have the orthogonal sum

$$\mathbb{R}^{n+4} = (M^\star)^{-1}\mathcal{A}^\star \oplus M^\star\mathcal{R}^\star, \tag{10.2.12}$$

so that any $x \in \mathbb{R}^{n+4}$ can be decomposed into

$$x = \xi^{\mathrm{a}} + \xi^{\mathrm{r}}, \tag{10.2.13}$$

with $\xi^{\mathrm{a}} \in (M^\star)^{-1}\mathcal{A}^\star$ and $\xi^{\mathrm{r}} \in M^\star\mathcal{R}^\star$.

We now introduce the linear operators $\Pi^{\mathrm{a}} = \mathbb{R}^{n^{\mathrm{a}}+4} \longrightarrow \mathbb{R}^{n+4}$ and $\Pi^{\mathrm{r}} = \mathbb{R}^{n-n^{\mathrm{a}}} \longrightarrow \mathbb{R}^{n+4}$ whose matrices in the canonical bases are given by

$$\Pi^{\mathrm{a}} = [a_1, \ldots, a_{n^{\mathrm{a}}}, a_{n+1}, a_{n+2}, a_{n+3}, a_{n+4}] \tag{10.2.14}$$

and

$$\Pi^{\mathrm{r}} = [a_{n^{\mathrm{a}}+1}, \ldots, a_n]. \tag{10.2.15}$$

We also introduce the metric matrices $\mathcal{J}^{\mathrm{a}}$ and $\mathcal{J}^{\mathrm{r}}$ of order $n^{\mathrm{a}}+4$ and $n-n^{\mathrm{a}}$, respectively, defined by

$$(\mathcal{J}^{\mathrm{a}})^{-1}_{i,j} = \langle a_i, a_j \rangle, \qquad i,j \in \mathfrak{A}^\star,$$

$$(\mathcal{J}^{\mathrm{r}})^{-1}_{i,j} = \langle a_i, a_j \rangle, \qquad i,j \in R.$$

After a little algebra, it is easily shown that, for any $x \in \mathbb{R}^{n+4}$, the vector $\mathcal{J}^{\mathrm{a}}(\Pi^{\mathrm{a}})^t x$ represents the coordinates of ξ^{a} with respect to the vectors $a_1, \ldots, a_{n^{\mathrm{a}}}$, a_{n+1}, a_{n+2}, a_{n+3}, and a_{n+4}, whereas the vector $\mathcal{J}^{\mathrm{r}}(\Pi^{\mathrm{r}})^t x$ represents the coordinates of ξ^{r} with respect to the vectors $a_{n^{\mathrm{a}}+1}, \ldots, a_n$. One may also easily check that we have the relations

$$\xi^{\mathrm{a}} = \Pi^{\mathrm{a}} \mathcal{J}^{\mathrm{a}} (\Pi^{\mathrm{a}})^t\, x, \qquad \xi^{\mathrm{r}} = \Pi^{\mathrm{r}} \mathcal{J}^{\mathrm{a}} (\Pi^{\mathrm{r}})^t\, x, \tag{10.2.16}$$

in such a way that we have in $\mathbb{R}^{n+4}$

$$I = \Pi^{\mathrm{a}} \mathcal{J}^{\mathrm{a}} (\Pi^{\mathrm{a}})^t + \Pi^{\mathrm{r}} \mathcal{J}^{\mathrm{r}} (\Pi^{\mathrm{r}})^t, \tag{10.2.17}$$

and, by definition of $\mathcal{J}^{\mathrm{a}}$, we also have in $\mathbb{R}^{n^{\mathrm{a}}+4}$

$$\mathcal{J}^{\mathrm{a}} (\Pi^{\mathrm{a}})^t \Pi^{\mathrm{a}} = I. \tag{10.2.18}$$

Finally, note that we have

$$\mathcal{U}^\star = (\mathcal{U}, 0, 0, 0, 0)^t = \sum_{l \in \mathfrak{A}} a_l, \tag{10.2.19}$$

in such a way that $\mathcal{U}^\star \in (M^\star)^{-1} \mathcal{A}^\star$. In particular, we have the relations

$$\rho = \sum_{k \in S} \rho_k = \sum_{l \in \mathfrak{A}} \widetilde{\rho}_l, \tag{10.2.20}$$

since $\rho = \langle \mathcal{U}^\star, U \rangle$ and $\widetilde{\rho}_l = \langle a_l, U \rangle = \langle \widetilde{\mathfrak{a}}_l, \varrho \rangle$, $l \in \mathfrak{A}$.

10.2.2. Atomic species and formation reactions

For investigating chemical equilibrium, we may choose the set of species large enough in such a way that the atomic species are present in the mixture. In order to simplify the presentation, we assume that these atomic species correspond to the first n^{a} species.

It is important then to distinguish between the atomic species mass density ρ_l, $l \in \mathfrak{A}$, and the atom mass density $\widetilde{\rho}_l = \langle \widetilde{\mathfrak{a}}_l, \varrho \rangle$, $l \in \mathfrak{A}$. For instance, in the mixture formed by the $n = 10$ species H, O, N, H_2, O_2, N_2,

OH, HO_2, H_2O, and H_2O_2, associated with the combustion of hydrogen in air, the atomic hydrogen species mass density is ρ_H, whereas the hydrogen atom mass density is

$$\widetilde{\rho}_H = \widetilde{m}_H \Big(\frac{\rho_H}{m_H} + 2\frac{\rho_{H_2}}{m_{H_2}} + \frac{\rho_{OH}}{m_{OH}} + \frac{\rho_{HO_2}}{m_{HO_2}} + 2\frac{\rho_{H_2O}}{m_{H_2O}} + 2\frac{\rho_{H_2O_2}}{m_{H_2O_2}} \Big).$$

Since the atomic species are chosen as the first n^a species, we can consider for $k \in S\backslash\mathfrak{A}$, that is, for $n^a + 1 \leq k \leq n$, the formation reaction vectors

$$\mathfrak{n}_k = (-\mathfrak{a}_{k1}, \ldots, -\mathfrak{a}_{kn^a}, 0, \ldots, 0, 1, 0, \ldots, 0)^t, \qquad k \in S\backslash\mathfrak{A}, \qquad (10.2.21)$$

where $\mathfrak{a}_{kl}$ is the number of lth atom in the kth species and the nonzero component 1 is at the kth place. The space $\mathcal{R}$ is easily shown to be spanned by the vectors $\mathfrak{n}_k$, $k \in S\backslash\mathfrak{A}$. We further assume that the mixture contains stable forms of the elements, as H_2 and O_2 for H and O, for instance. We denote by l' the species index of the stable version of the lth atom and by $\mathfrak{A}'$ the corresponding modified atom indexing set. We can then define similarly the formation vectors $\mathfrak{n}'_k$, $k \in S\backslash\mathfrak{A}'$, where the lth element is replaced by its stable version. We then assume that the energies of formation with respect to the stable elements at zero temperature $\langle (e_1^0, \ldots, e_n^0)^t, M\mathfrak{n}'_k \rangle$, $k \in S\backslash\mathfrak{A}'$, are positive.

10.2.3. Equations at chemical equilibrium

The equations governing chemical equilibrium flows are obtained from the nonequilibrium governing equation (8.2.3) by letting the chemical characteristic times go to zero. More specifically, we let the reaction constants K_i^s, $i \in R$, of (6.4.26) go to infinity, assuming that there are sufficient reactions in such a way that $\mathcal{R} = \mathcal{A}^\perp$. We only present here a formal analysis, the rigorous derivation being out of the scope of this book.

From the thermodynamics investigations of Chapter 6, the species and thermal variables needed to describe chemical equilibrium flows should be the atomic mass densities $\widetilde{\rho}_l = \langle \widetilde{\mathfrak{a}}_l, \varrho \rangle$, $l \in \mathfrak{A}$, and the temperature T, or, equivalently, the total energy per unit volume ρe^{tot}, which have to be completed by the momentum per unit volume in each spatial direction ρv_1, ρv_2, ρv_3. Therefore, the variable components for chemical equilibrium flows should be $\langle a_l, U \rangle$, $l \in \mathfrak{A}^\star$, and, regrouping these components into a single vector $U^a = (\Pi^a)^t U$, we have

$$U^a = (\Pi^a)^t U = \Big(\widetilde{\rho}_1, \ldots, \widetilde{\rho}_{n^a}, \rho v_1, \rho v_2, \rho v_3, \rho e^{\text{tot}}\Big)^t. \qquad (10.2.22)$$

In addition, multiplying the system (8.2.3) by any vector of $\mathcal{M}^{-1}\mathcal{A}^\star$ suppresses the source term and, thus, yields a conservation equation valid as the characteristic chemistry times go to infinity. As a consequence, multiplying (8.2.3) on the left by the linear operator $\Pi^{\rm a}$, we get

$$\partial_t (\Pi^{\rm a})^t U + \sum_{i\in C} \partial_i (\Pi^{\rm a})^t F_i + \sum_{i\in C} \partial_i (\Pi^{\rm a})^t F_i^{\rm dis} = 0. \tag{10.2.23}$$

On the other hand, letting the constants $K_i^{\rm s}$, $i \in R$, of (6.4.26) go to infinity, we obtain

$$\mu = (\mu_1, \dots, \mu_n)^t \in M^{-1}\mathcal{A}, \tag{10.2.24}$$

where $\mu = (\mu_1, \dots, \mu_n)^t$ and $\mu_k = \mathcal{G}_k/RT$, $k \in S$. The equations governing chemical equilibrium flows now appear as a system of partial differential equations (10.2.23) coupled to the algebraic constraints (10.2.24). In the next section, we use these constraints (10.2.24) in order to eliminate the reactive components $U^{\rm r} = (\Pi^{\rm r})^t U$ from U and obtain the set of partial differential equations involving the reduced atomic unknown $U^{\rm a} = (\Pi^{\rm a})^t U$.

10.2.4. Conservative and natural variables

We have already introduced $U^{\rm r} = (\Pi^{\rm r})^t U$, and we have to show that the equilibrium condition defines a smooth function $U^{\rm r} = U^{\rm r}(U^{\rm a})$. Using (10.2.17), a direct consequence will be that U is a function of the conservative variable $U^{\rm a}$ given by

$$U = \Pi^{\rm a} \mathcal{J}^{\rm a} U^{\rm a} + \Pi^{\rm r} \mathcal{J}^{\rm r} U^{\rm r}(U^{\rm a}). \tag{10.2.25}$$

To this purpose, we introduce the auxiliary variables $Z^{\rm a} = (\Pi^{\rm a})^t Z$ and have $Z^{\rm r} = (\Pi^{\rm r})^t Z = U^{\rm r}$. In the following, we first show that $U^{\rm r} = Z^{\rm r}$ is a smooth function of $Z^{\rm a}$ and we investigate the map $Z^{\rm a} \longrightarrow U^{\rm a}$.

Lemma 10.2.1. *The following property holds :*

$$\forall \widetilde{\varrho} \in (0,\infty)^{n^{\rm a}},\ \exists \varrho \in (0,\infty)^n,\ \forall l \in \mathfrak{A},\ \widetilde{\rho}_l = \langle \varrho, \widetilde{\mathfrak{a}}_l \rangle \tag{10.2.26}$$

and there exists a smooth map $Z^{\rm a} \longrightarrow Z^{\rm r}$ defined on $\mathcal{O}_{Z^{\rm a}} = (0,\infty)^{n^{\rm a}} \times \mathbb{R}^3 \times (0,\infty)$ such that $Z = \Pi^{\rm a}\mathcal{J}^{\rm a} Z^{\rm a} + \Pi^{\rm r}\mathcal{J}^{\rm r} Z^{\rm r}$ is a chemical equilibrium state, which satisfies, in particular, $\mu(Z) \in M^{-1}\mathcal{A}$, $V = \Pi^{\rm a}\mathcal{J}^{\rm a}(\Pi^{\rm a})^t V$, and

$$(\Pi^{\rm r})^t V = 0. \tag{10.2.27}$$

Proof. Let $\widetilde{\varrho} = (\widetilde{\rho}_1, \dots, \widetilde{\rho}_{n^{\rm a}})^t \in (0,\infty)^{n^{\rm a}}$ and consider the species vector

$$\varrho = \sum_{l\in\mathfrak{A}} \widetilde{\rho}_l e^l + \delta \sum_{k\in S\setminus\mathfrak{A}} M\mathfrak{n}_k, \tag{10.2.28}$$

for δ small and positive chosen such that all components of ϱ are positive. Since the vectors $M\mathfrak{n}_k$, $k \in S\backslash\mathfrak{A}$, are orthogonal to the mass atom vectors $\widetilde{\mathfrak{a}}_l$, $l \in \mathfrak{A}$, keeping in mind that $\widetilde{\mathfrak{a}}_l = \widetilde{m}_l M^{-1}\mathfrak{a}_l$, $l \in \mathfrak{A}$, and since $\widetilde{\mathfrak{a}}_{ll} = 1$, $l \in \mathfrak{A}$, it is readily seen that $\langle \varrho, \widetilde{\mathfrak{a}}_l \rangle = \widetilde{\rho}_l$, $l \in \mathfrak{A}$, as was to be shown. Using Proposition 6.5.3 we can now define uniquely $Z^{\mathrm{r}} = U^{\mathrm{r}}$ from $Z^{\mathrm{a}} \in \mathcal{O}_{Z^{\mathrm{a}}}$. The smoothness is then obtained by the implicit function theorem, as in Proposition 6.5.8. Finally, the property $\mu \in M^{-1}\mathcal{A}$ is also established in Proposition 6.5.3, and this directly implies (10.2.27) and $V = \Pi^{\mathrm{a}}\mathcal{J}^{\mathrm{a}}(\Pi^{\mathrm{a}})^t V$ from (10.2.17). ■

Remark 10.2.2. When the set of species does not contain the atomic species, it is necessary to reduce the domain of definition of $Z^{\mathrm{a}} \longrightarrow Z^{\mathrm{r}}$ to the set $\mathcal{O}_{Z^{\mathrm{a}}} = \mathfrak{p} \times \mathbb{R}^3 \times (0,\infty)$, where

$$\mathfrak{p} = \{\ \widetilde{\varrho} \in (0,\infty)^{n^{\mathrm{a}}},\ \exists \varrho \in (0,\infty)^n,\ \forall l \in \mathfrak{A},\ \widetilde{\rho}_l = \langle \varrho, \widetilde{\mathfrak{a}}_l \rangle\ \}$$

and $\mathcal{O}_{Z^{\mathrm{a}}}$ is open and convex. ■

Lemma 10.2.3. *The map $Z^{\mathrm{a}} \longrightarrow U^{\mathrm{a}}$ is a smooth diffeomorphism from $\mathcal{O}_{Z^{\mathrm{a}}} = (0,\infty)^{n^{\mathrm{a}}} \times \mathbb{R}^3 \times (0,\infty)$ onto the open convex set $\mathcal{O}_{U^{\mathrm{a}}}$ given by*

$$\mathcal{O}_{U^{\mathrm{a}}} = \{\ u \in \mathbb{R}^{n^{\mathrm{a}}+4};\, u_i > 0,\ 1 \le i \le n^{\mathrm{a}},\quad u_{n^{\mathrm{a}}+4} - \phi(u_1,\ldots,u_{n^{\mathrm{a}}+3}) > 0\ \}, \tag{10.2.29}$$

where

$$\phi(u_1,\ldots,u_{n^{\mathrm{a}}+3}) = \frac{1}{2}\frac{u_{n^{\mathrm{a}}+1}^2 + u_{n^{\mathrm{a}}+2}^2 + u_{n^{\mathrm{a}}+3}^2}{\sum_{l\in\mathfrak{A}} u_l} + \sum_{l\in\mathfrak{A}} u_l e_{l'}^0,$$

where $e_{l'}^0$ is the energy of formation at zero temperature of the stable version of the lth element.

Proof. From (10.2.20), (10.2.22), $\rho e^{\mathrm{tot}} = \frac{1}{2}\boldsymbol{v}\cdot\boldsymbol{v} + \mathcal{E}$, with

$$\mathcal{E}(T,\widetilde{\varrho}) = \sum_{k\in S} \rho_k(T,\widetilde{\varrho}) e_k(T)$$

and $\widetilde{\varrho} = (\widetilde{\rho}_1,\ldots,\widetilde{\rho}_{n^{\mathrm{a}}})^t$, it is readily seen that the map $Z^{\mathrm{a}} \longrightarrow U^{\mathrm{a}}$ defined on $\mathcal{O}_{Z^{\mathrm{a}}}$ is smooth.

In order to establish that this map is one to one, it is sufficient to establish that, at fixed atomic concentrations $\widetilde{\varrho}$, the map $T \longrightarrow \mathcal{E}(T,\widetilde{\varrho})$ is strictly increasing. From the definition of $\mathcal{E}$, we first obtain

$$\partial_T \mathcal{E} = \sum_{k\in S} \partial_T \rho_k(T,\widetilde{\varrho}) e_k(T) + \sum_{k\in S} \rho_k(T,\widetilde{\varrho}) c_{vk}(T). \tag{10.2.30}$$

However, differentiating with respect to the temperature T the relations $\widetilde{m}_l\langle \varrho, M^{-1}\mathfrak{a}_l\rangle = \widetilde{\rho}_l = Z_l^{\mathrm{a}}$, we find that

$$(\partial_T \rho_1, \ldots, \partial_T \rho_n)^t \in (M^{-1}\mathcal{A})^\perp = M\mathcal{R},$$

whereas differentiating the relation $(\Pi^{\mathrm{r}})^t V = 0$ with respect to T yields

$$(e_1 - r_1 T^2 \partial_T \log \rho_1, \ldots, e_n - r_n T^2 \partial_T \log \rho_n)^t \in M^{-1}\mathcal{A} = (M\mathcal{R})^\perp.$$

Upon multiplying scalarly these vectors, we obtain

$$\sum_{k\in S} \partial_T \rho_k(T, \widetilde{\varrho}) e_k(T) = \sum_{k\in S} r_k T^2 \rho_k \big(\partial_T \log \rho_k(T, \widetilde{\varrho})\big)^2,$$

so that, from the positivity of specific heats and (10.2.30), we now conclude that $T \longrightarrow \mathcal{E}(T, \widetilde{\varrho})$ is strictly increasing.

The Jacobian matrix of $Z^{\mathrm{a}} \longrightarrow U^{\mathrm{a}}$ is then easily shown to be nonsingular thanks to its triangular structure. From the inverse function theorem, this map is therefore a diffeomorphism from $\mathcal{O}_{Z^{\mathrm{a}}}$ onto its range which is an open set. In order to establish that its range is $\mathcal{O}_{U^{\mathrm{a}}}$, as given by (10.2.29), we simply have to show that $\mathcal{E}(T, \widetilde{\varrho}) \to \sum_{l\in\mathfrak{A}} \widetilde{\rho}_l e_{l'}^0$ as $T \to 0$ at fixed atom concentrations.

From $(\mathcal{G}_1, \ldots, \mathcal{G}_n)^t \in M^{-1}\mathcal{A}$, $\mathcal{G}_k(T, \widetilde{\varrho}) = e_k(T) + r_k T - T\mathcal{S}_k(T, \widetilde{\varrho})$, $k \in S$, and $M\mathfrak{n}_k' \in (M^{-1}\mathcal{A})^\perp$, we obtain that, for $k \in S\backslash\mathfrak{A}'$ and small T,

$$\log(\rho_k) = -\frac{\langle (e_1^0, \ldots, e_n^0)^t, M\mathfrak{n}_k'\rangle}{RT} + \sum_{l'\in\mathfrak{A}'} \mathfrak{a}_{kl'}' \log(\rho_{l'}) + O(|\log T|),$$

where $\mathfrak{a}_{kl'}'$ is the proportion of the stable version of the lth element in the kth species $\mathfrak{a}_{kl'}' = \mathfrak{a}_{kl}/\mathfrak{a}_{l'l}$ As a consequence, we have $\rho_k \leq \epsilon \prod_{l'\in\mathfrak{A}'} \rho_{l'}^{\mathfrak{a}_{kl'}'}$, $k \in S\backslash\mathfrak{A}'$, for small T, where ϵ is arbitrarily small, since the energies of formation $\langle (e_1^0, \ldots, e_n^0)^t, M\mathfrak{n}_k'\rangle$, $k \in S\backslash\mathfrak{A}'$ are positive. From the constraint $\widetilde{\rho}_l = \langle \varrho, \widetilde{\mathfrak{a}}_l\rangle$, we deduce that

$$\lim_{T\to 0} \rho_{l'} = \widetilde{\rho}_l, \quad l' \in \mathfrak{A}' \qquad \lim_{T\to 0} \rho_k = 0, \quad k \in S\backslash\mathfrak{A}',$$

so that $\mathcal{E}(T, \widetilde{\varrho}) \to \sum_{l\in\mathfrak{A}} \widetilde{\rho}_l e_{l'}^0$ as $T \to 0$ at fixed atom concentrations $\widetilde{\varrho}$.

The convexity of $\mathcal{O}_U$ is finally a direct consequence of the convexity of ϕ, which is established by evaluating its second derivative. ■

Remark 10.2.4. These results can be generalized to the situation in which the most stable species contain several elements, at the cost of a more complex formulation. ■

10.2.5. Fluxes at chemical equilibrium

In this section, we investigate the convective and diffusive fluxes in terms of the variable $U^{\rm a}$. The convective flux $F_i^{\rm a}$ in the $i^{\rm th}$ direction is easily rewritten in terms of $U^{\rm a}$ in the form

$$F_i^{\rm a}(U^{\rm a}) = (\Pi^{\rm a})^t F_i\big(\Pi^{\rm a}\mathcal{J}^{\rm a}U^{\rm a} + \Pi^{\rm r}\mathcal{J}^{\rm r}U^{\rm r}(U^{\rm a})\big), \tag{10.2.31}$$

and this yields by differentiation the Jacobian matrix $A_i^{\rm a}$

$$A_i^{\rm a} = \partial_{U^{\rm a}} F_i^{\rm a} = (\Pi^{\rm a})^t A_i(U)\big(\Pi^{\rm a}\mathcal{J}^{\rm a} + \Pi^{\rm r}\mathcal{J}^{\rm r}\partial_{U^{\rm a}}U^{\rm r}\big). \tag{10.2.32}$$

Similarly, the diffusive fluxes $F_i^{\rm dis,a} = (\Pi^{\rm a})^t F_i^{\rm dis}$ have now to be expressed in terms of dissipation matrices. Since $F_i^{\rm dis} = -\sum_{j\in C} B_{ij}(U)\partial_j U$, we have

$$F_i^{\rm dis,a} = -\sum_{j\in C} B_{ij}^{\rm a}(U^{\rm a})\partial_j U^{\rm a}, \tag{10.2.33}$$

where

$$B_{ij}^{\rm a}(U^{\rm a}) = (\Pi^{\rm a})^t B_{ij}(U)\Big(\Pi^{\rm a}\mathcal{J}^{\rm a} + \Pi^{\rm r}\mathcal{J}^{\rm r}\partial_{U^{\rm a}}U^{\rm r}\Big), \qquad i,j\in C, \tag{10.2.34}$$

are the dissipation matrices at chemical equilibrium.

10.2.6. Quasilinear form at chemical equilibrium

The equations at equilibrium can finally be rewritten as a quasilinear system of second-order equations in terms of the variable $U^{\rm a}$

$$\partial_t U^{\rm a} + \sum_{i\in C} A_i^{\rm a}(U^{\rm a})\partial_i U^{\rm a} = \sum_{i,j\in C} \partial_i\big(B_{ij}^{\rm a}(U^{\rm a})\partial_j U^{\rm a}\big), \tag{10.2.35}$$

in such a way that its structure can be investigated by using the tools developed in Chapters 8 and 9.

Remark 10.2.5. Gravity is easily added to the model since the momentum conservation equation is left unchanged. ∎

10.3. Entropy and symmetrization

In this section we investigate entropy at chemical equilibrium and subsequently obtain the corresponding symmetrized equations.

10.3.1. Entropy at chemical equilibrium

The mathematical entropy function σ^{a} at chemical equilibrium is taken to be the mathematical entropy σ out of equilibrium evaluated at $U(U^{\mathrm{a}})$, so that

$$\sigma^{\mathrm{a}}(U^{\mathrm{a}}) = \sigma\big(U(U^{\mathrm{a}})\big). \tag{10.3.1}$$

We check in this section that the assumptions of Definition 8.4.2 are satisfied. Note that we adopt here a mathematical point of view and that we directly use definition (10.3.1). From a kinetic theory point of view, one may also infer that the entropy at chemical equilibrium coincides with the nonequilibrium entropy evaluated at $U(U^{\mathrm{a}})$ [EG98a].

Theorem 10.3.1. *The function σ^{a} is a mathematical entropy in the sense of Definition* 8.4.2. *The corresponding entropy fluxes q_i^{a}, $i \in C$, are given by*

$$q_i^{\mathrm{a}} = q_i\big(U(U^{\mathrm{a}})\big), \tag{10.3.2}$$

and the symmetrizing variable $\mathcal{V}^{\mathrm{a}} = \big(\partial_{U^{\mathrm{a}}}\sigma^{\mathrm{a}}\big)^t$ is

$$\mathcal{V}^{\mathrm{a}} = \Big(\partial_{U^{\mathrm{a}}}\sigma^{\mathrm{a}}\Big)^t = \mathcal{J}^{\mathrm{a}}(\Pi^{\mathrm{a}})^t V,$$

so that we have $V = \Pi^{\mathrm{a}}\mathcal{V}^{\mathrm{a}}$ and, in general, $\mathcal{V}^{\mathrm{a}} \neq V^{\mathrm{a}} = (\Pi^{\mathrm{a}})^t V$.

Proof. Using $U = \Pi^{\mathrm{a}}\mathcal{J}^{\mathrm{a}}U^{\mathrm{a}} + \Pi^{\mathrm{r}}\mathcal{J}^{\mathrm{r}}U^{\mathrm{r}}$ and the definition of σ^{a}, we first get by differentiation

$$\partial_{U^{\mathrm{a}}}\sigma^{\mathrm{a}} = \partial_U\sigma\Big(\Pi^{\mathrm{a}}\mathcal{J}^{\mathrm{a}} + \Pi^{\mathrm{r}}\mathcal{J}^{\mathrm{r}}\partial_{U^{\mathrm{a}}}U^{\mathrm{r}}\Big). \tag{10.3.3}$$

However, at chemical equilibrium, we have (10.2.27) which yields by transposition that $(\partial_U\sigma)\,\Pi^{\mathrm{r}} = 0$. We thus obtain $\partial_{U^{\mathrm{a}}}\sigma^{\mathrm{a}} = (\partial_U\sigma)\,\Pi^{\mathrm{a}}\mathcal{J}^{\mathrm{a}}$ which yields by transposition

$$\mathcal{V}^{\mathrm{a}} = \mathcal{J}^{\mathrm{a}}(\Pi^{\mathrm{a}})^t V, \tag{10.3.4}$$

since $\mathcal{J}^{\mathrm{a}}$ is symmetric. Differentiating $(\Pi^{\mathrm{r}})^t V = 0$, we further deduce that

$$(\Pi^{\mathrm{r}})^t\partial_U V\Big(\Pi^{\mathrm{a}}\mathcal{J}^{\mathrm{a}} + \Pi^{\mathrm{r}}\mathcal{J}^{\mathrm{r}}\partial_{U^{\mathrm{a}}}U^{\mathrm{r}}\Big) = 0. \tag{10.3.5}$$

Differentiating then (10.3.4), we get

$$\partial_{U^{\mathrm{a}}}\mathcal{V}^{\mathrm{a}} = \mathcal{J}^{\mathrm{a}}(\Pi^{\mathrm{a}})^t\partial_U V\Big(\Pi^{\mathrm{a}}\mathcal{J}^{\mathrm{a}} + \Pi^{\mathrm{r}}\mathcal{J}^{\mathrm{r}}\partial_{U^{\mathrm{a}}}U^{\mathrm{r}}\Big), \tag{10.3.6}$$

since $\partial_{U^{\mathrm{a}}}U = \Pi^{\mathrm{a}}\mathcal{J}^{\mathrm{a}} + \Pi^{\mathrm{r}}\mathcal{J}^{\mathrm{r}}\partial_{U^{\mathrm{a}}}U^{\mathrm{r}}$, and, from (10.3.5), we obtain

$$\partial_{U^{\mathrm{a}}}\mathcal{V}^{\mathrm{a}} = \Big(\Pi^{\mathrm{a}}\mathcal{J}^{\mathrm{a}} + \Pi^{\mathrm{r}}\mathcal{J}^{\mathrm{r}}\partial_{U^{\mathrm{a}}}U^{\mathrm{r}}\Big)^t\partial_U V\Big(\Pi^{\mathrm{a}}\mathcal{J}^{\mathrm{a}} + \Pi^{\mathrm{r}}\mathcal{J}^{\mathrm{r}}\partial_{U^{\mathrm{a}}}U^{\mathrm{r}}\Big). \tag{10.3.7}$$

This shows that $\partial_{U^{\mathrm{a}}}\mathcal{V}^{\mathrm{a}} = \partial^2_{U^{\mathrm{a}}}\sigma^{\mathrm{a}}$ is positive definite and that σ^{a} is strictly convex over the convex open set $\mathcal{O}_{U^{\mathrm{a}}}$ so that Property (E_1) of Definition (8.4.2) is established.

For the entropy fluxes, we note from $(\partial_{U^{\mathrm{a}}}\sigma^{\mathrm{a}})^t = \mathcal{V}^{\mathrm{a}} = \mathcal{J}^{\mathrm{a}}(\Pi^{\mathrm{a}})^t V$ and (10.2.32) that

$$(\partial_{U^{\mathrm{a}}}\sigma^{\mathrm{a}})A_i^{\mathrm{a}} = V^t\Pi^{\mathrm{a}}\mathcal{J}^{\mathrm{a}}(\Pi^{\mathrm{a}})^t A_i\Big(\Pi^{\mathrm{a}}\mathcal{J}^{\mathrm{a}} + \Pi^{\mathrm{r}}\mathcal{J}^{\mathrm{r}}\partial_{U^{\mathrm{a}}}U^{\mathrm{r}}\Big).$$

Using now $V = \Pi^{\mathrm{a}}\mathcal{J}^{\mathrm{a}}(\Pi^{\mathrm{a}})^t V$ and $V^t A_i = \partial_U q_i$, we obtain

$$(\partial_{U^{\mathrm{a}}}\sigma^{\mathrm{a}})\,A_i^{\mathrm{a}} = \partial_U q_i\Big(\Pi^{\mathrm{a}}\mathcal{J}^{\mathrm{a}} + \Pi^{\mathrm{r}}\mathcal{J}^{\mathrm{r}}\partial_{U^{\mathrm{a}}}U^{\mathrm{r}}\Big) = \partial_U q_i \partial_{U^{\mathrm{a}}}U,$$

so that, finally,

$$(\partial_{U^{\mathrm{a}}}\sigma^{\mathrm{a}})\,A_i^{\mathrm{a}} = \partial_{U^{\mathrm{a}}}q_i^{\mathrm{a}}$$

and q_i^{a}, $i \in C$, are the entropy fluxes, and Property (E_2) of Definition (8.4.2) is established.

In order to establish that the products $B_{ij}^{\mathrm{a}}(U^{\mathrm{a}})\,(\partial^2_{U^{\mathrm{a}}}\sigma^{\mathrm{a}})^{-1}$, $i,j \in C$, satisfy Properties (E_3) and (E_4) of Definition (8.4.2), we first note that $(\partial^2_{U^{\mathrm{a}}}\sigma^{\mathrm{a}})^{-1} = \partial_{\mathcal{V}^{\mathrm{a}}}U^{\mathrm{a}}$. Furthermore, from (10.3.4) and $V = \Pi^{\mathrm{a}}\mathcal{J}^{\mathrm{a}}(\Pi^{\mathrm{a}})^t V$, we also have

$$V = \Pi^{\mathrm{a}}\mathcal{V}^{\mathrm{a}}. \tag{10.3.8}$$

By using (10.2.34), we can now evaluate

$$B_{ij}^{\mathrm{a}}(U^{\mathrm{a}})\,(\partial^2_{U^{\mathrm{a}}}\sigma^{\mathrm{a}})^{-1} = (\Pi^{\mathrm{a}})^t B_{ij}\partial_{U^{\mathrm{a}}}U\,\partial_{\mathcal{V}^{\mathrm{a}}}U^{\mathrm{a}}, \qquad i,j \in C,$$

and since

$$\partial_{U^{\mathrm{a}}}U\,\partial_{\mathcal{V}^{\mathrm{a}}}U^{\mathrm{a}} = \partial_{\mathcal{V}^{\mathrm{a}}}U = \partial_V U \partial_{\mathcal{V}^{\mathrm{a}}}V = (\partial_V U)\,\Pi^{\mathrm{a}},$$

we obtain

$$B_{ij}^{\mathrm{a}}(U^{\mathrm{a}})\,(\partial^2_{U^{\mathrm{a}}}\sigma^{\mathrm{a}})^{-1} = (\Pi^{\mathrm{a}})^t B_{ij}\partial_V U\,\Pi^{\mathrm{a}}, = (\Pi^{\mathrm{a}})^t \widetilde{B}_{ij}\,\Pi^{\mathrm{a}}, \qquad i,j \in C.$$

Properties (E_3) and (E_4) at chemical equilibrium are then a direct consequence of the same properties at chemical nonequilibrium, taking into account that Π^{a} is one to one. ∎

10.3.2. Symmetrized equations

In this section, we express the symmetrized equations at chemical equilibrium in terms of the symmetrized equations out of equilibrium. We first evaluate the symmetrizing variables $\mathcal{V}^{\mathrm{a}}$ from the variables V.

Proposition 10.3.2. *The symmetrizing variable $\mathcal{V}^{\mathrm{a}}$ reads as*

$$\mathcal{V}^{\mathrm{a}} = \frac{1}{T}\Big(\widetilde{\mathcal{G}}_1 - \tfrac{1}{2}\boldsymbol{v}\cdot\boldsymbol{v}, \ldots, \widetilde{\mathcal{G}}_{n^{\mathrm{a}}} - \tfrac{1}{2}\boldsymbol{v}\cdot\boldsymbol{v},\, v_1,\, v_2,\, v_3,\, -1\Big)^t, \tag{10.3.9}$$

where $\widetilde{\mathcal{G}}_l$, $l \in \mathfrak{A}$, are defined from

$$\big(\widetilde{\mathcal{G}}_1, \ldots, \widetilde{\mathcal{G}}_{n^{\mathrm{a}}}\big)^t = \mathcal{J}^{\mathrm{a}}_{n^{\mathrm{a}}\times n^{\mathrm{a}}}(\Pi^{\mathrm{a}}_{n\times n^{\mathrm{a}}})^t\,\big(\mathcal{G}_1, \ldots, \mathcal{G}_n\big)^t \tag{10.3.10}$$

and $\mathcal{J}^{\mathrm{a}}_{n^{\mathrm{a}}\times n^{\mathrm{a}}}$ and $\Pi^{\mathrm{a}}_{n\times n^{\mathrm{a}}}$ denotes the upper left blocks of size $n^{\mathrm{a}}\times n^{\mathrm{a}}$ of $\mathcal{J}^{\mathrm{a}}$ and of size $n\times n^{\mathrm{a}}$ of Π^{a}, respectively, and we also have the relations

$$\big(\mathcal{G}_1, \ldots, \mathcal{G}_n\big)^t = \Pi^{\mathrm{a}}_{n\times n^{\mathrm{a}}}\,\big(\widetilde{\mathcal{G}}_1, \ldots, \widetilde{\mathcal{G}}_{n^{\mathrm{a}}}\big)^t. \tag{10.3.11}$$

Finally, the map $U^{\mathrm{a}} \longrightarrow \mathcal{V}^{\mathrm{a}}$ from $\mathcal{O}_{U^{\mathrm{a}}}$ onto $\mathcal{O}_{\mathcal{V}^{\mathrm{a}}} = \mathbb{R}^{n^{\mathrm{a}}+3}\times(-\infty,0)$ is a C^∞ diffeomorphism.

Proof. This results from straightforward calculations. ■

Theorem 10.3.3. *The symmetrized equations at chemical equilibrium are given by*

$$\widetilde{A}^{\mathrm{a}}_0\partial_t\mathcal{V}^{\mathrm{a}} + \sum_{i\in C}\widetilde{A}^{\mathrm{a}}_i\partial_i\mathcal{V}^{\mathrm{a}} = \sum_{i,j\in C}\partial_i\big(\widetilde{B}^{\mathrm{a}}_{ij}(\mathcal{V}^{\mathrm{a}})\partial_j\mathcal{V}^{\mathrm{a}}\big), \tag{10.3.12}$$

where $\widetilde{A}^{\mathrm{a}}_0 = (\Pi^{\mathrm{a}})^t\widetilde{A}_0\,\Pi^{\mathrm{a}}$, $\widetilde{A}^{\mathrm{a}}_i = (\Pi^{\mathrm{a}})^t\widetilde{A}_i\,\Pi^{\mathrm{a}}$, $i\in C$, and $\widetilde{B}^{\mathrm{a}}_{ij} = (\Pi^{\mathrm{a}})^t\widetilde{B}_{ij}\,\Pi^{\mathrm{a}}$, $i,j\in C$, for $\mathcal{V}^{\mathrm{a}}\in\mathcal{O}_{\mathcal{V}^{\mathrm{a}}}$.

Proof. From Theorem 8.4.3, the system (10.2.35) rewritten in terms of the variable $\mathcal{V}^{\mathrm{a}}$ is symmetric. From Definition 8.4.1, we have $\widetilde{A}^{\mathrm{a}}_0 = \partial_{\mathcal{V}^{\mathrm{a}}}U^{\mathrm{a}}$, so that

$$\widetilde{A}^{\mathrm{a}}_0 = (\Pi^{\mathrm{a}})^t\partial_{\mathcal{V}^{\mathrm{a}}}U = (\Pi^{\mathrm{a}})^t(\partial_{\mathcal{V}}U)\,\partial_{\mathcal{V}^{\mathrm{a}}}\mathcal{V} = (\Pi^{\mathrm{a}})^t\partial_{\mathcal{V}}U\,\Pi^{\mathrm{a}}, \tag{10.3.13}$$

since $\mathcal{V} = \Pi^{\mathrm{a}}\mathcal{V}^{\mathrm{a}}$. Furthermore, from Definition 8.4.1, $\widetilde{A}^{\mathrm{a}}_i = A^{\mathrm{a}}_i\partial_{\mathcal{V}^{\mathrm{a}}}U^{\mathrm{a}} = \partial_{U^{\mathrm{a}}}F^{\mathrm{a}}_i\partial_{\mathcal{V}^{\mathrm{a}}}U^{\mathrm{a}} = \partial_{\mathcal{V}^{\mathrm{a}}}F^{\mathrm{a}}_i$, $i\in C$, so that, for $i\in C$, we have

$$\widetilde{A}^{\mathrm{a}}_i = (\Pi^{\mathrm{a}})^t\partial_{\mathcal{V}^{\mathrm{a}}}F_i = (\Pi^{\mathrm{a}})^t\partial_{\mathcal{V}}F_i\,\partial_{\mathcal{V}^{\mathrm{a}}}\mathcal{V}, = (\Pi^{\mathrm{a}})^t\widetilde{A}_i\,\Pi^{\mathrm{a}}, \tag{10.3.14}$$

since $\widetilde{A}_i = \partial_{\mathcal{V}}F_i$, $i\in C$. Similarly, for $i,j\in C$, we have

$$\widetilde{B}^{\mathrm{a}}_{ij} = (\Pi^{\mathrm{a}})^tB_{ij}\partial_{U^{\mathrm{a}}}U\partial_{\mathcal{V}^{\mathrm{a}}}U^{\mathrm{a}} = (\Pi^{\mathrm{a}})^tB_{ij}\partial_{\mathcal{V}}U\,\partial_{\mathcal{V}^{\mathrm{a}}}\mathcal{V} = (\Pi^{\mathrm{a}})^t\widetilde{B}_{ij}\,\Pi^{\mathrm{a}}, \tag{10.3.15}$$

which completes the proof. ■

The explicit expressions of the symmetric matrices $\widetilde{A}_0^{\mathrm{a}}$, $\widetilde{A}_i^{\mathrm{a}}$, $i \in C$, and $\widetilde{B}_{ij}^{\mathrm{a}}$, $i,j \in C$, are thus readily evaluated form the explicit forms of $\widetilde{A}_0$, $\widetilde{A}_i$, $i \in C$, and $\widetilde{B}_{ij}$, $i,j \in C$, already obtained in Theorem 8.6.2.

10.4. Normal forms

In this section, we express the normal forms in terms of those of nonequilibrium equations.

10.4.1. Nullspace invariance property

In this section, we establish that condition N is satisfied and introduce the linear operator Π'^{a} associated with the auxiliary variables.

Proposition 10.4.1. *The nullspace of the matrix*

$$\widetilde{B}^{\mathrm{a}} = \sum_{i,j \in C} \widetilde{B}_{ij}^{\mathrm{a}} w_i w_j \tag{10.4.1}$$

is one-dimensional and given by

$$N(\widetilde{B}^{\mathrm{a}}) = span(1, \dots, 1, 0, 0, 0, 0)^t, \tag{10.4.2}$$

and we have $\widetilde{B}_{ij}^{\mathrm{a}} N(\widetilde{B}^{\mathrm{a}}) = 0$*, for* $i,j \in C$ *.*

Proof. Since we have the relation

$$\widetilde{B}^{\mathrm{a}} = (\Pi^{\mathrm{a}})^t \widetilde{B}\, \Pi^{\mathrm{a}}, \tag{10.4.3}$$

we deduce that for $x^{\mathrm{a}} \in \mathbb{R}^{n^{\mathrm{a}}+4}$, $(x^{\mathrm{a}})^t \widetilde{B}^{\mathrm{a}} x^{\mathrm{a}} = 0$ if and only if $z^t \widetilde{B} z = 0$, where $z = \Pi^{\mathrm{a}} x^{\mathrm{a}}$, that is, if and only if we have $\Pi^{\mathrm{a}} x^{\mathrm{a}} \in \mathbb{R}\mathcal{U}^\star$.

However, $\Pi^{\mathrm{a}} x^{\mathrm{a}} \in \mathbb{R}\mathcal{U}^\star$ if and only if $\sum_{l \in \mathfrak{A}^\star} x_l^{\mathrm{a}} a_l \in \mathbb{R}\mathcal{U}^\star$, that is, from (10.2.19), if and only if there exists $\alpha \in \mathbb{R}$ such that

$$\sum_{l \in \mathfrak{A}^\star} x_l^{\mathrm{a}} a_l = \alpha \sum_{l \in \mathfrak{A}} a_l .$$

Since a_l, $l \in \mathfrak{A}^\star$, form a basis, the nullspace of $\widetilde{B}^{\mathrm{a}}(\mathcal{V}^{\mathrm{a}}, w)$ is one-dimensional and spanned by $(1, \dots, 1, 0, 0, 0, 0)^t \in \mathbb{R}^{n^{\mathrm{a}}+4}$ and thus independent of the variables $\mathcal{V}^{\mathrm{a}} \in \mathcal{O}_{\mathcal{V}^{\mathrm{a}}}$ and $w \in \Sigma^2$. Finally, one can easily check that $\widetilde{B}_{ij}^{\mathrm{a}} N(\widetilde{B}^{\mathrm{a}}) = 0$, $i,j \in C$. ▪

We now establish some useful relations that will be needed when investigating normal forms. The system of equations governing chemical equilibrium flows satisfies the nullspace invariance condition, and we can now introduce the auxiliary variables $U^{a\prime}$ and $V^{a\prime}$ of Lemma 8.5.3. From Proposition 10.4.1 and Lemma 8.5.3, the matrix P^{a} of size $n^{a}+4$ can be taken to be

$$P^{a} = \begin{pmatrix} 1 & 0 & \cdots & \cdots & 0 & 0 & 0 & 0 & 0 \\ 1 & 1 & \ddots & & \vdots & \vdots & \vdots & \vdots & \vdots \\ \vdots & 0 & \ddots & \ddots & \vdots & \vdots & \vdots & \vdots & \vdots \\ \vdots & \vdots & \ddots & \ddots & 0 & \vdots & \vdots & \vdots & \vdots \\ 1 & 0 & \cdots & 0 & 1 & 0 & 0 & 0 & 0 \\ 0 & \cdots & \cdots & \cdots & 0 & 1 & 0 & 0 & 0 \\ 0 & \cdots & \cdots & \cdots & 0 & 0 & 1 & 0 & 0 \\ 0 & \cdots & \cdots & \cdots & 0 & 0 & 0 & 1 & 0 \\ 0 & \cdots & \cdots & \cdots & 0 & 0 & 0 & 0 & 1 \end{pmatrix}, \tag{10.4.4}$$

and has the same structure as the matrix P of size $n+4$ introduced for nonequilibrium flows. We then have

$$U^{a\prime} = (P^{a})^{t} U^{a} = \Big(\rho,\, \widetilde{\rho}_2, \ldots, \widetilde{\rho}_{n^{a}},\, \rho v_1,\, \rho v_2,\, \rho v_3,\, \rho e^{\mathrm{tot}}\Big)^{t}, \tag{10.4.5}$$

and the associated entropic variable is then

$$V^{a\prime} = \frac{1}{T}\Big(\widetilde{\mathcal{G}}_1 - \tfrac{1}{2}\boldsymbol{v}\cdot\boldsymbol{v},\, \widetilde{\mathcal{G}}_2 - \widetilde{\mathcal{G}}_1, \ldots, \widetilde{\mathcal{G}}_{n^{a}} - \widetilde{\mathcal{G}}_1,\, v_1,\, v_2,\, v_3,\, -1\Big)^{t}. \tag{10.4.6}$$

After some algebra, we also have the relations

$$V' = P^{-1}\Pi^{a} P^{a} \mathcal{V}^{a\prime}, \qquad \mathcal{V}^{a\prime} = (P^{a})^{-1}\mathcal{J}^{a}(\Pi^{a})^{t} P V', \tag{10.4.7}$$

$$U' = P^{t}\Big(\Pi^{a}\mathcal{J}^{a}((P^{a})^{-1})^{t} U^{a\prime} + \Pi^{r}\mathcal{J}^{r} U^{r}\Big), \qquad U^{a\prime} = (P^{a})^{t}(\Pi^{a})^{t}(P^{-1})^{t} U'. \tag{10.4.8}$$

10.4.2. Intermediate normal form

From the expression of the auxiliary variables, we can now introduce the natural normal variable $\mathcal{W}^{a}$ at chemical equilibrium

$$\mathcal{W}^{a} = \Big(\rho,\, \frac{\widetilde{\mathcal{G}}_2 - \widetilde{\mathcal{G}}_1}{T}, \ldots, \frac{\widetilde{\mathcal{G}}_{n^{a}} - \widetilde{\mathcal{G}}_1}{T},\, \frac{v_1}{T},\, \frac{v_2}{T},\, \frac{v_3}{T},\, \frac{-1}{T}\Big)^{t} \tag{10.4.9}$$

or, taken into account the possibility of mixing parabolic variables, the intermediate normal variable W^{a}

$$\mathsf{W}^{a} = \Big(\rho,\, \frac{\widetilde{\mathcal{G}}_2 - \widetilde{\mathcal{G}}_1}{T}, \ldots, \frac{\widetilde{\mathcal{G}}_{n^{a}} - \widetilde{\mathcal{G}}_1}{T},\, v_1,\, v_2,\, v_3,\, T\Big)^{t}, \tag{10.4.10}$$

for which the velocity components are uncoupled from the other components and should not be confused with $\mathcal{W}^{\rm a}$.

These normal variables $\mathcal{W}^{\rm a}$ and $\mathsf{W}^{\rm a}$ are now related to the corresponding variables out of chemical equilibrium, that is, to the natural normal variable $\mathcal{W}$ and to the intermediate variable W introduced in Chapter 8. To this end, we introduce the new vectors of $\mathbb{R}^{n+4}$

$$a_1' = e^1, \qquad a_l' = (\widetilde{\mathfrak{a}}_l - \widetilde{\mathfrak{a}}_{1l}\mathcal{U}, 0, 0, 0, 0)^t = a_l - a_{1l}\mathcal{U}^\star, \qquad l \in \{2, \ldots, n^{\rm a}\},$$

and

$$a_l' = a_l = e^l, \qquad l \in \{n+1, \ldots, n+4\}.$$

We can then define the linear operator $\Pi^{\rm a\prime} = \mathbb{R}^{n^{\rm a}+4} \longrightarrow \mathbb{R}^{n+4}$ whose matrix in the canonical bases is given by

$$\Pi^{\rm a\prime} = \left[a_1', a_2', \ldots, a_{n^{\rm a}}', a_{n+1}', a_{n+2}', a_{n+3}', a_{n+4}'\right]. \tag{10.4.11}$$

We also introduce the metric matrix $\mathcal{J}^{\rm a\prime}$ of order $n^{\rm a}+4$ defined as for the matrix $\mathcal{J}^{\rm a}$ by

$$(\mathcal{J}^{\rm a\prime})^{-1}_{i,j} = \langle a_i', a_j' \rangle, \qquad i, j \in \mathfrak{A}^\star.$$

We also define

$$(M^\star)^{-1}\mathcal{A}^{\star\prime} = \mathrm{span}\{\ a_l',\ l \in \mathfrak{A}^\star\ \}, \tag{10.4.12}$$

and one may easily check that

$$\forall x \in (M^\star)^{-1}\mathcal{A}^{\star\prime}, \qquad x = \Pi^{\rm a\prime}\mathcal{J}^{\rm a\prime}(\Pi^{\rm a\prime})^t\, x, \tag{10.4.13}$$

and, from the definition of $\mathcal{J}^{\rm a\prime}$, we have in $\mathbb{R}^{n^{\rm a}+4}$

$$\mathcal{J}^{\rm a\prime}(\Pi^{\rm a\prime})^t\Pi^{\rm a\prime} = I. \tag{10.4.14}$$

Proposition 10.4.2. *We have the following relations between natural and intermediate normal variables*

$$\mathcal{W} = \Pi^{\rm a\prime}\mathcal{W}^{\rm a}, \qquad \mathcal{W}^{\rm a} = \mathcal{J}^{\rm a\prime}(\Pi^{\rm a\prime})^t\mathcal{W}, \tag{10.4.15}$$

$$\mathsf{W} = \Pi^{\rm a\prime}\mathsf{W}^{\rm a}, \qquad \mathsf{W}^{\rm a} = \mathcal{J}^{\rm a\prime}(\Pi^{\rm a\prime})^t\mathsf{W}. \tag{10.4.16}$$

Proof. Directly results from the relations between auxiliary variables. ■

We can now obtain normal forms for the system of partial differential equations governing flows at chemical equilibrium.

Theorem 10.4.3. *The system in the variable* $\mathsf{W}^{\rm a} = (\mathsf{W}^{\rm a}_{I}, \mathsf{W}^{\rm a}_{II})^t$, *on the open convex set* $\mathcal{O}_{\mathsf{W}^{\rm a}} = (0,\infty)\times\mathbb{R}^{n-1}\times\mathbb{R}^3\times(0,\infty)$, *with hyperbolic variable*

$$\mathsf{W}^{\rm a}_{I} = \rho \tag{10.4.17}$$

and parabolic variable

$$\mathcal{W}^{\mathrm{a}}_{II} = \Big(\frac{\widetilde{\mathcal{G}}_2 - \widetilde{\mathcal{G}}_1}{T}, \ldots, \frac{\widetilde{\mathcal{G}}_{n^{\mathrm{a}}} - \widetilde{\mathcal{G}}_1}{T}, v_1, v_2, v_3, T\Big)^t, \qquad (10.4.18)$$

can be written in the form

$$\overline{\mathcal{A}}_0^{\mathrm{a}I,I} \partial_t \mathcal{W}^{\mathrm{a}}_I + \sum_{i\in C} \overline{\mathcal{A}}_i^{\mathrm{a}I,I} \partial_i \mathcal{W}^{\mathrm{a}}_I + \sum_{i\in C} \overline{\mathcal{A}}_i^{\mathrm{a}I,II} \partial_i \mathcal{W}^{\mathrm{a}}_{II} = 0, \qquad (10.4.19)$$

$$\overline{\mathcal{A}}_0^{\mathrm{a}II,II} \partial_t \mathcal{W}^{\mathrm{a}}_{II} + \sum_{i\in C} \overline{\mathcal{A}}_i^{\mathrm{a}II,I} \partial_i \mathcal{W}^{\mathrm{a}}_I + \sum_{i\in C} \overline{\mathcal{A}}_i^{\mathrm{a}II,II} \partial_i \mathcal{W}^{\mathrm{a}}_{II} =$$

$$\sum_{i,j\in C} \partial_i\Big(\overline{\mathcal{B}}_{ij}^{\mathrm{a}II,II} \partial_j \mathcal{W}^{\mathrm{a}}_{II}\Big) + \overline{\mathcal{T}}^{\mathrm{a}}_{II}, \qquad (10.4.20)$$

and is of the normal form. Moreover, we have the relations

$$\overline{\mathcal{A}}_0^{\mathrm{a}} = (\Pi^{\mathrm{a}\prime})^t \overline{\mathcal{A}}_0 \Pi^{\mathrm{a}\prime}, \qquad \overline{\mathcal{A}}_i^{\mathrm{a}} = (\Pi^{\mathrm{a}\prime})^t \overline{\mathcal{A}}_i \Pi^{\mathrm{a}\prime}, \quad i \in C, \qquad (10.4.21)$$

and

$$\overline{\mathcal{B}}_{ij}^{\mathrm{a}} = (\Pi^{\mathrm{a}\prime})^t \overline{\mathcal{B}}_{ij} \Pi^{\mathrm{a}\prime}, \qquad i,j \in C. \qquad (10.4.22)$$

The proof of Theorem 10.4.3 is straightforward, and the explicit expressions of the symmetric matrices $\overline{\mathcal{A}}_0^{\mathrm{a}}$, $\overline{\mathcal{A}}_i^{\mathrm{a}}$, $i \in C$, and $\overline{\mathcal{B}}_{ij}^{\mathrm{a}}$, $i,j \in C$, are readily evaluated from the explicit forms of $\overline{\mathcal{A}}_0$, $\overline{\mathcal{A}}_i$, $i \in C$, and $\overline{\mathcal{B}}_{ij}$, $i,j \in C$, obtained in Theorem 8.7.6.

Remark 10.4.4. A similar result holds for the natural normal variables $\mathcal{W}^{\mathrm{a}}$ and $\mathcal{W}$. ∎

10.5. Global existence

In this section we investigate existence of solutions around constant states for the equations at chemical equilibrium.

10.5.1. Local dissipativity

In this section, we establish that Theorems 9.4.1 and 9.5.1 apply to the chemical equilibrium flow governing equations (10.2.35). We have already established in the previous sections that the system (10.2.35) is symmetrizable into (10.3.12) and have obtained a normal form in Theorem 10.4.3. In order to apply Theorems 9.4.1 and 9.5.1, around any constant state, it is necessary to investigate the local dissipative structure. Moreover, we only have to investigate the dissipative structure of the linearized normal form and establish that (Dis_2) holds, since all other required properties trivially hold, the source term being zero. We will use the normal variable $\mathcal{W}^{\mathrm{a}}$ introduced in the previous section and denote by $\mathcal{W}^{\mathrm{ae}}$ any constant state of $\mathcal{O}_{\mathcal{W}^{\mathrm{a}}}$. Note, however, that the natural normal variable $\mathcal{W}^{\mathrm{a}}$ could be used as well.

Proposition 10.5.1. *For a sufficiently small positive α, the matrices $\mathcal{K}^{\mathrm{a}j}$, $j \in C$, defined by*

$$\sum_{j\in C} \xi_j \mathcal{K}^{\mathrm{a}j} = \alpha \begin{pmatrix} 0 & 0_{1\times(n^{\mathrm{a}}-1)} & \boldsymbol{\xi}^t & 0 \\ 0_{(n^{\mathrm{a}}-1)\times 1} & 0_{(n^{\mathrm{a}}-1)\times(n^{\mathrm{a}}-1)} & 0_{(n^{\mathrm{a}}-1)\times 3} & 0_{(n^{\mathrm{a}}-1)\times 1} \\ -\boldsymbol{\xi} & 0_{3\times(n^{\mathrm{a}}-1)} & 0_{3\times 3} & 0_{3\times 1} \\ 0 & 0_{1\times(n^{\mathrm{a}}-1)} & 0_{1\times 3} & 0 \end{pmatrix} \overline{\mathcal{A}}_0^{\mathrm{a}}(\mathcal{W}^{\mathrm{ae}})^{-1}, \tag{10.5.1}$$

where $\boldsymbol{\xi} = (\xi_1, \xi_2, \xi_3)^t$, are compensating matrices. In particular, the products $\mathcal{K}^{\mathrm{a}j}\overline{\mathcal{A}}_0^{\mathrm{a}}(\mathcal{W}^{\mathrm{ae}})$ are skew-symmetric, and the matrix

$$\sum_{i,j\in C} \mathcal{K}^{\mathrm{a}j}\overline{\mathcal{A}}_i^{\mathrm{a}}(\mathcal{W}^{\mathrm{ae}}) w_i w_j + \overline{\mathcal{B}}^{\mathrm{a}}(\mathcal{W}^{\mathrm{ae}}, w)$$

is positive definite for $w \in \Sigma^2$.

Proof. It is obvious by construction that the products $\mathcal{K}^{\mathrm{a}j}\overline{\mathcal{A}}_0^{\mathrm{a}}(\mathcal{W}^{\mathrm{ae}})$, $j \in C$, are skew-symmetric. On the other hand, a direct calculation yields

$$\sum_{i,j\in C} \mathcal{K}^{\mathrm{a}j}\overline{\mathcal{A}}_i^{\mathrm{a}} \xi_i \xi_j = \begin{pmatrix} \dfrac{\alpha T^{\mathrm{e}}}{\Sigma_\rho^{\mathrm{e}}}|\boldsymbol{\xi}|^2 & \alpha O(|\boldsymbol{\xi}|^2) & \alpha O(|\boldsymbol{\xi}|^2) & \alpha O|\boldsymbol{\xi}|^2 \\ 0_{(n^{\mathrm{a}}-1)\times 1} & 0_{(n^{\mathrm{a}}-1)\times(n^{\mathrm{a}}-1)} & 0_{(n^{\mathrm{a}}-1)\times 3} & 0_{(n^{\mathrm{a}}-1)\times 1} \\ -\alpha(\boldsymbol{v}^{\mathrm{e}}\cdot\boldsymbol{\xi})\boldsymbol{\xi} & 0_{3\times(n^{\mathrm{a}}-1)} & -\alpha\rho^{\mathrm{e}}\boldsymbol{\xi}\otimes\boldsymbol{\xi} & 0_{3\times 1} \\ 0 & 0_{1\times(n^{\mathrm{a}}-1)} & 0_{1\times 3} & 0 \end{pmatrix}. \tag{10.5.2}$$

Proceeding now as in Proposition 9.7.1 we conclude that local dissipativity holds. ∎

10.5.2. Global existence

In the previous sections, we have established that Theorem 9.4.1 and Theorem 9.5.1 can be applied to the system (10.4.19) (10.4.20) governing multicomponent chemical equilibrium flows, written in the $\mathcal{W}^{\mathrm{a}} = (\mathcal{W}_I^{\mathrm{a}}, \mathcal{W}_{II}^{\mathrm{a}})^t$ variable, with the hyperbolic variable

$$\mathcal{W}_I^{\mathrm{a}} = \rho \tag{10.5.3}$$

and the parabolic variable

$$\mathcal{W}^{\mathrm{a}}_{II} = \Big(\frac{\widetilde{\mathcal{G}}_2 - \widetilde{\mathcal{G}}_1}{T}, \ldots, \frac{\widetilde{\mathcal{G}}_{n^{\mathrm{a}}} - \widetilde{\mathcal{G}}_1}{T}, v_1, v_2, v_3, T\Big)^t. \tag{10.5.4}$$

Theorem 10.5.2. *Consider the quasilinear system* (10.4.19) (10.4.20), *let* $d \in \{1,2,3\}$, $l \geq [d/2]+2$, *and let* $\mathcal{W}^{\mathrm{a}0}(x)$, *such that*

$$\mathcal{W}^{\mathrm{a}0} - \mathcal{W}^{\mathrm{ae}} \in W_2^l(\mathbb{R}^d).$$

Then, if $\|\mathcal{W}^{\mathrm{a}0} - \mathcal{W}^{\mathrm{ae}}\|_{l,2}$ *is small enough, there exists a unique global solution to the the Cauchy problem* (10.4.19) (10.4.20)

$$\overline{\mathcal{A}}^{\mathrm{a}}_0 \partial_t \mathcal{W}^{\mathrm{a}} + \sum_{i \in C} \overline{\mathcal{A}}^{\mathrm{a}}_i \partial_i \mathcal{W}^{\mathrm{a}} = \sum_{i,j \in C} \partial_i \Big(\overline{\mathcal{B}}^{\mathrm{a}}_{ij} \partial_j \mathcal{W}^{\mathrm{a}}\Big) + \overline{\mathcal{T}}^{\mathrm{a}},$$

with initial condition

$$\mathcal{W}^{\mathrm{a}}(0,x) = \mathcal{W}^{\mathrm{a}0}(x),$$

such that

$$\begin{cases} \mathcal{W}^{\mathrm{a}}_I - \mathcal{W}^{\mathrm{ae}}_I \in C^0\big([0,\infty); W_2^l(\mathbb{R}^d)\big) \cap C^1\big([0,\infty); W_2^{l-1}(\mathbb{R}^d)\big), \\ \mathcal{W}^{\mathrm{a}}_{II} - \mathcal{W}^{\mathrm{ae}}_{II} \in C^0\big([0,\infty); W_2^l(\mathbb{R}^d)\big) \cap C^1\big([0,\infty); W_2^{l-2}(\mathbb{R}^d)\big), \end{cases} \tag{10.5.5}$$

and

$$\begin{cases} \mathcal{W}^{\mathrm{a}}_I - \mathcal{W}^{\mathrm{ae}}_I \in L^2(0,\infty; W_2^l(\mathbb{R}^d)), \\ \mathcal{W}^{\mathrm{a}}_{II} - \mathcal{W}^{\mathrm{ae}}_{II} \in L^2(0,\infty; W_2^{l+1}(\mathbb{R}^d)). \end{cases} \tag{10.5.6}$$

Furthermore, $\mathcal{W}^{\mathrm{a}}$ *satisfies the estimate*

$$\begin{aligned} \|\mathcal{W}^{\mathrm{a}}(t) - \mathcal{W}^{\mathrm{ae}}\|^2_{l,2} &+ \int_0^t \Big(\|\boldsymbol{\partial_x}\rho(\tau)\|^2_{l-1,2} + \sum_{l \in \{2,\ldots,n^{\mathrm{a}}\}} \|\boldsymbol{\partial_x}(\widetilde{\mathcal{G}}_l - \widetilde{\mathcal{G}}_1)/T)(\tau)\|^2_{l,2} \\ &+ \|\boldsymbol{\partial_x} v(\tau)\|^2_{l,2} + \|\boldsymbol{\partial_x} T(\tau)\|^2_{l,2}\Big)\, d\tau \;\leq\; \beta \|\mathcal{W}^{\mathrm{a}}(0) - \mathcal{W}^{\mathrm{ae}}\|^2_{l,2}, \end{aligned}$$

where β *is a positive constant and* $\sup_{x \in \mathbb{R}^d} \big|\mathcal{W}^{\mathrm{a}}(t) - \mathcal{W}^{\mathrm{ae}}\big|$ *goes to zero as* $t \to \infty$.

Theorem 10.5.3. *Keeping the assumptions of the preceding theorem, assume that* $l \geq [d/2]+3$ *and* $\mathcal{W}^{\mathrm{a}}(0) - \mathcal{W}^{\mathrm{ae}} \in W_2^l(\mathbb{R}^d) \cap L^p(\mathbb{R}^d)$, *with* $p=1$, *if* $d=1$, *and* $p \in [1,2)$, *if* $d \geq 2$. *Then, if* $\|\mathcal{W}^{\mathrm{a}}(0) - \mathcal{W}^{\mathrm{ae}}\|_{l,2|p}$ *is small enough, the unique global solution to Cauchy problem also satisfies the decay estimate*

$$\|\mathcal{W}^{\mathrm{a}}(t) - \mathcal{W}^{\mathrm{ae}}\|_{l-2,2} \leq \beta(1+t)^{-\gamma} \|\mathcal{W}^{\mathrm{a}}(0) - \mathcal{W}^{\mathrm{ae}}\|_{l-2,2|p}, \qquad t \in [0,\infty),$$

where β *is a positive constant and* $\gamma = d \times (1/2p - 1/4)$.

10.6. Notes

10.1. The equations governing equilibrium flows have been derived from the equations governing nonequilibrium flows. It is also possible, however, to derive these equations directly from the kinetic theory by assuming that chemical equilibrium already arises at the Boltzmann level. The resulting macroscopic equations are similar, as are the expressions of the transport fluxes. Even the mathematical properties of the associated transport coefficients are identical. However, the linear systems defining the transport coefficients differ, even though they have similar mathematical properties, so that the corresponding transport coefficients differ.

10.2. Chemical equilibrium boundary layer flows and boundary layer edge flows have been investigated by Boillat [Boi95] and Pousin [Pou93].

10.7. References

[And89] J. D. Anderson, Jr., *Hypersonics and High Temperature Gas Dynamics,* McGraw-Hill Book Company, New York, (1989).

[Boi95] E. Boillat, *Existence and Uniqueness of the Solution to the Edge Problem in a Two-Dimensional Reactive Boundary Layer,* Math. Mod. Meth. Appl. Sci., **5**, (1995), pp. 1–27.

[EG98a] A. Ern and V. Giovangigli, *The Kinetic Equilibrium Regime,* Physica-A, **260**, (1998), pp. 49–72.

[Gok88] S. A. Gokoglu, *Significance of Vapor Phase Chemical Reactions on CVD Rates Predicted by Chemically Forzen and Local Thermochemical Equilibrium Boundary Layer Theories,* J. Electrochem. Soc., **135**, (1988), pp. 1562–1570.

[LH60] G. Ludwig and M. Heil, *Boundary Layer Theory with Dissociation and Ionization,* In Advances in Applied Mechanics, Volume **VI**, Academic Press, New York, (1960), pp. 39–118.

[MVZ97] L. Mottura, L. Vigevano, and M. Zaccanti, *An Evaluation of Roe's Scheme Generalizations for Equilibrium Real Gas Flows,* J. Comp. Phys., **138**, (1997), pp. 354–399.

[Pou93] J. Pousin, *Modélisation et Analyse Numérique de Couches Limites Réactives d'Air,* Doctorat es Sciences, Ecole Polytechnique Fédérale de Lausanne, 1112, (1993).

[Wil85] F. A. Williams, *Combustion Theory,* Second ed., The Benjamin/Cummings Publishing Company, Inc., Menlo Park, (1985).

11

Anchored Waves

11.1. Introduction

Traveling waves in inert or reactive flows can be classified into deflagration waves and detonation waves [Wil85]. The structure of reactive exothermic detonation waves will not be addressed in this book and we refer the reader to [Wil85]. On the other hand, in the context of combustion phenomena, weak deflagration waves typically correspond to plane flames and have been the object of intensive mathematical, physical, chemical, and asymptotic investigations over the last decade [Wil85].

From a mathematical point of view, numerous studies have recently been devoted to plane laminar flames with complex chemistry [Hei87] [VV90] [Bon92] [BeLa93] [VV94] [Bon94]. Important advances have been made in these papers that are essentially concerned with irreversible exothermic chemical reaction networks and elementary transport models.

In this chapter, we investigate the plane flame equations derived from the kinetic theory of dilute polyatomic reactive gas mixtures [Gio97] [Gio98] [Gio99]. These equations are one-dimensional steady models derived from the isobaric equations presented in Chapter 3. The corresponding thermochemistry and transport properties have been presented in Chapters 6 and 7. In particular, all species' second derivatives are coupled through diffusive processes and to the temperature second derivatives through thermal diffusion, and the reaction mechanism that we consider is composed of an arbitrary number of reversible chemical reactions.

A typical difficulty associated with traveling waves in infinite reactive media $(-\infty, +\infty)$ is that of boundary conditions, which need to be equilibrium points of the source terms. However, reactive source terms generally have a single equilibrium state in atom conservation manifolds. In the context of flames—which does not decrease the generality of the problem, but makes things more explicit—this is the cold boundary difficulty. More specifically, the hot boundary condition, say, at $+\infty$, corresponds to hot combustion products and is an equilibrium point, but source terms do not vanish at the cold boundary $-\infty$. For unsteady waves, the solution to the cold boundary difficulty is to use unsteady boundary conditions, as suggested by Zeldovitch in [AZK79] and investigated mathematically by Roquejoffre [Roq90]. For steady waves, however, the most satifying model is that of anchored waves.

The anchored flame model has been introduced by Hirschfelder et al. [HCC53] and represents an idealized adiabatic flame holder located at the origin. The corresponding domain is the half line $[0, \infty)$, and the boundary condition at $x = 0$ suppresses the cold boundary difficulty without artificially perturbing source terms. It modelizes an experimental configuration in which the flame is dynamically stabilized far away from the injection device by using laser tomography [Qal84] [Cla85]. This model also corresponds to boundary conditions often used in numerical simulations of complex chemistry flames [Ser86] [GS92]. A second type of boundary condition that we consider is that of an infinitely far cold boundary. The corresponding domain is then the full line $(-\infty, \infty)$. In this situation, we use a Heavyside cutoff function for the source term to remove the cold boundary difficulty, and both models are shown to be equivalent.

We first consider the problem on a bounded domain $[0, a]$. The existence theorem is obtained by using a fixed point formulation and the Leray–Schauder topological degree. The definition of the degree is obtained by estimating a priori the solutions. To this purpose, we use, in particular, the fundamental diffusion inequality established in Chapter 7 in conjunction with the entropy conservation equation. This inequality yields a natural entropy production norm for diffusive processes that is a solution weighted norm involving mass fractions at the denominator of mass fraction gradients squared.

We then let $a \to \infty$ and obtain an existence theorem as well as an exponential convergence towards equilibrium at infinity. We derive, in particular, a lower bound independent of a for the eigenvalue of the flame problem. To these purposes, an important tool is the exponential decay of entropy production residuals close to equilibrium, derived by using entropic estimates and thermodynamic stability inequalities.

The governing equations for steady plane flames are presented in Section 11.2. The equivalence of both problem formulations is obtained in Section 11.3. Existence on a bounded domain is obtained in Section 11.4, and, finally, existence of a solution is obtained in Section 11.5.

11.2. Governing equations

The governing equations for steady plane laminar flames with complex chemistry and detailed transport are obtained from the isobaric multicomponent reactive flow equations (3.3.9)–(3.3.13) presented in Chapter 3.

11.2.1. Conservation equations

We consider a steady plane laminar flame and denote by x the coordinate normal to the flame. The velocity components are $\boldsymbol{v} = (u, 0, 0)^t$, and we denote by "$\,'\,$" the space derivative operator with respect to the normal variable x. From the total mass conservation equation (3.3.9), we first observe that $(\rho u)' = 0$, so that ρu is a constant denoted by c

$$\rho u = c. \tag{11.2.1}$$

This constant c represents a mass flow rate per unit surface and is an eigenvalue of the problem. The normal momentum conservation equation (3.3.11) can then be discarded since it uncouples and would be only useful for evaluating the perturbed pressure $\widetilde{p}$.

Denoting by

$$\boldsymbol{\mathcal{F}}_k = (\mathcal{F}_k, 0, 0)^t$$

the diffusion flux of the kth species, we can write the species conservation equations (3.3.10) in the form

$$c\,Y_k' + \mathcal{F}_k' = m_k \omega_k, \qquad k \in S, \tag{11.2.2}$$

where Y_k is the mass fraction of the kth species, $\mathcal{F}_k$ is the mass flux of the kth species in the normal direction, m_k is the molar mass of the kth species, ω_k is the molar production of the kth species, $S = \{1, \dots, n\}$ is the set of species indices, and n is the number of species.

The energy conservation equation in the enthalpy form (3.3.12) yields

$$c\,h' + q' = 0, \tag{11.2.3}$$

where h is the specific enthalpy of the mixture and q is the normal heat flux in such a way that

$$\boldsymbol{\mathcal{Q}} = (q, 0, 0)^t.$$

By using the expressions of the mixture enthalpy h and of the heat flux q, a governing equation for the absolute temperature T will be obtained in Section 11.2.5.

The unknowns are the mass flow rate c, which is a nonlinear eigenvalue of the problem, the mass fraction vector $Y = (Y_1, \dots, Y_n)^t$, and the enthalpy h—or, equivalently, the absolute temperature T. It will be convenient to introduce the combined unknown ξ defined by

$$\xi = (h,\, Y_1, \dots,\, Y_n)^t. \tag{11.2.4}$$

The corresponding fluxes are denoted by

$$\phi = (q,\, \mathcal{F}_1, \dots,\, \mathcal{F}_n)^t, \tag{11.2.5}$$

and the corresponding sources are denoted by

$$w = (0, m_1\omega_1, \ldots, m_n\omega_n)^t. \tag{11.2.6}$$

The conservation equations can then be written in the compact form

$$c\,\xi' + \phi' = w, \tag{11.2.7}$$

and these equations have to be completed by the relations expressing the transport fluxes $\mathcal{F}_k$, $k \in S$, and q, the thermodynamic properties, such as the enthalpy h, and the chemical source terms ω_k, $k \in S$.

11.2.2. Thermodynamic properties

Since the mass fractions appear as the natural conservative variables in the species governing equations, we will use the thermodynamic formalism introduced in Section 6.3. The notation and assumptions (Th_1)–(Th_4) presented in Chapter 6 are used throughout the chapter. We only indicate here the simplifications arising from the isobaric approximation.

In agreement with the small Mach number approximation presented in Section 3.3, the pressure is taken to be a positive given constant

$$p = \mathrm{Cte}. \tag{11.2.8}$$

The entropy of the kth species can now be written as

$$s_k(T, Y) = s_k^{\mathrm{st},p} + \int_{T^{\mathrm{st}}}^{T} \frac{c_{pk}(t)}{t}\, dt - \frac{R}{m_k} \log(\mathcal{X}_k), \tag{11.2.9}$$

where $s_k^{\mathrm{st},p}$ is the formation entropy of the kth species at the positive standard temperature T^{st} and pressure p. It will be convenient to define

$$s_k^p(T) = s_k^{\mathrm{st},p} + \int_{T^{\mathrm{st}}}^{T} \frac{c_{pk}(t)}{t}\, dt, \tag{11.2.10}$$

which represents the entropy of the kth species at pressure p and the standard free enthalpy g_k^p at pressure p

$$g_k^p(T) = h_k(T) - s_k^p(T)\, T. \tag{11.2.11}$$

Finally, we also introduce for convenience the reduced quantities

$$\mu_k^p(T) = \frac{g_k^p(T)}{RT}, \qquad \mu_k(T, Y) = \frac{g_k}{RT} = \mu_k^p(T) + \frac{1}{m_k} \log(\mathcal{X}_k), \tag{11.2.12}$$

which naturally appear in various thermochemistry expressions.

11.2.3. Maxwellian chemistry

We consider a system of n^{r} reversible reactions for n species, as described in Section 6.4. The molar production rates have already been discussed in Chapter 6, and we only discuss here the simplifications arising from the constant pressure assumption. More specifically, we can now write the rate τ_i in the form

$$\tau_i = \overline{K}_i^{\mathrm{f}} \prod_{l\in S} \mathcal{X}_l^{\nu_{li}^{\mathrm{f}}} - \overline{K}_i^{\mathrm{b}} \prod_{l\in S} \mathcal{X}_l^{\nu_{li}^{\mathrm{b}}}, \qquad (11.2.13)$$

where $\mathcal{X}_k$ is given by (6.3.8) and $\overline{K}_i^{\mathrm{f}}$ and $\overline{K}_i^{\mathrm{b}}$ are modified direct and reverse rate constants of the ith reaction, respectively. These expressions are more suited to flame problems and are easily obtained, after a little algebra, making use of the state law and of the isobaric approximation. The equilibrium constant corresponding to the expressions (11.2.13) is now

$$\log \overline{K}_i^{\mathrm{e}}(T) = -\sum_{k\in S} \nu_{ki} m_k \,\frac{g_k^p(T)}{RT}, \qquad (11.2.14)$$

where $g_k^p(T)$ is defined by (11.2.11), so that $\overline{K}_i^{\mathrm{f}} = \overline{K}_i^{\mathrm{b}}\, \overline{K}_i^{\mathrm{e}}$, $i \in R$.

11.2.4. Transport fluxes

The mathematical properties of transport coefficients have already been investigated in Chapter 7. Assumptions (Tr_1)–(Tr_5) are assumed to hold throughout the chapter. In particular, we now make use of the assumptions (Tr_{4b}) and (Tr_5) since we have no a priori control of the mass fractions positivity, keeping in mind, for instance, that some of the cold state mass fractions may vanish. We indicate here the simplifications arising from the one-dimensional setting and the constant pressure assumption.

Under the isobaric approximation, the species fluxes $\mathcal{F}_k$, $k \in S$, and the heat flux q are taken in the form

$$\mathcal{F}_k = -\sum_{l\in S} C_{kl}\big(\mathcal{X}_l' + \mathcal{X}_l \widetilde{\chi}_l (T'/T)\big), \qquad k \in S, \qquad (11.2.15)$$

$$q = -\lambda T' + RT \sum_{l\in S} (\widetilde{\chi}_l/m_l)\mathcal{F}_l + \sum_{l\in S} h_l \mathcal{F}_l, \qquad (11.2.16)$$

where $C = (C_{kl})_{k,l\in S}$ is the flux diffusion matrix, $\mathcal{X}_k$ is the quantity given in (6.3.8), $\widetilde{\chi} = (\widetilde{\chi}_1, \dots, \widetilde{\chi}_n)^t$ is the rescaled thermal diffusion ratios vector, and λ is the thermal conductivity.

For positive mass fractions, one can further introduce the species diffusion velocities in the normal direction $\mathcal{V}_k$, $k \in S$, defined by

$$\mathcal{V}_k = \frac{\mathcal{F}_k}{\rho Y_k}, \qquad (11.2.17)$$

and from (11.2.15), one can express these velocities as

$$\mathcal{V}_k = -\sum_{l\in S} D_{kl}\big(\mathcal{X}_l' + \mathcal{X}_l\widetilde{\chi}_l(T'/T)\big), \qquad k\in S, \tag{11.2.18}$$

where $D = (D_{kl})_{k,l\in S}$ is the diffusion matrix defined by $D_{kl} = C_{kl}/(\rho Y_k)$, $k,l \in S$. The diffusion matrix D is symmetric positive semidefinite and associated with entropy production, as shown in Section 2.6.

Note that all species' second derivatives are coupled through the flux diffusion coefficients C and that multicomponent fluxes naturally involve mole fractions derivatives. In addition, species and temperature derivatives are coupled through the thermal diffusion coefficients $\widetilde{\chi}$. These effects were first considered in flame models by the author in [Gio97], [Gio98], and [Gio99].

11.2.5. The temperature equation

In order to derive a governing equation for the absolute temperature T, we can use equations (11.2.2) and (11.2.3) and the definition of the mixture enthalpy h. After a little algebra, we easily obtain

$$c\,c_pT' = -\Big(q - \sum_{k\in S} h_k\mathcal{F}_k\Big)' - \sum_{k\in S} c_{pk}T'\mathcal{F}_k - \sum_{k\in S} h_k m_k\omega_k. \tag{11.2.19}$$

Using now the expression (11.2.16) for the heat flux q, we finally obtain the temperature equation

$$c\,c_pT' = \Big(\lambda T' - RT\sum_{l\in S}(\widetilde{\chi}_l/m_l)\mathcal{F}_l\Big)' - \sum_{k\in S} c_{pk}T'\mathcal{F}_k - \sum_{k\in S} h_k m_k\omega_k. \tag{11.2.20}$$

It is worthwhile to point out several differences between the temperature equation (11.2.20) and the temperature equations considered in previous work. In all previous work, it has been assumed that the species specific heats are species independent, that is, $c_{pk} = \bar{c}_p$, $k \in S$. In this situation, one has

$$\sum_{k\in S} c_{pk}T'\mathcal{F}_k = \bar{c}_p\,T'\Big(\sum_{k\in S}\mathcal{F}_k\Big) = 0,$$

using the mass constraint $\sum_{k\in S}\mathcal{F}_k = 0$. In this situation, one also has

$$\sum_{k\in S} h_k m_k\omega_k = \sum_{k\in S} h_k^{\mathrm{st}} m_k\omega_k,$$

using the mass conservation relation $\sum_{k\in S} m_k\omega_k = 0$, since the temperature dependent part of the enthalpies $h_k - h_k^{\mathrm{st}} = \bar{c}_p(T - T^{\mathrm{st}})$ are identical.

Moreover, the transport and specific heat coefficients were generally assumed to be constant with a resulting simplified equation in the form

$$c\bar{c}_p T' = \lambda T'' - \sum_{k\in S} h_k^{\mathrm{st}} m_k \omega_k.$$

In comparison, we first note that in the temperature equation (11.2.20) the source term $\sum_{k\in S} h_k m_k \omega_k$ is not bounded a priori even when assuming that the rates ω_k, $k \in S$, are bounded, since the enthalpies are linearly increasing functions of temperature. In addition, there is a quadratic derivative term $\sum_{k\in S} c_{pk} T' \mathcal{F}_k$, and the coefficients λ and c_p are not constants. Finally, maximum principles cannot be used when thermal diffusion is included, that is, when $\widetilde{\chi}$ is nonzero.

11.2.6. Boundary conditions

We now specify the flame boundary conditions at the cold and hot boundaries. We assume that the cold boundary is on the left side and the hot equilibrium boundary is on the right side.

The right boundary condition is in the form

$$\xi(\infty) = \xi^{\mathrm{e}}, \qquad (11.2.21),$$

where ξ^{e} is an equilibrium point where source terms vanish $w(\xi^{\mathrm{e}}) = 0$. Existence and uniqueness of the proper equilibrium ξ^{e} is discussed in Section 11.2.7. In particular, this equilibrium state is such that $Y^{\mathrm{e}} > 0$, that is, satisfies $Y_k^{\mathrm{e}} > 0$, $k \in S$, where the superscript ^{e} is used to denote the value at $\xi = \xi^{\mathrm{e}}$ of any function of the state variables.

On the other hand, we consider two types of cold boundary conditions. The first type corresponds to the anchored flame model with an unknown function ξ defined on the the half line $[0, \infty)$ and with the boundary conditions

$$c(\xi(0) - \xi^{\mathrm{f}}) + \phi(0) = 0 \qquad (11.2.22)$$

and

$$T(0) = T^{\mathrm{i}}, \qquad (11.2.23)$$

where ξ^{f} is a given nonequilibrium state and T^{i} is a temperature such that $T^{\mathrm{f}} < T^{\mathrm{i}}$. The state ξ^{f} is such that $Y^{\mathrm{f}} \geq 0$, $\sum_{k\in S} Y_k^{\mathrm{f}} = 1$, and each species of the mixture is reachable from Y^{f} by the chemical network. The superscript ^{f} is used to denote the value at $\xi = \xi^{\mathrm{f}}$ of any function of the state variables.

The anchored flame model has been introduced by Hirschfelder et al. [HCC53]. This model corresponds to an idealized adiabatic flame holder located at the origin. The anchored flame model suppresses the cold boundary difficulty without any artificial modification of the source term [HCC53]

[BuLu82] [Wil85]. This model also corresponds to practical experimental configurations. Indeed, it is possible to inject a reactive mixture through a porous plate in a tube and stabilize a flame far from the injection device by using laser tomography, which triggers the injection velocity [Qal84] [Cla85]. For such flames, by integrating the conservation equations through the adiabatic porous burner, we obtain Equation (11.2.22). By choosing a temperature T^{i} slightly above the cold mixture temperature, we then recover the model. For exothermic systems, the temperature T^{i} can also be interpreted as an ignition temperature [Wil85]. Finally, the boundary conditions (11.2.22) (11.2.23) are also used in numerical modeling of complex chemistry flames [Ser86] [GS92].

It is also possible to consider an unknown function ξ defined on the real line $(-\infty, \infty)$ and replace (11.2.22) by the condition

$$\xi(-\infty) = \xi^{\mathrm{f}}. \tag{11.2.24}$$

The relation (11.2.23) is still needed to remove the translational invariance of the model. However, it is well-known that the chemistry source terms only vanish at equilibrium points, so that we have $w(\xi^{\mathrm{f}}) \neq 0$ and (11.2.24) cannot be satisfied. This problem is the well known "cold boundary difficulty." Various cutoff functions ψ have then been used to modify the source terms w into ψw, in order to satisfy the boundary condition (11.2.24). In this paper, we consider a cutoff function in the form $\psi(x) = 1_{[0,\infty)}(x)$ and, thus, source terms in the form $1_{[0,\infty)}w$. For this cutoff function, we establish rigorously in Section 11.3 that both formulations on $(-\infty, \infty)$ and $[0, \infty)$ are equivalent. As a consequence, it will be sufficient to investigate the anchored flame problem on $[0, \infty)$.

In both situations, the adiabatic plane flame velocity is defined by c/ρ^{f}, where c is the mass flow rate eigenvalue. This plane flame velocity is a characteristic of the flowing mixture and of fundamental importance in flame theory [Wil85].

Note that, when a continuous function of the state variables is used as a cutoff function, it introduces artificial equilibrium points in the phase space, and the dynamics of the reactive system has to guarantee that such artificial points are not further reached in the flame. For a priori exothermic systems, any increasing function of temperature can be used as a cutoff function. However, this is no longer the case for arbitrary reversible chemical networks, due to the lack of a priori monotonicity of the temperature and, more generally, of state functions when multicomponent transport is considered.

Remark 11.2.1. Note the difference between anchored flames and burner stabilized flames for which the mass flux is imposed. Denoting by c^{f} this prescribed mass flux, burner stabilized flames satisfy the boundary conditions

$$c^{\mathrm{f}}\left(Y(0) - Y^{\mathrm{f}}\right) + \mathcal{F}(0) = 0, \tag{11.2.25}$$

$$T(0) = T^{\mathrm{f}}, \tag{11.2.26}$$

and the imposed mass flux c^{f} is such that $0 < c^{\mathrm{f}} < c$, where c is the eigenvalue of the anchored flame problem. The anchored flame problem thus corresponds to an adiabatic burner with $c\big(h(0) - h^{\mathrm{f}}\big) + q(0) = 0$, whereas heat losses arise for burner stabilized flames with $T(0) = T^{\mathrm{f}}$. ■

11.2.7. Equilibrium limit

In order to characterize the equilibrium point ξ^{e}, we multiply (11.2.2) by any vector $u \in (M\mathcal{R})^{\perp}$ and obtain

$$c\,\langle Y, u\rangle' + \langle \mathcal{F}, u\rangle' = 0. \qquad (11.2.27),$$

Integrating formally over $[0,\infty)$—or, equivalently, over $(-\infty,\infty)$—we deduce that

$$\langle Y^{\mathrm{e}} - Y^{\mathrm{f}}, u\rangle = 0, \tag{11.2.28}$$

keeping in mind that we only seek solutions with a positive mass flow rate c. Note that such formal integrations can easily be justified rigorously. As a consequence, the equilibrium point ξ^{e} must be such that $Y^{\mathrm{e}} \in Y^{\mathrm{f}} + M\mathcal{R}$. Similarly, integrating (11.2.3) yields

$$h^{\mathrm{e}} = h^{\mathrm{f}}, \tag{11.2.29}$$

and, from the isobaric approximation, we also have $p^{\mathrm{e}} = p^{\mathrm{f}} = p$.

As a consequence, we have to investigate existence and uniqueness of the equilibrium point with a given specific enthalpy $h^{\mathrm{e}} = h^{\mathrm{f}}$, a given pressure $p^{\mathrm{e}} = p^{\mathrm{f}}$, and such that Y^{e} is in the conservation simplex

$$\mathfrak{Y} = (Y^{\mathrm{f}} + M\mathcal{R}) \cap (0,\infty)^n \cap \{\, Y, \sum_{k\in S} Y_k h_k^0 < h^{\mathrm{f}} \,\}. \tag{11.2.30}$$

Assuming that $T^{\mathrm{f}} > 0$, $Y^{\mathrm{f}} \geq 0$, $\langle Y^{\mathrm{f}}, \mathcal{U}\rangle = 1$, and $(Y^{\mathrm{f}} + M\mathcal{R}) \cap (0,\infty)^n \neq \emptyset$, and under Assumptions (Th_1)–(Th_4), we have seen in Proposition 6.5.5 that there exists a unique equilibrium vector Y^{e} in the simplex $\mathfrak{Y}$ where the source term ω vanishes. At Y^{e}, the reaction rates of progress also vanish and $\mu^{\mathrm{e}} \in (M\mathcal{R})^{\perp}$.

We have also seen in Proposition 6.5.8 that the equilibrium point ξ^{e} only depends on $\mathcal{P}(\xi^{\mathrm{f}})$ and is a smooth function of $\mathcal{P}(\xi^{\mathrm{f}})$, where $\mathcal{P}$ denotes the orthogonal projector onto the linear space $(0, M\mathcal{R})^{\perp}$. We have also established the local stability inequality of Corollary 6.7.4 between the difference of entropy with equilibrium and the chemical dissipation rate. Further assuming that there are no boundary equilibrium points, we have also established the global inequality of Proposition 6.8.1.

These extra assumptions concerning the cold state ξ^{f} and the equilibrium temperature T^{e} can now be summarized as follows.

(Th_5) The cold state ξ^{f} is such that $T^{\mathrm{f}} > 0$, $Y^{\mathrm{f}} \geq 0$, $\langle Y^{\mathrm{f}}, \mathcal{U} \rangle = 1$,

$$(Y^{\mathrm{f}} + M\mathcal{R}) \cap (0,\infty)^n \neq \emptyset,$$

and we have $T^{\mathrm{f}} < T^{\mathrm{i}} < T^{\mathrm{e}}$.

(Th_6) There are no boundary equilibrium points in the set $\overline{\mathfrak{Y}}$ defined by

$$\overline{\mathfrak{Y}} = (Y^{\mathrm{f}} + M\mathcal{R}) \cap [0,\infty)^n \cap \{\, Y,\ \sum_{k\in S} Y_k h_k^0 < h^{\mathrm{f}} \,\}. \tag{11.2.31}$$

Note that, for flame problems, the local inequality of Proposition 6.7.3 and Corollary 6.7.4 only involve high temperatures and small characteristic times. The global inequality of Proposition 6.8.1, however, involves low temperatures and large characteristic times depending on the lower temperature bound.

11.2.8. The matrix $\mathcal{L}$

The transport fluxes ϕ are naturally expressed in terms of the gradients of the state variables T and Y. In order to manipulate the conservation equations, we need to express ϕ in terms of the gradients of the conservative variables h and Y. Note, however, that symmetrization of the flame equations is not needed since we investigate one-dimensional waves.

Definition 11.2.2. *Define the matrix $\mathcal{L}$ by*

$$\mathcal{L} = \begin{pmatrix} 1 & \widehat{h}_1,\ldots,\widehat{h}_n \\ 0 & I \end{pmatrix} \begin{pmatrix} \lambda/c_p & 0 \\ C\chi/c_pT & C\,\partial_Y\mathcal{X} \end{pmatrix} \begin{pmatrix} 1 & -h_1,\ldots,-h_n \\ 0 & I \end{pmatrix}, \tag{11.2.32}$$

where $\widehat{h}_k = h_k + RT\widetilde{\chi}_k/m_k$, $k \in S$. Then we have

$$\phi = -\mathcal{L}\,\xi', \tag{11.2.33}$$

and $\mathcal{L}$ is such that $R(\mathcal{L}) = \mathfrak{U}^\perp$, where $\mathfrak{U} = (0,\mathcal{U})^t$ and $N(\mathcal{L}) = \mathbb{R}\xi$.

Proof. A straightforward calculation directly yields the expression of $\mathcal{L}$ and the identity $\phi = -\mathcal{L}\xi'$. From this expression, the relation (6.3.37)

$$\partial_Y\mathcal{X} = \frac{1}{\langle Y,\mathcal{U}\rangle}\Big(I - \frac{X\otimes\mathcal{U}}{\langle Y,\mathcal{U}\rangle}\Big)\,E,$$

where $X = (X_1,\ldots,X_n)^t$, and the identity $\langle Y,\mathcal{U}\rangle = \langle X,\mathcal{U}\rangle$, we then obtain that $R(\mathcal{L}) = \mathfrak{U}^\perp$ and $N(\mathcal{L}) = \mathbb{R}\xi$. Indeed, we have $C(\partial_Y\mathcal{X})\,x = 0$ if and only if $(\partial_Y \mathcal{X})x = 0$ since $R(\partial_Y \mathcal{X}) = \mathcal{U}^\perp$ and $N(C) \cap \mathcal{U}^\perp = \{0\}$. On the other hand, thanks to $EY = X$, we have $(\partial_Y \mathcal{X})x = 0$ if and only if $Ex \in \mathbb{R}X$, that is, if and only if $x \in \mathbb{R}Y$. ∎

The following proposition is essentially a direct consequence of classical results concerning generalized inverses with prescribed range and nullspace obtained in Proposition 7.3.5 [BG74]. The smoothness results from the inversion formula $\mathcal{L}^\sharp = (\mathcal{L} + \alpha\xi \otimes \mathfrak{U})^{-1} - \beta\xi \otimes \mathfrak{U}$, valid for positive α and β with $\alpha\beta\langle Y, \mathcal{U}\rangle^2 = 1$.

Proposition 11.2.3. *Assume that $T > 0$ and $Y \geq 0$, $Y \neq 0$. Then there exists a unique matrix $\mathcal{L}^\sharp$ such that $\mathcal{L}\mathcal{L}^\sharp\mathcal{L} = \mathcal{L}$, $\mathcal{L}^\sharp\mathcal{L}\mathcal{L}^\sharp = \mathcal{L}^\sharp$, $N(\mathcal{L}^\sharp) = N(\mathcal{L})$, and $R(\mathcal{L}^\sharp) = R(\mathcal{L})$. The matrix $\mathcal{L}^\sharp$ is a smooth function of $T > 0$ and $Y \geq 0$, $Y \neq 0$, and $\mathcal{L}\mathcal{L}^\sharp = \mathcal{L}^\sharp\mathcal{L}$ is the oblique projector onto $R(\mathcal{L})$ parallel to $N(\mathcal{L})$.*

Finally, we specialize the flux decomposition of Lemma 7.3.4 to the case of one-dimensional flows under the isobaric approximation, assuming that the only force acting on the species is gravity.

Lemma 11.2.4. *The diffusion flux $\mathcal{F}_k$ of the kth species can we written in the form*

$$\mathcal{F}_k = -\mathfrak{d}_k Y_k' - Y_k \mathfrak{f}_k, \tag{11.2.34}$$

where

$$\mathfrak{d}_k = \frac{C_{kk}}{m_k} \frac{\overline{m}}{\langle Y, \mathcal{U}\rangle} \tag{11.2.35}$$

and

$$\begin{aligned} \mathfrak{f}_k = & -\frac{C_{kk}}{m_k} \frac{\overline{m}^2}{\langle Y, \mathcal{U}\rangle^2} \sum_{l \in S} \frac{Y_l'}{m_l} \\ & + \frac{C_{kk}}{m_k} \widetilde{\chi}_k \frac{\overline{m}}{\langle Y, \mathcal{U}\rangle} \frac{T'}{T} + \sum_{\substack{l \in S \\ l \neq k}} \rho D_{kl} \big(\mathcal{X}_l' + \mathcal{X}_l \widetilde{\chi}_l \frac{T'}{T}\big), \end{aligned} \tag{11.2.36}$$

so that the nondiagonal part of the multicomponent flux $\mathcal{F}_k$ vanishes for $Y_k = 0$. Moreover, we always have $\mathfrak{d}_k \geq 0$, and $\mathfrak{d}_k = 0$ if and only if $Y_l = 0$ for $l \neq k$ and $Y_k > 0$. In particular, whenever $\langle Y, \mathcal{U}\rangle = 1$, $\mathfrak{d}_k/\varphi$ is positive and bounded away from zero when $0 \leq Y_k \leq \delta$ for any $\delta < 1$.

11.2.9. Entropy conservation equation

The entropy conservation equation will play a fundamental role in the analysis. In this section, we specialize the results of Section 2.6 to the one-dimensional setting and the constant pressure assumption.

Lemma 11.2.5. *Assume that (T, Y) is a regular solution of the flame equations and that T and Y are positive. Then the entropy function $s(h, Y)$ satisfies the equation*

$$c\, s' + \langle \partial_\xi s, \phi \rangle' = \mathfrak{v}, \tag{11.2.37}$$

where the dissipation rate $\mathfrak{v}$ is given by

$$\mathfrak{v} = \lambda \Big(\frac{T'}{T}\Big)^2 + \frac{p}{T} \sum_{k,l \in S} D_{kl} \big(\mathcal{X}_k' + \chi_k \frac{T'}{T}\big) \big(\mathcal{X}_l' + \chi_l \frac{T'}{T}\big) - \sum_{k \in S} \frac{g_k m_k \omega_k}{T}. \tag{11.2.38}$$

Proof. We have $s' = \langle \partial_\xi s, \xi' \rangle$, so that, from the conservation equations, we obtain

$$c\, s' = \langle \partial_\xi s, -\phi' + w \rangle = -\langle \partial_\xi s, \phi \rangle' + \mathfrak{v},$$

where $\mathfrak{v}$ is defined by $\mathfrak{v} = \big\langle (\partial_\xi s)', \phi \big\rangle + \langle \partial_\xi s, w \rangle$. The term $\langle \partial_\xi s, w \rangle$ is the entropy production rate due to chemistry $\langle \partial_\xi s, w \rangle = R\,\zeta$, and we have

$$\big\langle (\partial_\xi^2 s) \xi', \phi \big\rangle = \big\langle (\partial_\xi s)', \phi \big\rangle = \Big(\frac{1}{T}\Big)' q - \sum_{k \in S} \Big(\frac{g_k}{T}\Big)' \mathcal{F}_k,$$

since $\partial_h s = 1/T$ and $\partial_{Y_k} s = -g_k / T$. Using

$$\Big(\frac{g_k}{T}\Big)' = -\frac{h_k}{T^2} T' + \frac{R}{m_k} \frac{\mathcal{X}_k'}{\mathcal{X}_k},$$

we obtain, after a little algebra,

$$\big\langle (\partial_\xi^2 s) \xi', \phi \big\rangle = -\frac{(q - \sum_{k \in S} h_k \mathcal{F}_k) T'}{T^2} - \sum_{k \in S} \frac{R}{m_k} \frac{\mathcal{X}_k'}{\mathcal{X}_k} \mathcal{F}_k,$$

which can be rewritten, by using the expression of the heat flux, in the form

$$\big\langle (\partial_\xi^2 s) \xi', \phi \big\rangle = \frac{(\lambda T' - RT \sum_{k \in S} (\widetilde{\chi}_k / m_k) \mathcal{F}_k) T'}{T^2} - \sum_{k \in S} \frac{R}{m_k} \frac{\mathcal{X}_k'}{\mathcal{X}_k} \mathcal{F}_k,$$

and, finally,

$$\big\langle (\partial_\xi^2 s) \xi', \phi \big\rangle = \lambda (T'/T)^2 + \frac{p}{T} \sum_{k \in S} D_{kl} \big(\mathcal{X}_k' + \chi_k (T'/T)\big) \big(\mathcal{X}_l' + \chi_l (T'/T)\big),$$

and the proof is complete. ∎

Note that Equation (11.2.37) is only valid when the mass fractions Y_k, $k \in S$, are positive. Since $\langle (\partial_\xi^2 s) \xi', \phi \rangle$ is the nonreactive part of the dissipation rate, the Hessian matrix may also be seen as a metric that correlates the flux ϕ and the gradient ξ', but this metric is singular for zero mass fractions. The entropy equation also shows that the natural norm for multicomponent diffusion is not the W_2^1 Sobolev norm, but involves mass fractions at the denominator from (11.2.38) and (7.6.1).

11.3. First properties

In this section, we establish that each solution of the anchored flame problem can be extended uniquely into a solution on the full line satisfying (11.2.24). Conversely, we establish that the reduction to $[0, \infty)$ of any solution on $(-\infty, \infty)$ is a solution of the anchored flame problem.

11.3.1. Preliminaries

A triplet (T, Y, c) such that T and Y are $C^2[0, \infty)$, such that $T > 0$, $Y \geq 0$, $Y \neq 0$, and $c > 0$, and which satisfies pointwise the governing equations and boundary conditions (11.2.21)–(11.2.23), will be said to be a solution of the anchored flame problem. Similarly, a triplet (T, Y, c) such that T and Y are $C^2[0, \infty) \cap C^2[-\infty, 0]$ and $C^1(-\infty, \infty)$, such that $T > 0$, $Y \geq 0$, $Y \neq 0$, and $c > 0$, and which satisfies pointwise the governing equations and boundary conditions (11.2.21), (11.2.24), and (11.2.23), will be said to be a solution on the full line.

We first establish some a priori estimates for any solution of the anchored flame problem, which will be further needed.

Lemma 11.3.1. *Let (T, Y, c) be a solution of the anchored flame problem. Then the mass fractions are positive $Y > 0$ and sum to unity $\langle \mathcal{U}, Y \rangle = 1$.*

Proof. Let (T, Y, c) be a solution of the anchored flame problem. We first deduce from the species equations that $c\langle \mathcal{U}, Y \rangle' = 0$ since $R(C) = \mathcal{U}^\perp$ and $M\omega \in \mathcal{U}^\perp$. This shows that $\langle \mathcal{U}, Y \rangle$ is constant, and this constant is equal to unity since the boundary condition (11.2.22) implies $\langle \mathcal{U}, Y \rangle = \langle \mathcal{U}, Y^{\mathrm{f}} \rangle = 1$.

Arguing by contradiction, we now establish that $Y(0) > 0$. We assume that there exists $k \in S$ such that $Y_k(0) = 0$. Then, using Lemma 11.2.4 and the boundary condition $c(Y_k(0) - Y_k^{\mathrm{f}}) + \mathcal{F}_k(0) = 0$, we deduce that

$$Y_k'(0) = -\frac{cY_k^{\mathrm{f}}}{\mathfrak{d}_k(0)}.$$

If $Y_k^{\mathrm{f}} > 0$, then Y is negative over an interval $(0, \epsilon)$ with $\epsilon > 0$, which contradicts the assumptions that (T, Y, c) is a solution of the anchored flame problem. Hence, if $Y_k(0) = 0$, we must have $Y_k^{\mathrm{f}} = 0$, and, consequently, we have $Y_k'(0) = 0$. By expanding the derivative $\mathcal{F}_k'$ in (11.2.2) and then letting $x = 0$, we further obtain

$$\mathfrak{d}_k Y_k'' = -m_k \mathcal{C}_k,$$

since $\mathcal{D}_k = 0$ for $Y_k = 0$. If $\mathcal{C}_k > 0$, then $Y_k''(0) < 0$, which is impossible since Y is assumed to be a solution of the anchored flame problem. Hence, we have $\mathcal{C}_k = 0$ at $x = 0$ and $Y_k''(0) = 0$.

We now introduce the set $\mathcal{I} = \{\, k \in S,\ Y_k(0) = 0 \,\}$, and, from the preceding discussion, we know that

$$\forall k \in \mathcal{I}, \qquad Y_k(0) = 0,\ Y_k'(0) = 0,\ Y_k''(0) = 0,\ \mathcal{C}_k(0) = 0. \tag{11.3.1}$$

As a consequence, any reaction creating a species $\mathfrak{M}_k$ with $k \in \mathcal{I}$ needs at least one species $\mathfrak{M}_i$ with $i \in \mathcal{I}$ as a reactant, otherwise, one would get a positive production term $\mathcal{C}_k$. This shows that the mass fractions Y_k, $k \in \mathcal{I}$, are factorized in the production terms ω_l, $l \in \mathcal{I}$. Now the mass fractions Y_k, $k \in \mathcal{I}$, are solutions of the system

$$cY_k' = \Big(\mathfrak{d}_k(x) Y_k' + Y_k \mathfrak{f}_k(x)\Big)' + m_k \omega_k, \qquad k \in \mathcal{I},$$

where $\mathfrak{d}_k$ and $\mathfrak{f}_k$ are defined in Lemma 11.2.4, and the mass fractions vanish at $x = 0$ with their first derivatives. However, since $\mathfrak{d}_k$, $k \in \mathcal{I}$, is locally positive, this system has a locally unique solution by virtue of the Cauchy–Lipschitz theorem. Since this system also admits the trivial solution, we conclude that locally, we have $Y_k(x) = 0$, for $x \in [0, \epsilon]$, and $k \in \mathcal{I}$. An easy argument then yields that $Y_k(x) = 0$, for $x \in [0, \infty)$, and $k \in \mathcal{I}$, contradicting $Y_k(\infty) = Y_k^{\mathrm{e}} > 0$.

We now have to show that $Y > 0$ over $(0, \infty)$. Arguing by contradiction, assume that there exists a point $x_0 > 0$ and $k \in S$ such that $Y_k(x_0) = 0$. We may also assume that x_0 is the smallest point where a species vanishes. We then have $Y_k'(x_0) \le 0$ by construction. If $Y_k'(x_0) < 0$, we obtain a contradiction with the fact that Y is nonnegative. On the other hand, if $Y_k'(x_0) = 0$, we can then argue as above and show that Y_k is zero in the neighborhood of x_0, contradicting the definition of x_0, and the proof is complete. ∎

11.3.2. Reduction to a problem on $[0, \infty)$

Proposition 11.3.2. *Let* (T, Y, c) *be a solution on the full line. Then we have*

$$c\big(\xi(x) - \xi^{\mathrm{f}}\big) + \phi(x) = 0, \qquad x \le 0, \tag{11.3.2}$$

and the reduction of (T, Y, c) *to* $[0, \infty)$ *is a solution of the anchored flame problem.*

Proof. From the governing equations, we have $\langle Y', \mathcal{U} \rangle = 0$, so that $\langle Y, \mathcal{U} \rangle$ is a constant, and this constant is unity from $\langle Y^{\mathrm{f}}, \mathcal{U} \rangle = 1$. From the governing equations, the quantity $c\xi + \phi = c\xi - \mathcal{L}\xi'$ is a constant over $(-\infty, 0]$, say, $c\xi - \mathcal{L}\xi' = u$. Since $\xi' \in \mathfrak{U}^{\perp}$, we can write that $\xi' = \mathcal{L}^{\sharp}(c\xi - u)$. This shows that ξ' admits a limit as $x \to -\infty$ since $\xi(-\infty) = \xi^{\mathrm{f}}$, and this limit can only be zero. As a consequence, we must have $u = c\,\xi^{\mathrm{f}}$ and conclude that $c\xi + \phi = c\xi^{\mathrm{f}}$ over $(\infty, 0]$, so that the restriction to $[0, \infty)$ of (T, Y, c) is a solution of the anchored flame problem. ∎

11.3.3. Extension to $(-\infty, 0)$

In this section, we establish that any solution of the anchored flame problem can be extended over $(-\infty, 0)$. However, for a priori estimates, we will further need to extend any solution obtained on a bounded domain. As a consequence, we investigate the extension starting from any positive species at the origin and apply this result to both situations.

Proposition 11.3.3. *Let* $Y^0 > 0$ *and* $c > 0$ *be given such that* $\langle Y^0, \mathcal{U} \rangle = 1$, *and let* $\xi^0 = (h(T^{\mathrm{i}}, Y^0), Y^0)$. *Then there exists a unique solution* (T, Y) *to the flame equations over* $(-\infty, 0]$ *such that* $\xi(0) = \xi^0$ *and* $\xi(-\infty) = \xi^{\mathrm{f}}$. *Furthermore, this solution satisfies* (11.3.2) *and there exist positive constants* α *and* β *such that* $0 < \alpha \leq T \leq \beta$. *In addition, the mass fractions are positive, satisfy* $\langle Y, \mathcal{U} \rangle = 1$, *and the integral over* $(-\infty, 0]$ *of the dissipation rate* $\mathfrak{v}$ *is finite and given by*

$$\int_{-\infty}^{0} \mathfrak{v}\, dx = c\Big(s(0) - s^{\mathrm{f}} - \big\langle \partial_{\xi} s(0), \xi(0) - \xi^{\mathrm{f}} \big\rangle\Big). \tag{11.3.3}$$

Proof. We have established in the proof of Proposition 11.3.2 that any solution of the flame equations, such that (11.2.24) holds, must satisfy (11.3.2). As a consequence, any such solution satisfies the ordinary differential equation

$$z' = c\mathcal{L}^{\sharp}(z - \xi^{\mathrm{f}}), \tag{11.3.4}$$

with initial value $\xi(0) = \xi^0$. Now such a solution locally exists and is unique from the Cauchy–Lipschitz theorem, so that we only have to extend this solution over $(-\infty, 0]$.

Consider such a solution over $[x_0, 0]$ and, locally, such that the species are positive. It is easily checked that $\langle Y, \mathcal{U} \rangle = 1$, so that the mass fractions are bounded. From the governing equations, we obtain, after a little algebra,

$$c \sum_{k \in S} Y_k^{\mathrm{f}} (h_k - h_k^{\mathrm{f}}) - \lambda T' + c\, RT \sum_{k \in S} \frac{\widetilde{\chi}_k}{m_k} Y_k^{\mathrm{f}} = 0.$$

Consider any x_1 where T reaches its minimum over $[x_0, 0]$. If $T(x_1) \leq T^{\rm f}$, then we must either have $T'(x_1) = 0$ when $x_1 \in (x_0, 0)$ or $T'(x_1) \geq 0$ when $x_0 = x_1$, $x_1 = 0$ being excluded since $T^{\rm f} < T^{\rm i}$. As a consequence, if $T(x_1) \leq T^{\rm f}$, then

$$\sum_{k \in S} Y_k^{\rm f} \Big(h_k(T(x_1)) - h_k(T^{\rm f}) \Big) + RT(x_1) \sum_{k \in S} \frac{\widetilde{\chi}_k}{m_k} Y_k^{\rm f} \geq 0,$$

where we have used $c > 0$. Keeping in mind that $T(x_1) \leq T^{\rm f}$, we get

$$\underline{c}_p (T(x_1) - T^{\rm f}) + T(x_1)\overline{\chi} \geq 0,$$

so that $T \geq \alpha = \underline{c}_p\, T^{\rm f} / (\underline{c}_p + \overline{\chi})$ where $R \sum_{k \in S} |\widetilde{\chi}_k| Y_k^{\rm f} / m_k \leq \overline{\chi}$, and T is uniformly bounded from below.

This method cannot be used, however, to derive an upper bound for T, unless the thermal diffusion factors are small quantities. In particular, when $\widetilde{\chi} = 0$, one easily shows that $T^{\rm f} \leq T \leq T^{\rm i}$. For the general case $\widetilde{\chi} \neq 0$, we integrate the entropy conservation equation over $[x, 0]$

$$c(s(0) - s) + \langle \partial_\xi s, \phi \rangle(0) - \langle \partial_\xi s, \phi \rangle = \int_x^0 \mathfrak{v} \, d\hat{x},$$

so that, from $c(\xi - \xi^{\rm f}) + \phi = 0$, we obtain

$$c \left(s - s^{\rm f} - \langle \partial_\xi s, \xi - \xi^{\rm f} \rangle \right) = c \left(s(0) - s^{\rm f} - \langle \partial_\xi s(0), \xi(0) - \xi^{\rm f} \rangle \right) - \int_x^0 \mathfrak{v} \, d\hat{x}. \quad (11.3.5)$$

As a consequence, the functional $s - s^{\rm f} - \langle \partial_\xi s, \xi - \xi^{\rm f} \rangle$ is increasing over $[x_0, 0]$ and nonnegative by concavity. After a little algebra, we easily evaluate this functional and deduce that

$$\frac{h^{\rm f}}{T} - \sum_{k \in S} Y_k^{\rm f} \frac{h_k(T)}{T} + \sum_{k \in S} Y_k^{\rm f} s_k(T, Y) \leq s(0) - \langle \partial_\xi s(0), \xi(0) - \xi^{\rm f} \rangle.$$

Using now the lower bound for T and the expression (11.2.9), we get that

$$\sum_{k \in S} Y_k^{\rm f} \Big(-\frac{R}{m_k} \log X_k + \int_\alpha^T \frac{c_{pk}(t)}{t} dt \Big) \leq b,$$

where b is a positive constant that depends only on the data and on ξ^0. This implies that $T \leq \beta$ over $[x_0, 0]$, where β is independent of x_0, and that Y_k is bounded away from 0 when k is such that $Y_k^{\rm f} > 0$.

We now consider a maximal solution defined on $(x_1, 0]$ and argue by contradiction by assuming that x_1 is finite. In this situation, from the preceding estimates, and the smoothness of the coefficients, ξ' remains finite in the neighborhood of x_1, so that the limit $\xi(x_1)$ exists as well as $\xi'(x_1)$. As a consequence, if x_1 is finite, we have either $Y(x_1) > 0$ and the solution can be extended, contradicting the definition of x_1, or there exists $k \in S$ such that $Y_k(x_1) = 0$. Since $\forall x \in (x_1, 0]$, $\forall k \in S$, $Y_k(x) > 0$, we have $Y_k'(x_1) \geq 0$ by construction and, from (11.3.2) and (11.2.3), $Y_k'(x_1) = -cY_k^{\mathrm{f}}/\rho\mathfrak{d}_k$. When $Y_k^{\mathrm{f}} > 0$, this yields $Y_k'(x_1) < 0$, and we have obtained a contradiction. On the other hand, if $Y_k^{\mathrm{f}} = 0$, this yields $Y_k'(x_1) = 0$ and, thus, $Y_k = 0$ by the Cauchy–Lipschitz theorem applied to $z = Y_k$ solution of

$$cz = \mathfrak{d}_k(x)z' + z\mathfrak{f}_k(x),$$

where $\mathfrak{d}_k$ and $\mathfrak{f}_k$ are defined in Lemma 11.2.4 . We then obtain $Y_k = 0$ in the neighborhood of x_1, contradicting the definition of x_1. As a consequence, the solution can be extended over the half line $(-\infty, 0]$ and the temperature and mass fractions remain positive.

From the identity (11.3.5), we also obtain that the integral of the dissipation rate $\mathfrak{v}$ remains finite. Using now the estimate $\delta\langle\phi, \phi\rangle \leq \mathfrak{v}$, where δ is a positive constant and $\phi = -c(\xi - \xi^{\mathrm{f}})$, we obtain $\xi - \xi^{\mathrm{f}} \in W_2^1(-\infty, 0)$ and, thus, $\xi(-\infty) = \xi^{\mathrm{f}}$. This now shows that

$$c\Big(s(0) - s^{\mathrm{f}} - \langle\partial_\xi s(0), \xi(0) - \xi^{\mathrm{f}}\rangle\Big) = \int_{-\infty}^{0} \mathfrak{v}\, dx,$$

and that the upper bound for T is finally controlled by the data and the integral of the reduced dissipation rate $\mathfrak{v}/c$. ∎

We can now apply this result to any solution of the anchored flame problem since we know that $Y(0) > 0$.

Corollary 11.3.4. *Let* (T, Y, c) *be a solution of the anchored flame problem. Then this solution has a unique extension over* $(-\infty, 0)$, *such that* (11.2.24) *holds. Moreover, this solution satisfies* (11.3.2), *the mass fractions are positive over* $(-\infty, 0]$, *and the identity* (11.3.3) *holds.*

11.4. Existence on a bounded domain

In this section, we establish an existence theorem on a bounded domain $[0, a]$. In the next section, we will let $a \to \infty$ and obtain a solution of the anchored flame problem.

11.4.1. Preliminaries

For technical reasons, we need to extend the domain of definition of the equation coefficients. For this purpose, we introduce the definition

$$\widetilde{Y} = \begin{cases} Y^+, & \text{if } \langle Y, \mathcal{U} \rangle \geq 1, \\ Y^+ + \dfrac{1 - \langle Y, \mathcal{U} \rangle}{n} \mathcal{U}, & \text{if } \langle Y, \mathcal{U} \rangle \leq 1. \end{cases} \tag{11.4.1}$$

Note that we have $\widetilde{Y} \geq 0$ and $\langle \widetilde{Y}, \mathcal{U} \rangle \geq 1$ for any $Y \in \mathbb{R}^n$. In addition, whenever $\langle Y, \mathcal{U} \rangle = 1$, we have $\widetilde{Y} = Y^+$, so that both properties $Y \geq 0$ and $\langle Y, \mathcal{U} \rangle = 1$ imply that $\widetilde{Y} = Y$. For a fixed point formulation, the transport coefficients, thermodynamic properties, and chemical production rates can be taken to be functions of $(T, \widetilde{Y})$ and are thus defined for $T > 0$ and any $Y \in \mathbb{R}^n$.

In order to obtain a suitable fixed point formulation, it is preferable to control the temperature and thus to use (T, Y, c) as an unknown, rather than to use the specific enthalpy h. The enthalpy equation, however, is simpler than the temperature equation and can be integrated once, making use of the flux boundary conditions. In order to keep both advantages, we will solve the equations in the form

$$\xi' = \mathcal{L}^\sharp \Big(c(\xi - \xi^{\mathrm{f}}) - \int_0^x w \, d\hat{x} \Big),$$

with h' replaced by $c_p T' + \sum_{k \in S} h_k(T) Y_k'$ and h replaced by $\langle c_p \rangle T + \sum_{k \in S} h_k^0 Y_k$, where $\langle c_p \rangle = \sum_{k \in S} Y_k \int_0^1 c_{pk}(\theta T) \, d\theta$. That is, we will solve the problem in the form

$$\Big(c_p T' + \sum_{k \in S} h_k(T) Y_k', \, Y' \Big)^t =$$
$$\mathcal{L}^\sharp \bigg(c \Big(\langle c_p \rangle T + \sum_{k \in S} h_k^0 Y_k, \, Y \Big)^t - c\xi^{\mathrm{f}} - \int_0^x w \, d\hat{x} \bigg),$$

with the backward initial conditions $\xi(a) = \xi^{\mathrm{e}}$ and the extra relation $T(0) = T^{\mathrm{i}}$ used as an equation for the eigenvalue c.

In order to establish the existence of a solution, we will use the Leray–Schauder topological degree theory and the homotopy invariance of the degree. We refer the reader to [LS34] [Sch69] and [Dei88] for more details about the Leray–Schauder topological degree theory. We will first eliminate the chemistry in a first homotopy path and then simplify thermodynamic properties and transport fluxes along a second homotopy path. Evaluation of the resulting degree will conclude the existence proof.

11.4.2. Fixed point formulation

We consider the Banach space

$$\mathcal{B} = \big(C^1[0,a]\big)^{n+1} \times \mathbb{R} \tag{11.4.2}$$

and the open set $\mathcal{O} \subset \mathcal{B}$

$$\mathcal{O} = \{\ (T, Y, c) \in \mathcal{B},\ T > 0,\ c > 0\ \}. \tag{11.4.3}$$

We introduce the operator $\mathcal{K}_\tau\ :\ \mathcal{O} \longrightarrow \mathcal{B}$ defined by

$$\mathcal{K}_\tau(T, Y, c) = \Big(\mathfrak{t}, \mathfrak{y}, c + \mathfrak{t}(0) - T^{\mathrm{i}}\Big), \tag{11.4.4}$$

where $(\mathfrak{t}, \mathfrak{y})$ are solutions of the system

$$\begin{aligned}&\Big(c_p(T,\widetilde{Y})\,\mathfrak{t}' + \sum_{k\in S} h_k(T)\,\mathfrak{y}_k',\ \mathfrak{y}'\Big)^t = \\ &\mathcal{L}^\sharp(T,\widetilde{Y})\Big(c\Big(\langle c_p(T,\widetilde{Y})\rangle\,\mathfrak{t} + \sum_{k\in S} h_k^0 \mathfrak{y}_k,\ \mathfrak{y}\Big)^t - c\xi^{\mathrm{f}} - \tau\int_0^x w(T,\widetilde{Y})\,d\hat{x}\Big),\end{aligned} \tag{11.4.5}$$

with the backward initial conditions

$$\mathfrak{t}(a) = T^{\mathrm{e}}, \qquad \mathfrak{y}(a) = Y^{\mathrm{e}}. \tag{11.4.6}$$

Note that nonlinearities are taken as functions of $(T, \widetilde{Y})$ in the homotopy path and that for $\tau = 0$ the chemistry source terms vanish. The extra left boundary condition $T(0) = T^{\mathrm{i}}$ is also used as an equation for the eigenvalue c.

Proposition 11.4.1. *The operator $\mathcal{K}_\tau$ from $\mathcal{O}$ to $\mathcal{B}$ is well defined.*

Proof. Consider the backward ordinary differential equation (11.4.5) with initial condition $(\mathfrak{t}, \mathfrak{y})(a) = (T^{\mathrm{e}}, Y^{\mathrm{e}})$. This system of ordinary differential equations admits a unique solution from the Cauchy–Lipschitz theorem, since the right member is an affine function of $(\mathfrak{t}, \mathfrak{y})$ with coefficients that are Lipschitz continuous with respect to x [Bou76]. ■

We also have the following property, which allows us to use the homotopy invariance of the degree.

Lemma 11.4.2. *Let B be a closed bounded set in $\mathcal{O}$ and $\mathcal{K} :\ [0,1]\times B \to \mathcal{B}$ be the map defined by $\mathcal{K}\big(\tau, (T, Y, c)\big) = \mathcal{K}_\tau(T, Y, c)$. Then the map $\mathcal{K}$ is compact.*

Proof. We can easily establish from the definition of $(\mathfrak{t}, \mathfrak{y})$ that, for any $\big(\tau, (T, Y, c)\big) \in [0,1]\times B$, the functions $\mathfrak{t}$, $\mathfrak{y}$, $\mathfrak{t}'$, and $\mathfrak{y}'$ are uniformly bounded and that $\mathfrak{t}'$ and $\mathfrak{y}'$ are uniformly Lipschitzian. This implies that the range of $\mathcal{K}$ is relatively compact and that $\mathcal{K}$ is continuous by uniqueness of the solution of (11.4.5) and (11.4.6). ∎

We now introduce the open set Ω defined by

$$\Omega = \{ \ (T, Y, c) \in \mathcal{O}, \ \|(T, Y)\|_{C^1[0,a]} < \beta, \ \alpha < T < \beta, \ \alpha < c < \beta \ \}, \tag{11.4.7}$$

where α and β are positive constants, and we assume that $a \geq 1$.

Theorem 11.4.3. *There exist constants α and β such that, for all $\tau \in [0, 1]$, the Leray–Schauder degree $d(I - \mathcal{K}_\tau, \Omega, 0)$ is well defined.*

In order to establish this theorem, we have to show that, for α small enough and β large enough, we have

$$\forall \tau \in [0, 1], \qquad 0 \notin (I - \mathcal{K}_\tau)(\partial \Omega). \tag{11.4.8}$$

This will be obtained in the next section by estimating fixed points of $\mathcal{K}_\tau$.

11.4.3. Existence of the degree

We estimate in this section the fixed points (T, Y, c) of $\mathcal{K}_\tau$ for $\tau \in [0, 1]$. By definition, we have $T > 0$, $c > 0$, and $(T, Y, c) = (\mathfrak{t}, \mathfrak{y}, c + \mathfrak{t}(0) - T^{\mathrm{i}})$, so that the equations obtained from (11.4.5) with $\mathfrak{t} = T$ and $\mathfrak{y} = Y$ hold and $T(0) = T^{\mathrm{i}}$. A priori estimates are successively obtained in the following lemmas. Note that these estimates could also be conducted over $[0, \infty)$.

Lemma 11.4.4. *Let (T, Y, c) be a fixed point of $\mathcal{K}_\tau$. Then the mass fractions are positive $Y > 0$ and sum to unity $\langle \mathcal{U}, Y \rangle = 1$.*

Proof. We deduce from the species equations (11.4.5) that $c\langle \mathcal{U}, Y \rangle' = 0$ since we have $R(\mathcal{L}^\sharp) = \mathfrak{U}^\perp$. As a consequence, $\langle \mathcal{U}, Y \rangle$ is a constant that is unity from (11.4.6), and this shows that $\widetilde{Y} = Y^+$

Step 1. We first establish that $Y(0) \geq 0$. Arguing by contradiction, assume that there exists $k \in S$ such that $Y_k(0) < 0$. Then, from the species equations written at $x = 0$, we have $c(Y_k(0) - \xi_k^{\mathrm{f}}) - \sum_{l \in S} \mathcal{L}_{kl}\big((T(0), Y^+(0)\big) Y_l'(0) = 0$, and we also have $Y_k^+(0) = 0$, so that, from Lemma 11.2.4, we deduce that

$$Y_k'(0) = \frac{c}{\mathfrak{d}_k(0)} (Y_k(0) - Y_k^{\mathrm{f}}) < 0.$$

Then Y_k is negative over an interval of the form $(0, \epsilon)$ with $\epsilon > 0$. Over this interval $(0, \epsilon)$, we now have

$$\mathfrak{d}_k Y_k' = c(Y_k - Y_k^{\mathrm{f}}) - \tau \int_0^x m_k \mathcal{C}_k(T, \widetilde{Y}) \, d\hat{x},$$

since $\omega_k(T,\widetilde{Y}) = \mathcal{C}_k(T,\widetilde{Y})$, for $\widetilde{Y}_k = Y_k^+ = 0$. This shows that Y_k' is negative and Y_k is decreasing. By using a forward shooting argument, we deduce that Y_k remains negative, contradicting the backward initial condition $Y_k(a) = Y_k^{\mathrm{e}} > 0$.

Step 2. We now claim that $Y \geq 0$. Arguing by contradiction, assume that there exist a point x_1 and a species $k \in S$ such that $Y_k(x_1) < 0$. From the preceding discussion and since $Y(a) = Y^{\mathrm{e}}$, we necessarily have $x_1 \in (0,a)$. Define now

$$x_0 = \sup\{\, t \in (0,x_1],\ Y_k(t) = 0 \,\}.$$

This set is nonempty since $Y_k(0) \geq 0$ and $Y_k(x_1) < 0$. In addition, x_0 is different from x_1 since Y_k is Lipschitzian. Indeed, we have $|Y_k(x_1)| \leq M(x_1 - t)$ for any $t < x_1$ with $Y_k(t) = 0$, where M is a Lipschitz constant for Y_k. Since $x_0 \neq x_1$, we now consider the interval (x_0, x_1) where Y_k is negative since $Y_k(x_1) < 0$ and Y_k has no variation in sign over (x_0, x_1) by construction. There exists $\bar{x} \in (x_0, x_1)$ with $Y_k'(\bar{x}) < 0$ and a forward shooting argument, using the identity

$$\mathfrak{d}_k Y_k' = c\big(Y_k - Y_k(\bar{x})\big) - \tau \int_{\bar{x}}^{x} m_k \mathcal{C}_k(T,\widetilde{Y})\, d\hat{x} + \mathfrak{d}_k(\bar{x}) Y_k'(\bar{x}),$$

shows that Y_k' and Y_k are negative over $[\bar{x}, a]$, an obvious contradiction with $Y_k(a) = Y_k^{\mathrm{e}}$.

Step 3. We now show that $Y(0) > 0$. Arguing by contradiction, assume that $Y_k(0) = 0$ for some $k \in S$. Then, from the species equation, we obtain

$$Y_k'(0) = \frac{c}{\mathfrak{d}_k(0)}(Y_k(0) - Y_k^{\mathrm{f}}) \leq 0,$$

and, since $Y \geq 0$, we cannot have $Y_k'(0) < 0$ so that necessarily $Y_k^{\mathrm{f}} = 0$. From the preceding steps, we know that $\langle Y, \mathcal{U}\rangle = 1$ and that $Y \geq 0$, so that $\widetilde{Y} = Y$ and from (11.4.5) the solution is then twice differentiable. Multiplying (11.4.5) by $\mathcal{L}$, deriving the species components and substituting $x = 0$, we deduce that

$$\mathfrak{d}_k(0) Y_k''(0) = -\tau m_k \mathcal{C}_k(0) \leq 0.$$

Since $Y \geq 0$, $Y_k''(0)$ cannot be negative, so that necessarily $\mathcal{C}_k(0) = 0$. We now introduce the set $\mathcal{I} = \{\, k \in S,\ Y_k(0) = 0 \,\}$ and, from the preceding discussion, we have

$$\forall k \in \mathcal{I}, \qquad Y_k(0) = 0,\ Y_k'(0) = 0,\ Y_k''(0) = 0,\ \tau \mathcal{C}_k(0) = 0. \qquad (11.4.9)$$

Now, if $\tau > 0$, any reaction creating a species $\mathfrak{M}_k$, with $k \in \mathcal{I}$, needs at least one species $\mathfrak{M}_i$ with $i \in \mathcal{I}$ as a reactant, otherwise, one would get a positive production term. This shows that the mass fractions Y_k, $k \in \mathcal{I}$, are factorized in the production terms ω_l, $l \in \mathcal{I}$. Now the mass fractions Y_k, $k \in \mathcal{I}$, are a solution of the system

$$cY_k' = \Big(\mathfrak{d}_k(x)Y_k' + Y_k\mathfrak{f}_k(x)\Big)' + \tau m_k\omega_k,$$

where $\mathfrak{d}_k$ and $\mathfrak{f}_k$ are defined in Lemma 11.2.4 . This system, however, has a locally unique solution from the Cauchy–Lipschitz theorem, but it also admits the trivial solution for any $\tau \in [0,1]$. A shooting argument now shows that $Y_k = 0$ over $[0,a]$ for any $k \in \mathcal{I}$, an obvious contradiction.

Step 4. We finally establish that $Y > 0$. Still arguing by contradiction, we consider the first point x_0 such that there exists $l \in S$ with $Y_l(x_0) = 0$. Of course, we have $x_0 \neq a$, since $Y_l^{\mathrm{e}} > 0$. At this point, since $Y_l > 0$ on $[0,x_0)$, we have $Y_l'(x_0) \leq 0$. We cannot have $Y_l'(x_0) < 0$, since $Y \geq 0$, so that $Y_l'(x_0) = 0$ and $Y_l(x_0) = 0$. We can now proceed as in the preceding step to conclude that Y_l is locally zero, which contradicts the definition of x_0, and the proof is complete. ■

Lemma 11.4.5. *Let (T,Y,c) be a fixed point of $\mathcal{K}_\tau$ and assume that is has been extended over $(-\infty,0)$ by using Proposition 11.3.3. We know from Lemma 11.4.4 and 11.3.3 that the species are positive, so that the entropy and the dissipation rate are well defined. We then have the entropic estimates*

$$\int_{-\infty}^{a} \mathfrak{v}\, dx = c(s^{\mathrm{e}} - s^{\mathrm{f}}), \tag{11.4.10}$$

where

$$\mathfrak{v} = \lambda\Big(\frac{T'}{T}\Big)^2 + \frac{p}{T}\sum_{k,l\in S} D_{kl}\big(\mathcal{X}_k' + \chi_k(T'/T)\big)\,\big(\mathcal{X}_l' + \chi_l(T'/T)\big) - \tau\sum_{k\in S}\frac{g_k m_k\omega_k}{T}, \tag{11.4.11}$$

and we use the notation $\omega_k(x) = 0$ for $x < 0$ and $k \in S$.

Proof. Since the species are positive and we may thus use the entropy governing equation with $s = s(h,Y)$

$$c\,s' + \langle\partial_\xi s, \phi\rangle' = \mathfrak{v},$$

where the dissipation rates $\mathfrak{v}$ is given by (11.4.11). We also know from Proposition 11.3.3 that the integral of $\mathfrak{v}$ over $(-\infty,0)$ is given by (11.3.3). Integrating the entropy conservation equation over $[0,a]$, thus, yields

$$c(s^{\mathrm{e}} - s^{\mathrm{f}}) + \big\langle(\partial_\xi s)^{\mathrm{e}}, \phi(a)\big\rangle = \int_{-\infty}^{a} \mathfrak{v}\, dx,$$

where $(\partial_\xi s)^{\rm e} = (\partial_\xi s)(\xi^{\rm e})$, and we only have to show that $\big\langle (\partial_\xi s)^{\rm e}, \phi(a)\big\rangle = 0$. Moreover, integrating the conservation equations over $[0,a]$—or equivalently over $(-\infty, a]$—we obtain

$$c(\xi^{\rm e} - \xi^{\rm f}) + \phi(a) = \int_0^a w\, dx, \tag{11.4.12}$$

in such a way that $\phi(a) \in (0, M\mathcal{R})$, since $\xi^{\rm e} - \xi^{\rm f} \in (0, M\mathcal{R})$ by definition of $\xi^{\rm e}$ and $w \in (0, M\mathcal{R})$. On the other hand, we know from Proposition 6.5.5 that $-(g_1^{\rm e}, \ldots, g_n^{\rm e})^t / T^{\rm e} = (\partial_Y s)^{\rm e} \in (M\mathcal{R})^\perp$, so that $(\partial_\xi s)^{\rm e} \in (0, M\mathcal{R})^\perp$ and, therefore, $\big\langle (\partial_\xi s)^{\rm e}, \phi(a)\big\rangle = 0$, and the proof is complete. ∎

Lemma 11.4.6. *Let (T, Y, c) be a fixed point of $\mathcal{K}_\tau$. Then there exist positive constants α and β independent of a such that $\alpha < T < \beta$.*

Proof. We start from the entropy conservation equation

$$c\, s' + \langle \partial_\xi s, \phi\rangle' = \mathfrak{v}$$

that we multiply by $\langle \partial_\xi s, \phi\rangle$ and integrate over $[0,x]$. This yields the identity

$$\tfrac{1}{2}\langle \partial_\xi s, \phi\rangle^2 = \tfrac{1}{2}\langle \partial_\xi s, \phi\rangle^2(0) + \int_0^x \Big(\mathfrak{v}\, \langle \partial_\xi s, \phi\rangle - c s' \langle \partial_\xi s, \phi\rangle\Big)\, d\hat{x}.$$

The first integral term can be bounded by

$$\int_0^x \mathfrak{v}\, \langle \partial_\xi s, \phi\rangle\, d\hat{x} \;\le\; \int_0^x \big(c + \frac{1}{c}\langle \partial_\xi s, \phi\rangle^2\big)\mathfrak{v}\, d\hat{x} \;\le\; bc^2 + \frac{1}{c}\int_0^x \langle \partial_\xi s, \phi\rangle^2 \mathfrak{v}\, d\hat{x}.$$

On the other hand, the second integral term can be estimated with

$$\big|s'\langle \partial_\xi s, \phi\rangle\big| = \Big| c_p \frac{T'}{T} + \sum_{k\in S} s_k Y_k' \Big|\, \Big| -\lambda \frac{T'}{T} + \sum_{k\in S} (s_k + R\frac{\widetilde{\mathcal{X}}_k}{m_k})\mathcal{F}_k \Big|,$$

which yields, by using the definition of the species entropies s_k, $k \in S$,

$$\begin{aligned}\big|s'\langle \partial_\xi s, \phi\rangle\big| \;&\le\; b\, \varphi(T)^{1/2}\Big(\big|\frac{T'}{T}\big| + \sum_{k\in S}(1 + |\log Y_k| + |\log T|)|Y_k'| \Big)\\ &\qquad \varphi(T)^{-1/2}\Big(\lambda\big|\frac{T'}{T}\big| + \sum_{k\in S}(1 + |\log Y_k| + |\log T|)|\mathcal{F}_k| \Big).\end{aligned}$$

We now use the estimate (11.4.10) and the inequality $2xy \le x^2 + y^2$ to estimate the integral of $\big|s'\langle \partial_\xi s, \phi\rangle\big|$. By using $Y_k |\log Y_k|^2 \le 1$, valid for

$0 < Y_k \leq 1$, and the expressions (11.2.34) for the multicomponent fluxes, we deduce that

$$\int_0^x \left|s'\langle \partial_\xi s, \phi\rangle\right| d\hat{x} \;\leq\; b\Big(c + \int_0^x |\log T|^2\, \mathfrak{v}\, d\hat{x}\Big).$$

On the other hand, by integrating the entropy conservation equation over $[0, x]$, we have

$$c(s - s^{\mathrm{f}}) + \langle \partial_\xi s, \phi\rangle = \int_{-\infty}^x \mathfrak{v}\, d\hat{x},$$

which implies that

$$c\,|\log T| \;\leq\; bc + \left|\langle \partial_\xi s, \phi\rangle\right| \qquad (11.4.13)$$

and that $|\langle \partial_\xi s, \phi\rangle(0)| \leq bc$, since the specific heats are bounded by positive constants and $T(0) = T^{\mathrm{i}}$. Combining these results yields

$$\int_0^x \left|s'\langle \partial_\xi s, \phi\rangle\right| d\hat{x} \;\leq\; bc + \frac{b}{c^2}\int_0^x \langle \partial_\xi s, \phi\rangle^2 \mathfrak{v}\, d\hat{x},$$

so that finally

$$\langle \partial_\xi s, \phi\rangle^2 \;\leq\; bc^2 + \frac{b}{c}\int_0^x \langle \partial_\xi s, \phi\rangle^2\, \mathfrak{v}\, d\hat{x}.$$

From the Gronwall Lemma and the entropic estimate (11.4.10), we then obtain $\langle \partial_\xi s, \phi\rangle^2 \leq bc^2$, and, using (11.4.13), we conclude that $|\log T| \leq b$, and the proof is complete. ∎

Lemma 11.4.6 is the first temperature upper bound obtained in a flame with nontrivial transport [Gio97] [Gio98] [Gio99]. Note also that the natural entropy production weighted norm has been used in the proof of this lemma.

We now estimate the eigenvalue c of the flame problem. In the following lemma, we obtain an upper bound independent of a and a lower bound that depends on a. The existence of a lower bound independent of a will only be needed when passing to the limit $a \to \infty$, and it is postponed to Section 11.7.

Lemma 11.4.7. *Let (T, Y, c) be a fixed point of $\mathcal{K}_\tau$. Then there exist positive constants α and β such that $\alpha < c < \beta$ and β is independent of a.*

Proof. When thermal diffusion effects are not included, there exists a simple way to derive an upper bound for the eigenvalue c. Indeed, using the identity $c(\xi - \xi^{\mathrm{f}}) + \phi = \tau \int_0^x w d\hat{x}$, we have

$$c\sum_{k\in S} Y_k^{\mathrm{f}}(h_k - h_k^{\mathrm{f}}) - \lambda T' = \tau \sum_{k\in S} h_k \int_0^x m_k\omega_k d\hat{x}.$$

Keeping in mind that $a \geq 1$, we denote by b_0 an upper bound—independent of a—of the sum appearing in the right member for $x \in [0,1]$. Then, either $c\underline{c}_p(T^{\mathrm{i}} - T^{\mathrm{f}}) \leq b_0$, and there is nothing to prove, or $c\underline{c}_p(T^{\mathrm{i}} - T^{\mathrm{f}}) \geq b_0$, and a shooting argument shows that $T \geq T^{\mathrm{i}}$ over $[0,1]$. In this situation, we may write the above equality at any point $x \in [0,1]$, such that $T'(x) = T(1) - T(0)$ to obtain an upper bound for c.

When thermal diffusion is included, we first note that, for $x \geq 0$,

$$c(s - s^{\mathrm{f}}) + \langle \partial_\xi s, \phi \rangle = \int_{-\infty}^{x} \mathfrak{v}\, d\hat{x} \geq \int_{-\infty}^{0} \mathfrak{v}\, d\hat{x} = c\Big(s(0) - s^{\mathrm{f}} - \big\langle \partial_\xi s(0), \xi(0) - \xi^{\mathrm{f}} \big\rangle \Big).$$

Since $T(0) = T^{\mathrm{i}} > T^{\mathrm{f}}$, we deduce that there exists $\delta > 0$ such that, for any $x \geq 0$,

$$s - s^{\mathrm{f}} + \frac{1}{c} \langle \partial_\xi s, \phi \rangle \geq \delta > 0.$$

In addition, we have $\langle \phi, \phi \rangle \leq b\mathfrak{v}$ and $\langle \partial_\xi s, \phi \rangle^2 \leq b\mathfrak{v}$ since the temperature is uniformly bounded and $(\log \xi)^2 \xi \leq 1$ for $\xi \leq 1$. As a consequence, we have

$$\int_0^1 \big(\langle \phi, \phi \rangle + \langle \partial_\xi s, \phi \rangle^2 \big)\, dx \;\leq\; bc,$$

and, from the governing equations $c(\xi - \xi^{\mathrm{f}}) + \phi = \mathcal{T} \int_0^x w\, d\hat{x}$, we finally obtain

$$c^2 \int_0^1 \Big(\big\| \xi - \xi^{\mathrm{f}} - \frac{\mathcal{T}}{c} \int_0^x w\, d\hat{x} \big\|^2 + \big| \frac{1}{c} \langle \partial_\xi s, \phi \rangle \big|^2 \Big)\, dx \;\leq\; bc.$$

We now use the uniform continuity of s over the compact set $Y \geq 0$, $\langle Y, \mathcal{U} \rangle = 1$, and $\alpha \leq T \leq \beta$. More specifically, there exists η such that $\|\xi_1 - \xi_0\| < \eta$ implies that $|s(\xi_1) - s(\xi_0)| < \delta/2$.

Now for $x \in (0,1)$ we have the following alternative. Either $\|\xi - \xi^{\mathrm{f}}\| < \eta$ and then $|s(\xi) - s(\xi^{\mathrm{f}})| < \delta/2$ and the above estimates yields

$$\big| \frac{1}{c} \langle \partial_\xi s, \phi \rangle \big| \geq \delta/2$$

or $\|\xi - \xi^{\mathrm{f}}\| \geq \eta$, so that, for $b_w \leq \frac{1}{2}\eta c$, where b_w is a uniform bound of $\| \int_0^1 w\, dx \|$, we then have

$$\| \xi - \xi^{\mathrm{f}} - \frac{\mathcal{T}}{c} \int_0^x w\, d\hat{x} \| \geq \tfrac{1}{2}\eta.$$

We have thus shown that $c \geq 2b_w/\eta$ implies that $c \inf\big(\eta^2, \delta^2\big) \leq 4b$, so that c is bounded from above. For the lower bound, we remark that

$$|T(a) - T(0)| \;\leq\; \int_0^a |T'(x)|\, dx \;\leq\; \sqrt{a} \Big(\int_0^a |T'(x)|^2\, dx \Big)^{1/2},$$

and we also have $|T'|^2 \leq b\mathfrak{v}$ uniformly from the temperature bounds. As a consequence, we obtain

$$T^{\mathrm{e}} - T^{\mathrm{i}} \;\leq\; b\sqrt{a}\left(\int_0^a \mathfrak{v}\,dx\right)^{1/2} \;\leq\; b\sqrt{a}\sqrt{c},$$

and this yields a lower bound for c, which depends on a. ∎

We now estimate the derivatives of fixed points of $\mathcal{K}_\tau$.

Lemma 11.4.8. *Let (T, Y, c) be a fixed point of $\mathcal{K}_\tau$. Then there exists a positive constant β independent of a such that $\|(T,Y)\|_{C^3[0,a]} < \beta$ and $\|(T',Y')\|_{W_2^2[0,a]} < \beta$*

Proof. We use the variable ξ for convenience, and, from the entropic estimate and the upper bounds derived in the previous lemmas, we obtain

$$\int_{-\infty}^{a} \|\xi'\|^2\,dx \;\leq\; \beta. \tag{11.4.14}$$

On the other hand, we may use the expression (6.5.4) of the chemical dissipation rate

$$\zeta = \sum_{i\in R} K_i^{\mathrm{s}}\big(\langle M\nu_i^{\mathrm{f}},\mu\rangle - \langle M\nu_i^{\mathrm{b}},\mu\rangle\big)\,\big(\exp\langle M\nu_i^{\mathrm{f}},\mu\rangle - \exp\langle M\nu_i^{\mathrm{b}},\mu\rangle\big).$$

Since $\mathcal{X}_k \leq 1$, $k \in S$, we know from (11.2.12) that $\langle\mu, M\nu_i^{\mathrm{f}}\rangle$ and $\langle\mu, M\nu_i^{\mathrm{b}}\rangle$ are bounded from above. As a consequence, there exists b such that

$$\big|\exp\langle\mu, M\nu_i^{\mathrm{f}}\rangle - \exp\langle\mu, M\nu_i^{\mathrm{b}}\rangle\big| \;\leq\; b\big|\langle\mu, M\nu_i^{\mathrm{f}}\rangle - \langle\mu, M\nu_i^{\mathrm{b}}\rangle\big|,$$

and, using now the positivity of K_i^{s}, $i \in R$, and the uniform estimates for the temperature, we conclude that

$$\sum_{i\in R} \tau_i^2 \;\leq\; b\zeta,$$

so that, finally, using the entropic estimates, there exists b such that

$$\tau\int_0^a \|w\|^2\,dx \leq b.$$

From the governing equations $c\xi' + \phi' = \tau w$, where $w = 0$ for $x < 0$, and from $0 \leq \tau \leq 1$, we now deduce that

$$\int_{-\infty}^{a} \|\phi'\|^2\,dx \leq \beta.$$

From Sobolev injection applied to $\phi = -\mathcal{L}\xi'$, we obtain $\|\phi\| \leq \beta$ uniformly over $(-\infty, a]$, and, from $\xi' = -\mathcal{L}^\sharp\phi$, we further obtain $\|\xi'\| \leq \beta$ uniformly over $(-\infty, a]$. Finally, from the governing equations, we deduce that $\|\phi'\|$ is uniformly bounded, and, upon expanding the derivatives, we conclude that $\|\xi''\|$ is also uniformly bounded. The L^∞ estimates for ϕ'' and ξ''' and the $W_2^2[0,a]$ estimates of derivatives are similar. ∎

11.4.4. Calculation of the degree

From the homotopy invariance of the degree, we have

$$d(I - \mathcal{K}_1, \Omega, 0) = d(I - \mathcal{K}_0, \Omega, 0), \tag{11.4.15}$$

and the map $\mathcal{K}_0$ no longer involves chemistry source terms. In order to evaluate this degree, we need a second homotopy in order to simplify transport properties and thermodynamics.

We introduce the operator $\mathcal{H}_\tau \ : \ \mathcal{O} \longrightarrow \mathcal{B}$ defined by

$$\mathcal{H}_\tau(T, Y, c) = \Big(\mathfrak{t}, \mathfrak{y}, c + \mathfrak{t}(0) - T^{\mathrm{i}}\Big), \tag{11.4.16}$$

where $(\mathfrak{t}, \mathfrak{y})$ are solutions of the system

$$\begin{aligned} &\Big(c_p^\tau(T,\widetilde{Y})\,\mathfrak{t}' + \sum_{k\in S} h_k^\tau(T)\,\mathfrak{y}_k',\ \mathfrak{y}'\Big)^t = \\ &\Big(\mathcal{L}^\tau(T,\widetilde{Y})\Big)^\sharp \Big(c\Big(\langle c_p^\tau(T,\widetilde{Y})\rangle \mathfrak{t} + \sum_{k\in S} h_k^0\,\mathfrak{y}_k,\ \mathfrak{y}\Big)^t - c\,\xi^{\mathrm{f},\tau}\Big), \end{aligned} \tag{11.4.17}$$

with the initial conditions

$$\mathfrak{t}(a) = T^{\mathrm{e}}, \qquad \mathfrak{y}(a) = Y^{\mathrm{e}}. \tag{11.4.18}$$

In these expressions, we have used the notation $c_{pk}^\tau = \tau c_{pk} + (1-\tau)\overline{c}_p$, $h_k^\tau = \langle c_{pk}^\tau\rangle T + h_k^0$, $\lambda^\tau = \tau\lambda + (1-\tau)\overline{\lambda}$, $C^\tau = \tau C + (1-\tau)\overline{C}\Big(I\langle Y, \mathcal{U}\rangle - Y\otimes\mathcal{U}\Big)$, $\chi_k^\tau = \tau\chi_k$, $m_k^\tau = \tau m_k + (1-\tau)m_0$, and

$$\xi^{\mathrm{f},\tau} = \Big(\sum_{k\in S} Y_k^{\mathrm{f}}\big(h_k^0 + \int_0^{T^{\mathrm{f}}} c_{pk}^\tau(\theta)d\theta\big),\ Y^{\mathrm{f}}\Big)^t,$$

where $\overline{c}_p$, $\overline{\lambda}$, $\overline{C}$, and m_0 are fixed positive constants with $\overline{\lambda}/\overline{c}_p\overline{C} = 1$ and $\mathcal{L}^\tau$ is taken as in Definition 11.2.2 in terms of the modified coefficients C^τ, χ^τ, λ^τ, c_p^τ, and m^τ.

The idea behind this new homotopy is to obtain trivial thermodynamics and trivial transport fluxes for $\tau = 0$. One could also simultaneously modify the chemistry source terms, thermodynamic properties, and transport properties in a single homotopy. However, it requires changing the equilibrium temperature boundary condition—which depends on thermodynamics—and, simultaneously, the cold temperature boundary condition, in order to prevent any temperature crossing associated with modified equilibrium temperatures. Using two distinct homotopies somewhat simplifies the presentation.

Proposition 11.4.9. *The operator $\mathcal{H}_\tau : \mathcal{O} \to \mathcal{B}$ is well posed, and the degree $d(I - \mathcal{H}_\tau, \Omega, 0)$ is well defined for $\tau \in [0,1]$ and $a \geq 1$. Moreover, for any closed bounded set B in $\mathcal{O}$, the map $\mathcal{H} : [0,1] \times B \to \mathcal{B}$ defined by $\mathcal{H}\big(\tau, (T, Y, c)\big) = \mathcal{H}_\tau(T, Y, c)$ is compact.*

Proof. It is exactly similar to the preceding case up to minor modifications. For instance, the entropic estimates now read as

$$\int_{-\infty}^{a} \mathfrak{v}\, dx = c\Big(s(\xi^{\mathrm{e},\tau}) - s(\xi^{\mathrm{f},\tau}) - \big\langle \partial_\xi s(\xi^{\mathrm{e},\tau}), \xi^{\mathrm{e},\tau} - \xi^{\mathrm{f},\tau} \big\rangle\Big),$$

where

$$\xi^{\mathrm{e},\tau} = \Big(\sum_{k\in S} Y_k^{\mathrm{e}} \big(h_k^0 + \int_0^{T^{\mathrm{e}}} c_{pk}^{\tau}(\theta) d\theta\big),\, Y^{\mathrm{e}}\Big)^t,$$

since $\xi^{\mathrm{e},\tau}$ no longer coincides with the maximum of s on the conservation simplex, the specific heats being modified along the homotopy path. ■

From the homotopy invariance of the degree, we now obtain

$$d(I - \mathcal{K}_1, \Omega, 0) = d(I - \mathcal{H}_0, \Omega, 0), \tag{11.4.19}$$

and a straightforward calculation—making use of $\langle \mathfrak{y}', \mathcal{U} \rangle = 0$—yields that the map $\mathcal{H}_0$ has the simple structure

$$\mathcal{H}_0(T, Y, c) = \Big(\mathfrak{t}_c, \mathfrak{y}_c, c + \mathfrak{t}_c(0) - T^{\mathrm{i}}\Big), \tag{11.4.20}$$

where $\mathfrak{t}_c$ and $\mathfrak{y}_c$ are given by

$$\mathfrak{t}_c(x) = T^{\mathrm{f}} + (T^{\mathrm{e}} - T^{\mathrm{f}}) \exp(c\,(x-a)\overline{c}_p/\overline{\lambda}), \tag{11.4.21}$$

$$\mathfrak{y}_c(x) = Y^{\mathrm{f}} + (Y^{\mathrm{e}} - Y^{\mathrm{f}}) \exp(c\,(x-a)\overline{c}_p/\overline{\lambda}), \tag{11.4.22}$$

respectively, and only depend on c.

From the multiplicative properties of the degree, by using an auxiliary homotopy [Gio93], one can then easily establish that

$$d(I - \mathcal{K}_1, \Omega, 0) = d(T^{\mathrm{i}} - \mathfrak{t}_c(0), (\alpha, \beta), 0).$$

From (11.4.21), we have

$$\mathfrak{t}_c(0) = T^{\mathrm{f}} + (T^{\mathrm{e}} - T^{\mathrm{f}}) \exp(-c\, a\overline{c}_p/\overline{\lambda}),$$

so that, finally,

$$d(I - \mathcal{K}_1, \Omega, 0) = 1,$$

since $\alpha < T^{\rm f} < T^{\rm i} < T^{\rm e} < \beta$ and there exists a solution to the problem posed on a bounded domain.

11.5. Existence of solutions

In this section, we pass to the limit $a \to \infty$ and obtain a solution of the anchored flame problem. A first step is to derive a lower bound for the flame eigenvalue c that is independent of a. Another important point is to investigate the behavior of the solution near the equilibrium state $\xi^{\rm e}$.

11.5.1. Uniform estimates for c

Lemma 11.5.1. *For* $0 < \delta < s^{\rm e} - s^{\rm f}$ *there exists a unique point* x_δ *such that*

$$\int_{x_\delta}^{a} \mathfrak{v}\, dx = \delta c.$$

Proof. The function Ψ defined on $(-\infty, a]$ by $\Psi(x) = \int_{-\infty}^{x} \mathfrak{v}\, d\hat{x}$ is strictly increasing since $\mathfrak{v}$ is strictly positive over $(-\infty, a]$. Arguing by contradiction, we indeed note that, for $x_0 < 0$, $\mathfrak{v}(x_0) = 0$ implies that $\phi(x_0) = 0$ and thus that $\xi(x_0) = \xi^{\rm f}$ from (11.3.2) and the Cauchy–Lipschitz theorem yields $\xi(x) = \xi^{\rm f}$ for any $x \le 0$, contradicting $T^{\rm f} \ne T^{\rm i}$. Similarly, for $x_0 \ge 0$, $\mathfrak{v}(x_0) = 0$ implies that $\xi'(x_0) = 0$ and $\zeta(x_0) = 0$. Since $Y(x_0) > 0$, we deduce from $\zeta(x_0) = 0$ that $\xi(x_0)$ is an equilibrium point, and, from the Cauchy–Lipschitz theorem, we obtain $\xi(x) = \xi(x_0)$ for $0 \le x \le a$, contradicting $T(0) = T^{\rm i} \ne T^{\rm e} = T(a)$. ∎

We now show that $\xi(x)$ is close to $\xi^{\rm e}$ when the integral of the dissipation rate $\mathfrak{v}$ over the interval $[x, a]$ is small, in particular, for $x \in [x_\delta, a]$.

Lemma 11.5.2. *There exists a constant* β_0 *such that, for any* $a \ge 1$, *we have*

$$\forall x \in [0,a], \qquad \|\xi(x) - \xi^{\rm e}\|^2 \le \beta_0 \Big(\frac{1}{c}\int_x^a \mathfrak{v}\, d\hat{x}\Big)^{1/2}. \tag{11.5.1}$$

Proof. We first note that there exists β such that we have uniformly

$$\|\xi - \xi^{\rm e}\|^2 \le \beta\Big(s^{\rm e} - s - \big\langle (\partial_\xi s)^{\rm e}, \xi^{\rm e} - \xi \big\rangle\Big),$$

since T is uniformly bounded and $\langle Y, \mathcal{U}\rangle = 1$. Since $(\partial_\xi s)^{\mathrm{e}} \in (0, M\mathcal{R})^\perp$, we easily obtain, by combining the conservation equations and the entropy equation,

$$c\Big(s' - \langle(\partial_\xi s)^{\mathrm{e}}, \xi\rangle'\Big) + \Big(\langle\partial_\xi s, \phi\rangle - \langle(\partial_\xi s)^{\mathrm{e}}, \phi\rangle\Big)' = \mathfrak{v}. \tag{11.5.2}$$

We multiply this equation by $\psi = \langle\partial_\xi s, \phi\rangle - \langle(\partial_\xi s)^{\mathrm{e}}, \phi\rangle$ and integrate over $[x, a]$. This yields

$$-\tfrac{1}{2}\psi^2 = \int_x^a \mathfrak{v}\psi\, d\hat{x} - c\int_x^a \big(s' - \langle(\partial_\xi s)^{\mathrm{e}}, \xi'\rangle\big)\, \psi\, d\hat{x}.$$

However, we can now estimate

$$\int_x^a \big|s' - \langle(\partial_\xi s)^{\mathrm{e}}, \xi'\rangle\big|\, |\psi|\, d\hat{x}, \;\leq\; \beta\int_x^a \mathfrak{v}\, d\hat{x},$$

so that, by using $|\psi| \leq c + (1/c)\psi^2$, we obtain

$$\psi^2 \;\leq\; \beta c\int_x^a \mathfrak{v}\, d\hat{x} + \frac{\beta}{c}\int_x^a \psi^2\, \mathfrak{v}\, d\hat{x}.$$

Using a (backward) generalized Gronwall Lemma [Bou76], we now obtain

$$\psi^2 = \Big|\langle\partial_\xi s, \phi\rangle - \langle(\partial_\xi s)^{\mathrm{e}}, \phi\rangle\Big|^2 \;\leq\; \beta c\int_x^a \mathfrak{v}\, d\hat{x}.$$

On the other hand, by integrating (11.5.2) over $[x, a]$, we get

$$c\Big(s^{\mathrm{e}} - s - \langle(\partial_\xi s)^{\mathrm{e}}, \xi^{\mathrm{e}} - \xi\rangle\Big) - \psi = \int_x^a \mathfrak{v}\, d\hat{x},$$

which yields

$$s^{\mathrm{e}} - s - \langle(\partial_\xi s)^{\mathrm{e}}, \xi^{\mathrm{e}} - \xi\rangle \;\leq\; \beta\Big(\frac{1}{c}\int_x^a \mathfrak{v}\, d\hat{x}\Big)^{1/2}$$

and completes the proof. ∎

We now estimate more closely the integral of the dissipation rate $\mathfrak{v}$ when ξ is close to ξ^{e}. This result will then be applied over the interval $[x_\delta, a]$.

Proposition 11.5.3. *There exist positive constants $\epsilon_1 < T^{\mathrm{e}} - T^{\mathrm{f}}$ and β_1 such that, for any $a \geq 1$,*

$$\forall x \in (-\infty, a], \qquad \|\xi(x) - \xi^{\mathrm{e}}\| \leq \epsilon_1 \quad\Longrightarrow\quad \int_x^a \mathfrak{v}\, d\hat{x} \;\leq\; \beta_1 \mathfrak{v}(x).$$

Proof. We define $\pi^{\rm e}(\xi)$ as the unique equilibrium point in the affine manifold $\xi+(0,M\mathcal{R})$. By construction, we have $(\partial_\xi s)(\pi^{\rm e}(\xi)) \in (0,M\mathcal{R})^\perp$. Integrating now the governing equations over $[x,a]$, multiplying by $(\partial_\xi s)(\pi^{\rm e}(\xi))$, and substracting the result from the entropy governing equation integrated over $[x,a]$, we easily obtain

$$c\Big(s^{\rm e} - s - \langle(\partial_\xi s)(\pi^{\rm e}), \xi^{\rm e} - \xi\rangle\Big) + \Big\langle(\partial_\xi s)(\pi^{\rm e}) - \partial_\xi s, \phi\Big\rangle = \int_x^a \mathfrak{v}\, d\hat{x},$$

keeping in mind that $\phi(a) \in (0,M\mathcal{R})$ from (11.4.12). This implies that

$$\begin{aligned} c\Big(s(\pi^{\rm e}) - s - \langle(\partial_\xi s)(\pi^{\rm e}), \pi^{\rm e} - \xi\rangle\Big) + c\Big(s^{\rm e} - s(\pi^{\rm e}) - \langle(\partial_\xi s)(\pi^{\rm e}), \xi^{\rm e} - \pi^{\rm e}\rangle\Big) \\ + \Big\langle(\partial_\xi s)(\pi^{\rm e}) - \partial_\xi s, \phi\Big\rangle = \int_x^a \mathfrak{v}\, d\hat{x}, \end{aligned} \tag{11.5.3}$$

so that, by using $(\partial_\xi s)(\pi^{\rm e}) \in (0,M\mathcal{R})^\perp$ and $\xi - \pi^{\rm e} \in (0,M\mathcal{R})$, and taking into account the concavity of s, we obtain

$$c\left(s(\pi^{\rm e}) - s\right) + \Big\langle(\partial_\xi s)(\pi^{\rm e}) - \partial_\xi s, \phi\Big\rangle \ \geq\ \int_x^a \mathfrak{v}\, d\hat{x}. \tag{11.5.4}$$

On the other hand, when ξ is close enough to $\xi^{\rm e}$, say, $\|\xi(x) - \xi^{\rm e}\| \leq \epsilon_1$, we can write that

$$\alpha\|\pi^{\rm e} - \xi\|^2 \ \leq\ s(\pi^{\rm e}) - s(\xi) \ \leq\ \beta\|\pi^{\rm e} - \xi\|^2$$

and

$$\Big|\langle\partial_\xi s(\pi^{\rm e}) - \partial_\xi s, \phi\rangle\Big| \ \leq\ \beta\Big(\|\pi^{\rm e} - \xi\|^2 + \|\phi\|^2\Big),$$

so that

$$\Big|\langle\partial_\xi s(\pi^{\rm e}) - \partial_\xi s, \phi\rangle\Big| \ \leq\ \beta\Big(s(\pi^{\rm e}) - s(\xi) + \langle(\partial_\xi^2 s)\xi', \phi\rangle\Big).$$

Finally, using the local stability inequality of Corollary 6.7.4, we deduce that there exists β_1 such that

$$\int_x^a \mathfrak{v}\, d\hat{x} \ \leq\ \beta_1 \mathfrak{v}(x),$$

for $\|\xi(x) - \xi^{\rm e}\| \leq \epsilon_1$, since c is bounded by above independently of a for $a \geq 1$, and the proof is complete. ∎

By using Proposition 11.5.3, we directly obtain the exponential decay of the integral of the dissipation rate $\mathfrak{v}$ near equilibrium.

Corollary 11.5.4. *For ϵ, such that $0 < \epsilon < T^{\mathrm{e}} - T^{\mathrm{i}}$, we define $z_\epsilon \geq 0$ to be the smallest x, such that $\|\xi(x) - \xi^{\mathrm{e}}\| = \epsilon$ and $\|\xi(t) - \xi^{\mathrm{e}}\| \leq \epsilon$ for all $t \in [x, a]$. Then, there exists a constant d such that, for any $a \geq 1$,*

$$\forall \epsilon \in (0, \epsilon_1],\ \forall x \in [z_\epsilon, a], \qquad \int_x^a \mathfrak{v}\, d\hat{x} \;\leq\; \Big(\int_{z_\epsilon}^a \mathfrak{v}\, d\hat{x}\Big) \exp\big(-d(x - z_\epsilon)\big), \tag{11.5.5}$$

where ϵ_1 is defined in Proposition 11.5.3.

We now obtain a positive lower bound independent of a for the flame eigenvalue c.

Theorem 11.5.5. *There exists a positive constant α independent of a such that*

$$\alpha \;\leq\; c. \tag{11.5.6}$$

Proof. We consider δ_1 small enough, such that

$$\sqrt{\beta_0}\sqrt[4]{\delta_1} \leq \epsilon_1,$$

so that, for $\delta \leq \delta_1$, we have $z_{\epsilon_1} \leq x_{\delta_1}$ from Lemma 11.5.2. We then consider the point $x_{\delta_1/2}$, which necessarily belongs to the interval (x_{δ_1}, a). Using the exponential estimates of Corollary 11.5.4, valid over $[z_{\epsilon_1}, a]$, we obtain

$$c\delta_1/2 = \int_{x_{\delta_1/2}}^a \mathfrak{v}\, dx \;\leq\; (s^{\mathrm{e}} - s^{\mathrm{f}}) c \exp\big(-d(x_{\delta_1/2} - z_{\epsilon_1})\big). \tag{11.5.7}$$

We also have

$$\sqrt{\beta_0}(\sqrt[4]{\delta_1} - \sqrt[4]{\delta_1/2}) \;\leq\; \|\xi(z_{\epsilon_1}) - \xi^{\mathrm{e}}\| - \|\xi(x_{\delta_1/2}) - \xi^{\mathrm{e}}\| \;\leq\; \|\xi(z_{\epsilon_1}) - \xi(x_{\delta_1/2})\|,$$

and, letting $\alpha_1 = \beta_0\sqrt{\delta_1}\,(1 - 1/\sqrt[4]{2})^2$, we obtain from the entropic estimates

$$\alpha_1 \leq \Big(\int_{z_{\epsilon_1}}^{x_{\delta_1/2}} \|\xi'\|\, dx\Big)^2 \leq (x_{\delta_1/2} - z_{\epsilon_1}) \int_{z_{\epsilon_1}}^{x_{\delta_1/2}} \|\xi'\|^2\, dx \leq \beta_2 c (x_{\delta_1/2} - z_{\epsilon_1}),$$

where β_2 is independent of a. This now yields

$$\delta_1 \leq 2(s^{\mathrm{e}} - s^{\mathrm{f}}) \exp\big(-\alpha_2/c\big),$$

where $\alpha_2 = d\alpha_1/\beta_2$, and the proof is complete. ∎

Note that, in Proposition 11.5.3 and Theorem 11.5.5, we have only used the local version of the stability inequality concerning the chemical dissipation rate ζ.

11.5.2. Convergence towards equilibrium

Theorem 11.5.6. *There exist $\overline{a}$, d, and $\mathcal{C}$ independent of a such that*

$$\forall a \geq \overline{a} \qquad \forall x \in [0,a], \qquad \|\xi(x) - \xi^{\mathrm{e}}\| \leq \mathcal{C}\exp(-dx). \tag{11.5.8}$$

Proof. In order to establish exponential convergence towards equilibrium, we only have to establish that, for a given $\epsilon > 0$, z_ϵ remains bounded independently of a. Let $0 < \epsilon < \epsilon_1$—where ϵ is determined later—and define $a_0\epsilon = \overline{c}(s^{\mathrm{e}} - s^{\mathrm{f}}) + 1$, where $\overline{c}$ is an upper bound for the eigenvalue c. Then, for $a \geq a_0$, there exists necessarily $x_0 \in [0, a_0]$ such that $\mathfrak{v}(x_0) < \epsilon$. Since we have

$$\langle \phi, \phi \rangle \leq b\,\mathfrak{v},$$

uniformly over $(-\infty, a]$ with b independent of a, we get $\|\phi(x_0)\|^2 \leq b\epsilon$. Since $\mathcal{P}(\xi - \xi^{\mathrm{e}}) = \mathcal{P}(\xi - \xi^{\mathrm{f}}) = -(1/c)\mathcal{P}(\phi)$, where $\mathcal{P}$ is the orthogonal projector onto $(0, M\mathcal{R})^{\perp}$, we can use the global stability inequality of Proposition 6.8.1 and assume that $\epsilon \leq \underline{c}\epsilon_0$, so that $s(\pi^{\mathrm{e}}) - s \leq \beta\epsilon$, and thus

$$\|\pi^{\mathrm{e}} - \xi\|^2 \leq \beta\epsilon.$$

In addition, we have $\|\pi^{\mathrm{e}} - \xi^{\mathrm{e}}\| \leq \beta\sqrt{\epsilon}$ since $\mathcal{P}(\xi - \xi^{\mathrm{e}}) = -(1/c)\mathcal{P}(\phi)$ and $\pi^{\mathrm{e}}(\xi)$ is a smooth function of $\mathcal{P}(\xi)$. As a consequence, we obtain

$$\left\|\xi(x_0) - \xi^{\mathrm{e}}\right\|^2 \leq \beta_3\epsilon,$$

where β_3 is independent of ϵ and a. Assuming now $\epsilon < \epsilon_1^2/(1+\beta_3)$, we obtain

$$\int_{x_0}^{a} \mathfrak{v}\, d\hat{x} \leq \beta_1\mathfrak{v}(x_0) \leq \beta_1\epsilon.$$

Further assuming $\beta_1\epsilon < \underline{c}(s^{\mathrm{e}} - s^{\mathrm{f}})$, then $x_{\beta_1\epsilon/\underline{c}}$ can be defined as in Lemma 11.5.1 and we have the estimate $x_{\beta_1\epsilon/\underline{c}} \leq a_0$. Letting

$$\bar{\epsilon} = \tfrac{1}{2}\inf\Big(\underline{c}\epsilon_0, \frac{\epsilon_1^2}{1+\beta_3}, \frac{\underline{c}\,(s^{\mathrm{e}} - s^{\mathrm{f}})}{\beta_1}, \frac{\underline{c}\,\epsilon_1^4}{\beta_0^2\beta_1}\Big),$$

we finally obtain from Lemma 11.5.2 that $z_{\bar{\epsilon}}$ is well defined with $\bar{\epsilon} < \epsilon_1$ and that

$$z_{\bar{\epsilon}} \leq a_0 = \big(\overline{c}\,(s^{\mathrm{e}} - s^{\mathrm{f}}) + 1\big)/\bar{\epsilon}.$$

Since $z_{\bar{\epsilon}}$ is bounded independently of a, we now obtain from Corollary 11.5.4

$$\forall a \geq a_0, \qquad \forall x \in [0,a], \qquad \int_x^a \mathfrak{v}\, d\hat{x} \leq \beta\exp(-dx),$$

and the proof is complete, again using Lemma 11.5.2 ∎

Theorem 11.5.6 is the only place where the global inequality of Proposition 6.8.1 is used.

11.5.3. Passage to the limit $a \to \infty$

Theorem 11.5.7. *There exists a* C^∞ *solution to the anchored flame problem.*

Proof. We consider a sequence of solutions (T^i, Y^i, c^i) of the anchored flame problem over the domains $[0, i]$ for $i \in \mathbb{N}\backslash\{0\}$. From the a priori estimates derived in Sections 11.3, 11.4, and 11.5, we can extract a subsequence converging towards (T, Y, c) on every compact in the C^2 topology. We then know that $\alpha < T < \beta$ and $\alpha < c < \beta$, where α and β are positive and independent of i, and that $Y \geq 0$, $\langle Y, \mathcal{U}\rangle = 1$. From the uniform L^2 estimates of derivatives, we easily deduce that T', Y', T'', and Y'' are in L^2, and, from the exponential estimates, we also deduce that $\xi(\infty) = \xi^{\mathrm{e}}$. As a consequence, we conclude that (T, Y, c) is a solution of the anchored flame problem, and, from Lemma 11.3.1, we also obtain $Y > 0$. Finally, the C^∞ regularity follows from the C^2 regularity and the species and energy governing equations. ▪

Note that the estimates obtained in Section 11.4 on a bounded domain $[0, a]$ could also be conducted directly over $[0, \infty)$.

11.6. Notes

11.1. For mathematical investigations of detonation flames, we refer the reader to Majda [Maj84], Roquejoffre and Vila [RV98], and to the references therein. For various mathematical problems from combustion theory, we refer the reader to the monographs of Volpert and Hudjaev [VH85], Bebernes and Eberly [BE89], Volpert *et al.* [VV94], and to the references therein.

11.7. References

[AZK79] A. P. Aldushin, Ya. B. Zeldovitch, and S. I. Khudyaev, *Numerical Investigation of the Flame Propagation in a Mixture Reacting at the Initial Temperature,* Fiz. Goreniya Vzryva (Comb. Explos. Shock Waves), **15**, (1979), pp. 20–27.

[BE89] J. Bebernes and D. Eberly, *Mathematical Problems from Combustion Theory,* Springer-Verlag, New York, (1989).

[BG74] A. Ben-Israel and T. N. E. Greville, *Generalized Inverses, Theory and Applications,* John Wiley& Sons, Inc., New York, (1974).

[BeLa93] H. Berestycki and B. Larrouturou, *Sur les Modèles de Flammes en Chimie Complexe,* C. R. Acad. Sci. Paris, Série I, **317**, (1993), pp. 173–176.

[Bon92] A. Bonnet, *Travelling Wave for Planar Flames with Complex Chemistry Reaction Network,* Comm. Pure Appl. Math., **XLV**, (1992), pp. 1271–1302.

[Bon94] A. Bonnet, *Contribution à l'Étude des Problèmes Elliptiques et à la Modélisation Mathématique de la Combustion, Mémoire d'Habilitation à Diriger des Recherches,* Université Paris 6, (1994).

[Bou76] N. Bourbaki, *Fonctions d'une Variable Réelle,* Éléments de Mathématiques, Diffusion C. C. L. S., (1976).

[BuLu82] J. D. Buckmaster and G. S. S. Ludford, *Theory of Laminar Flames,* Cambridge University Press, Cambridge, (1982).

[Cla85] P. Clavin, *Dynamic Behavior of Premixed Flame Fronts in Laminar and Turbulent Flows,* Prog. Ener. Comb. Sci., **11**, (1985), pp. 1–59.

[Dei88] K. Deimling, *Nonlinear Functional Analysis,* Springer-Verlag, Berlin, (1988).

[Gio93] V. Giovangigli, *An Existence Theorem for a Free-Boundary Problem of Hypersonic Flow Theory,* SIAM J. Math. Anal., **24**, (1993), pp. 571–582.

[Gio97] V. Giovangigli, *Plane Flames with Multicomponent Transport and Complex Chemistry,* Internal Report, **366**, Centre de Mathématiques Appliquées de l'Ecole Polytechnique, (1997).

[Gio98] V. Giovangigli, *Flames Planes avec Transport Multi-espece et Chimie Complexe,* C. R. Acad. Sci. Paris, **326**, Série I, (1998), pp. 775–780.

[Gio99] V. Giovangigli, *Plane Flames with Multicomponent Transport and Complex Chemistry,* Math. Mod. Meth. Appl. Sci., **9**, (1999), pp. 337–378.

[GS92] V. Giovangigli and M. Smooke, *Application of Continuation Techniques to Plane Premixed Laminar Flames,* Comb. Sci. Tech., **87**, (1992), pp. 241–256.

[Hei87] S. Heinze, *Travelling Waves in Combustion Processes with Complex Chemical Networks,* Trans. Amer. Soc., **304**, (1987), pp. 405–416.

[HCC53] J. O. Hirschfelder, C. F. Curtiss, and D. E. Campbell, *The Theory of Flames and Detonations,* Fourth International Symposium on Combustion, (1953), pp. 190–210.

[LS34] J. Leray and J. Schauder, *Topologie et Équations Fonctionnelles,* Ann. Sci. École Norm. Sup., **51**, (1934), pp. 45–78.

[Maj84] A. Majda, *Compressible Fluid Flow and Systems of Conservation Laws in Several Space Variables,* Appl. Math. Sci., **53**, Springer-Verlag, New York, (1984).

[Q*al*84] J. Quinard, G. Searby, and L. Boyer, *Cellular Structures on Premixed Flames in a Uniform Laminar Flow,* Lect. Notes Phys., **210**, Springer-Verlag, Berlin, (1984), pp. 331–341.

[Ser86] M. Sermange, *Mathematical and Numerical Aspects of One-Dimensional Laminar Flame Simulation,* Appl. Math. Opt., **14**, (1986), pp. 131–153.

[Roq90] J. M. Roquejoffre, *Étude Mathématique d'un Modèle de Flamme Plane,* C. R. Acad. Sci. Paris, **311**, Série I, (1990), pp. 593–596.

[RV98] J. M. Roquejoffre and J. P. Vila, *Stability of ZND Detonation Waves in the Majda Combustion Model,* Preprint, (1998).

[Sch69] J. T. Schwartz, *Nonlinear Functional Analysis,* Gordon and Breach, New York, (1969).

[VH85] A. Vol'pert and S Hudjaev, *Analysis in Classes of Discontinuous Functions and Equations of Mathematical Physics,* Mechanics: Analysis 8, Martinus Nijhoff Publishers, Dordrecht, (1985).

[VV90] V. A. Volpert and V. A. Volpert, *Travelling Wave Solutions of Monotone Parabolic Systems,* Preprint, Inst. Chem. Phys., Chernogolovka, (in Russian), (1990), (Also, Preprint **146**, Université de Lyon 1, 1993).

[VV94] A. I. Volpert, V. A. Volpert, and V. A. Volpert, *Travelling Wave Solutions of Parabolic Systems,* Trans. Math. Mono., AMS, **140**, (1994).

[Wil85] F. A. Williams, *Combustion Theory,* Second ed., The Benjamin/Cummings Publishing Company, Inc., Menlo Park, (1985).

12

Numerical Simulations

12.1. Introduction

Numerical simulation of compressible flows is a very difficult task that has been the subject of numerous textbooks and requires a solid background in fluid mechanics and numerical analysis [PT83] [GiRa86] [Hug87] [OB87] [Joh90] [FP96] [GoRa96]. The nature of compressible flows may be very complex, with features such as shock fronts, boundary layers, turbulence, acoustic waves, or instabilities. Taking into account chemical reactions dramatically increases the difficulties, especially when detailed chemical and transport models are considered. Interactions between chemistry and fluid mechanics are especially complex in reentry problems, combustion phenomena, or chemical vapor deposition reactors. As a consequence, it would be unthinkable to try to address these issues and discuss the development of reactive flow solvers in a single chapter or even in a single book.

More simply, in this chapter, we only want to illustrate the previous developments and indicate some of the difficulties that are specific to complex chemistry and multicomponent transport numerical models. As a typical illustration, we consider in this chapter a hydrogen-air laminar flame with complex chemistry and detailed transport.

We first derive the Bunsen laminar flame model from the previous chapters. These laminar flames can be modeled equivalently, either by the complete equations of Chapter 2 or by using the small Mach number limit presented in Chapter 3. We then present a typical hydrogen-air combustion mechanism composed by reversible elementary reactions involving active radicals. We also briefly address the numerical algorithms that have been used for the numerical simulation of Bunsen flames and discuss the numerical implementation of chemistry and transport software that evaluates thermochemistry expressions or transport coefficients. Finally, we discuss a typical lean hydrogen-air flame structure.

12.2. Laminar flame model

In this section we present a Bunsen laminar flame model and a typical hydrogen-air combustion mechanism. Bunsen flames are axisymmetric flames of conical shape that are obtained by flowing premixed mixtures of fuel and oxidant through cylindrical tubes, as schematically illustrated in Figure 1. These flames are of conical shape and sit at the mouth of the cylindrical burner when the exit velocity exceeds the plane flame speed [Wil85].

12.2.1. Governing equations

The governing equations are obtained from the fundamental equations presented in Chapter 2 specialized to the steady axisymmetric setting. We denote by (r, θ, z) the cylindrical corrdinates so that r is the radial coordinate and z the vertical coordinate. We denote by u and v the corresponding velocity coordinates so that $\boldsymbol{v} = (u, 0, v)^t$. The gravity vector is assumed to be oriented in the (negative) vertical direction so that $\boldsymbol{g} = (0, 0, g)^t$.

The total mass conservation is in the form

$$\frac{1}{r}\partial_r(\rho u r) + \partial_z(\rho v) = 0. \tag{12.2.1}$$

The radial momentum conservation equation reads as

$$\begin{aligned} \rho u \partial_r u + \rho v \partial_z u = &-\partial_r p + \frac{1}{r}\partial_r\Big(2\eta r \partial_r u + (\kappa - \tfrac{2}{3}\eta) r \big(\frac{\partial_r(ru)}{r} + \partial_z v\big)\Big) \\ &+ \partial_z\Big(\eta \partial_z u + \eta \partial_r v\Big) - \frac{1}{r}\Big(2\eta \frac{u}{r} + (\kappa - \tfrac{2}{3}\eta)\big(\frac{\partial_r(ru)}{r} + \partial_z v\big)\Big), \end{aligned} \tag{12.2.2}$$

whereas the vertical momentum conservation equation yields

$$\begin{aligned} \rho u \partial_r v + \rho v \partial_z v = &- \partial_z p + \frac{1}{r}\partial_r\Big(\eta r \partial_z u + \eta r \partial_r v\Big) \\ &+ \partial_z\Big(2\eta \partial_z v + (\kappa - \tfrac{2}{3}\eta)\big(\frac{\partial_r(ru)}{r} + \partial_z v\big)\Big) + \rho g, \end{aligned} \tag{12.2.3}$$

where ρ is the mass density, κ is the volume viscosity and η is the shear viscosity. On the other hand, the conservation of energy is in the form

$$\begin{aligned} \rho u \partial_r h + \rho v \partial_z h = &- \frac{1}{r}\partial_r(r q_r) - \partial_z q_z + u \partial_r p + v \partial_z p \\ &+ 2\eta\Big(\big(\partial_r u\big)^2 + \big(\frac{u}{r}\big)^2 + \big(\partial_z v\big)^2\Big) \\ &+ (\kappa - \tfrac{2}{3}\eta)\big(\frac{\partial_r(ru)}{r} + \partial_z v\big)^2 + \eta\big(\partial_z u + \partial_r v\big)^2, \end{aligned} \tag{12.2.4}$$

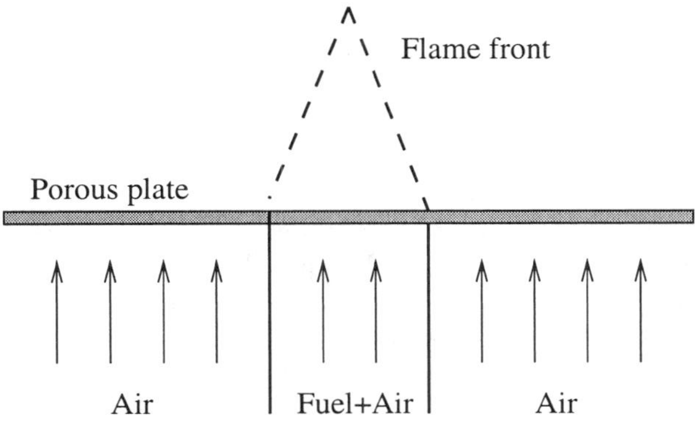

Figure 1. Schematic diagram of the flame.

and the species conservation equations are

$$\rho u \partial_r Y_k + \rho v \partial_z Y_k = -\frac{1}{r}\partial_r\left(r\rho Y_k \mathcal{U}_k\right) - \partial_z\left(\rho Y_k \mathcal{V}_k\right) + m_k\omega_k, \qquad k \in S, \tag{12.2.5}$$

where Y_k is the mass fraction of the kth species, $\boldsymbol{\mathcal{V}}_k = (\mathcal{U}_k, 0, \mathcal{V}_k)^t$ is the radial and vertical components of the diffusion velocity $\boldsymbol{\mathcal{V}}_k$ of the kth species, m_k is the molar weight of the kth species, ω_k is the molar production rate of the kth species, n is the number of species, p is the thermodynamic pressure, κ is the volume viscosity, η is the shear viscosity, h is the mixture specific enthalpy, and $\boldsymbol{\mathcal{Q}} = (q_r, 0, q_z)^t$ are the radial and vertical components of the heat flux $\boldsymbol{\mathcal{Q}}$, respectively.

The radial and vertical components of the species diffusion velocities $\boldsymbol{\mathcal{V}}_k = (\mathcal{U}_k, 0, \mathcal{V}_k)^t$, $k \in S$, are taken in the form

$$\mathcal{U}_k = -\sum_{l\in S} D_{kl}(\partial_r X_l + \chi_l \partial_r \log T), \qquad k \in S, \tag{12.2.6}$$

$$\mathcal{V}_k = -\sum_{l\in S} D_{kl}(\partial_z X_l + \chi_l \partial_z \log T), \qquad k \in S, \tag{12.2.7}$$

where D_{kl}, $k, l \in S$, are the multicomponent diffusion coefficients, X_l is the mole fraction of the lth species, T is the absolute temperature, and $\chi_l = X_l\widetilde{\chi}_l$ is the thermal diffusion ratio of the lth species. Furthermore, the radial and vertical components of the heat flux $\boldsymbol{\mathcal{Q}} = (q_r, 0, q_z)^t$ are in the form

$$q_r = -\lambda\partial_r T + \sum_{k\in S}(\rho Y_k h_k + p\chi_k)\mathcal{U}_k, \tag{12.2.8}$$

$$q_z = -\lambda\partial_z T + \sum_{k\in S}(\rho Y_k h_k + p\chi_k)\mathcal{V}_k, \tag{12.2.9}$$

where λ is the thermal conductivity and h_k is the enthalpy per unit mass of the kth species.

These equations have to be completed by the relations expressing thermodynamic properties, chemical production rates, and transport coefficients. These relations have been presented in the previous chapters and directly apply to this flame problem.

An alternative to the preceding model is that of the isobaric approximation, which is fully justified for laminar flames, since typical gas velocities are at most a few meters per seconds. The corresponding equations are straightforward to derive and are omitted. In this context, it is also possible to use a vorticity–velocity formulation, as described in Chapter 3 [ES93] [EGS96] [EG98b].

12.2.2. Boundary conditions

The computational domain is taken in the form $[0, R] \times [0, Z]$, the tube inner radius is denoted by r_{i}, the tube outer diameter is denoted by r_{o}, with $0 < r_{\text{i}} < r_{\text{o}} < R$, and $r_{\text{o}} - r_{\text{i}}$ is the tube width.

On the lower inflow boundary $z = 0$, $0 \leq r \leq R$, we impose the radial velocity, vertical velocity, and temperature

$$u(r, 0) = 0, \qquad v(r, 0) = v^{\text{in}}(r), \qquad T(r, 0) = T^{\text{f}}, \tag{12.2.10}$$

where $v^{\text{in}}(r)$ is a given function of r and T^{f} is the ambient temperature. The inflow velocity $v^{\text{in}}(r)$ is zero above the tube width $r_{\text{i}} \leq r \leq r_{\text{o}}$ and positive inside the tube $0 \leq r < r_{\text{i}}$ and in the air coflow jet $r_{\text{o}} < 0 \leq R$.

Denoting by Y_k^{f}, $k \in S$, the mass fractions of the premixed mixture flowing in the tube and by Y_k^{air}, $k \in S$, the air coflow mass fractions, we impose flux boundary conditions for the species in the form

$$\rho(r, 0) Y_k(r, 0) \big(v(r, 0) + \mathcal{V}_k(r, 0) \big) = \rho(r, 0) Y_{k0}(r) v(r, 0), \qquad k \in S, \tag{12.2.11}$$

where $Y_{k0}(r) = Y_k^{\text{f}}$, $k \in S$, inside the tube $0 \leq r \leq r_{\text{i}}$ and $Y_{k0}(r) = Y_k^{\text{air}}$, $k \in S$, in the coflow $r_{\text{o}} \leq r \leq R$.

At the outflow boundary $z = Z$, $0 \leq r \leq R$, we impose

$$u(r, Z) = 0, \qquad \partial_z v(r, Z) = 0, \qquad p(r, Z) = p^{\text{atm}}, \tag{12.2.12}$$

where p^{atm} is the atmospheric pressure, and

$$\partial_z T(r, Z) = 0, \qquad \partial_z Y_k(r, Z) = 0, \quad k \in S. \tag{12.2.13}$$

On the axis of symmetry, $r = 0$, $0 \leq z \leq Z$, we also have

$$u(0, z) = 0, \qquad \partial_r v(0, z) = 0, \qquad \partial_r p(0, z) = 0, \tag{12.2.14}$$

TABLE 1: Hydrogen-Air Reaction Mechanism
Coefficients in the form $K = \mathfrak{A}_i T_i^{\mathfrak{b}} \exp(-\mathfrak{E}_i/RT)$.

i	REACTION	$\mathfrak{A}_i$	$\mathfrak{b}_i$	$\mathfrak{E}_i$
1.	$H_2 + O_2 \rightleftharpoons 2OH$	1.70E+13	0.00	47780.
2.	$OH + H_2 \rightleftharpoons H_2O + H$	1.17E+09	1.30	3626.
3.	$H + O_2 \rightleftharpoons OH + O$	2.00E+14	0.00	16800.
4.	$O + H_2 \rightleftharpoons OH + H$	1.80E+10	1.00	8826.
5.	$H + O_2 + M \rightleftharpoons HO_2 + M^a$	2.10E+18	-1.00	0.
6.	$H + O_2 + O_2 \rightleftharpoons HO_2 + O_2$	6.70E+19	-1.42	0.
7.	$H + O_2 + N_2 \rightleftharpoons HO_2 + N_2$	6.70E+19	-1.42	0.
8.	$OH + HO_2 \rightleftharpoons H_2O + O_2$	5.00E+13	0.00	1000.
9.	$H + HO_2 \rightleftharpoons 2OH$	2.50E+14	0.00	1900.
10.	$O + HO_2 \rightleftharpoons O_2 + OH$	4.80E+13	0.00	1000.
11.	$2OH \rightleftharpoons O + H_2O$	6.00E+08	1.30	0.
12.	$H_2 + M \rightleftharpoons H + H + M^b$	2.23E+12	0.50	92600.
13.	$O_2 + M \rightleftharpoons O + O + M$	1.85E+11	0.50	95560.
14.	$H + OH + M \rightleftharpoons H_2O + M^c$	7.50E+23	-2.60	0.
15.	$H + HO_2 \rightleftharpoons H_2 + O_2$	2.50E+13	0.00	700.
16.	$HO_2 + HO_2 \rightleftharpoons H_2O_2 + O_2$	2.00E+12	0.00	0.
17.	$H_2O_2 + M \rightleftharpoons OH + OH + M$	1.30E+17	0.00	45500.
18.	$H_2O_2 + H \rightleftharpoons HO_2 + H_2$	1.60E+12	0.00	3800.
19.	$H_2O_2 + OH \rightleftharpoons H_2O + HO_2$	1.00E+13	0.00	1800.

Units are moles, centimeters, seconds, Kelvins, and calories.
[a] Third-body efficiencies : $\alpha_{5\,H_2O} = 21$, $\alpha_{5\,H_2} = 3.3$, $\alpha_{5\,N_2} = 0$, $\alpha_{5\,O_2} = 0$.
[b] Third-body efficiencies : $\alpha_{12\,H_2O} = 6$, $\alpha_{12\,H} = 2$, $\alpha_{12\,H_2} = 3$.
[c] Third-body efficiencies : $\alpha_{14\,H_2O} = 20$.

$$\partial_r T(0,z) = 0, \qquad \partial_r Y_k(0,z) = 0, \quad k \in S, \tag{12.2.15}$$

and, at the outer boundary $r = R$, $0 \le z \le Z$, we write

$$u(R,z) = 0, \qquad \partial_r v(R,z) = 0, \tag{12.2.16}$$

$$T(R,z) = T^{\mathrm{f}}, \qquad Y_k(R,z) = Y_k^{\mathrm{air}}, \quad k \in S. \tag{12.2.17}$$

12.2.3. Chemical mechanism

We consider the chemical reaction mechanism presented in Table 1, which is often used in hydrogen combustion applications. This mechanism contains $n = 9$ reactive species H, O, H_2, O_2, N_2, OH, HO_2, H_2O, and H_2O_2, participating in n^{r} reversible elementary reactions. The third body can be any species present in the mixture and is denoted by the symbol M.

If the reactions are rewritten explicitly with all third bodies, $n^{\mathrm{r}} = 57$ elementary reversible reactions are obtained, whereas, by using the usual notational shortcut M, only $n^{\mathrm{r}} = 19$ formal reactions are needed to describe the chemical mechanism.

12.3. Computational considerations

In this section we discuss some issues concerning computational techniques for complex chemistry laminar flows [Dix84] [OB87] [WMD96] [BPB96] [EGS96] [BS98] [EG98b] [BBR99].

12.3.1. Discretized equations

In order to simulate numerically multicomponent flows with complex chemistry, it is first necessary to discretize spatially the equations. Numerous discretization techniques have been introduced in the literature, such as the finite difference method, finite element method, or finite volume method. Detailed discussions of these techniques are given in the monographs of Peyret and Taylor [PT83], Girault and Raviart [GiRa86], Oran and Boris [OB87], Ferziger and Perić [FP96], Godlewski and Raviart [GoRa96], and Lucquin and Pironneau [LP96].

Assuming that a spatial discretization technique has been selected, the continuous equations are transformed into a set of semidiscrete equations in the form of an algebraic-differential system

$$\mathcal{A}(\mathcal{Z})\partial_t \mathcal{Z} \;+\; \Phi(\mathcal{Z}) = 0,$$

where $\mathcal{Z}(t)$ is the discrete unknown function and $\mathcal{A}(\mathcal{Z})$ is a matrix. These algebraic-differential equations can either be solved globally or by using operator splitting techniques between convection, diffusion, and chemical sources with fractional steps [OB87] [GoRa96] [FP96].

Assuming a coupled resolution and a first-order scheme for illustration, we obtain discrete equations in the form

$$f(\mathcal{Z}^{i+1}) = \mathcal{A}(\mathcal{Z}^{i+\tau})\frac{\mathcal{Z}^{i+1} - \mathcal{Z}^{i}}{t_{i+1} - t_i} \;+\; \Phi(\mathcal{Z}_{i+\tau}) = 0, \qquad i \geq 0,$$

where $\mathcal{Z}^i$ is an approximation of $\mathcal{Z}(t_i)$, $\tau = 1$ for an implicit method, and $\tau = 0$ for an explicit method when $\mathcal{A}$ is easy to invert.

12.3.2. Multiple time scales

An important aspect of complex chemistry flows is the presence of multiple time scales. For compressible flows, we already know that the presence of acoustic waves introduces small characteristic times for small Mach number flows. However, chemistry times are typically distributed between very short times and fairly long times. For combustion applications, for instance, chemical characteristic times can range typically from 10^{-8} seconds up to several seconds. In the presence of multiple time scales, implicit methods are advantageous, since otherwise explicit schemes would be limited by the smallest time scale. In complex chemistry flows, these multiple time scales typically arise from the reaction rate constants, which can differ by several orders of magnitude.

12.3.3. Multiple space scales

A second potential difficulty associated with the multicomponent aspect is the presence of multiple space scales. In combustion applications, for instance, the flame fronts are very thin and typically require space steps of 10^{-3} cm whereas typical flow scales may be of 10 cm. The multiple scales can only be solved by using adaptive grids obtained by successive refinements or by moving grids for unsteady problems [Smo82] [OB87] [Joh90] [BS98].

12.3.4. Nonlinear solvers

Nonlinear discrete equations can be solved by using Newton's method or any generalization. Starting from an initial estimate $\mathcal{Z}_0$ sufficiently close from the solution x^*, the system of discretized equations denoted symbolically

$$f(\mathcal{Z}) = 0,$$

can be solved by using a damped Newton's method

$$J(\mathcal{Z}_k)\,(\mathcal{Z}_{k+1} - \mathcal{Z}_k) = -\alpha_k f(\mathcal{Z}_k), \qquad k = 0, 1, \ldots,$$

where $\mathcal{Z}_k$ denotes the kth iterate, $J(\mathcal{Z}_k) = \partial_{\mathcal{Z}} f(\mathcal{Z}_k)$ denotes the Jacobian matrix, and α_k denotes the damping parameter.

The evaluation of Jacobian matrices is an expensive task, so that a modified Newton's method should be used whenever it is possible. However, the rate of convergence must then be controlled. The Jacobian matrices can be evaluated by finite differences or analytically. The resulting large sparse linear systems must then be solved by using a Krylov-type method, such as GMRES [SS86] [EGKS94]. More sophisticated methods involve coupled Newton–Krylov techniques [KMK96].

Evaluating aerothermochemistry quantities is computationally expensive since they involve multiple sums and products. Optimal evaluation requires a low-level parallelization, e.g., by using vector capabilities of computers, depending on the problem granularity [GD88].

12.3.5. Pseudo-unsteady iterations

Steady solutions are usually obtained by first bringing the initial estimates into the domain of convergence of Newton's method for steady iterations. For this purpose, one can use pseudo-unsteady iterations starting from polynomial estimates or simplified chemistry estimates [Smo82].

12.3.6. Thermochemistry and transport software

It is preferable, when writing numerical software, to clearly separate the numerical tools from the special type of equations that are under concern. In the context of multicomponent flows, it is therefore a good idea to write codes for general mixtures and use libraries that automatically evaluate thermochemistry properties and transport properties.

For thermodynamic properties and chemistry source terms, application independent softwares have been written in the eighties. The most well-known libraries are the CHEMKIN I and CHEMKIN II softwares, which have been used by numerous developers [KMJ80] [KRM87] [KRM89]. For transport properties, a very powerful library EGLIB [EG94] is available at the author's web site [EG96d].

12.4. Hydrogen-air Bunsen flame

In this section we present a typical Bunsen laminar flame structure.

12.4.1. Burner geometry

We consider a lean hydrogen-air Bunsen flame obtained by flowing a lean mixture of 20% hydrogen in air in mole fraction at 300 K and at atmospheric pressure. The tube inner diameter is $r_i = 4$ mm, the tube width is $w = 0.5$ mm, and the burner temperature is kept at 300 K. The flow is of plug type, and the flame is surrounded by a coflow of air. The maximum velocity in both flows is $v_{\rm inj} = 300$ cm/sec. More specifically, denoting by $r_o = r_i + w$ the outer tube diameter, the inflow velocity $v^{\rm in}(r)$ is given by $v^{\rm in}(r)/v_{\rm inj} = 1 - \exp\bigl(-(r_i - r)/\delta\bigr)$ for $0 \leq r \leq r_i$, $v^{\rm in}(r) = 0$ for $r_i \leq r \leq r_o$, and $v^{\rm in}(r)/v_{\rm inj} = 1 - \exp\bigl(-(r - r_o)/\delta\bigr)$ for $r_o \leq r$, where δ is the gradient parameter taken to be $\delta = 0.5$ mm. The computational

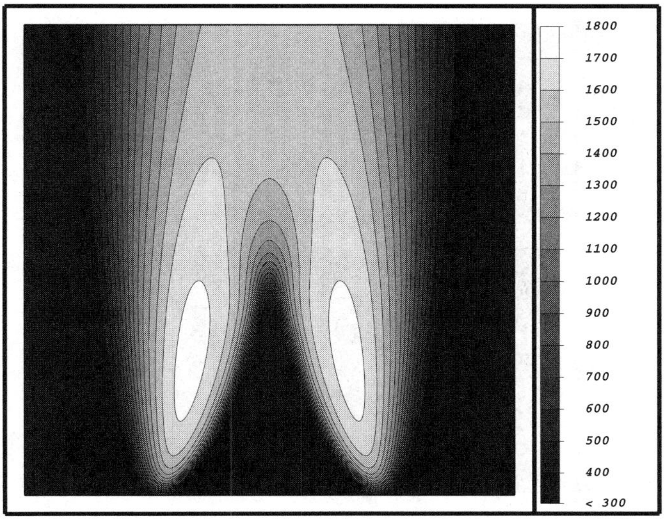

Figure 2. Temperature distribution in a hydrogen-air flame.

domain is [0,1.5]×[0,30] in centimeters and approximately 200 points are adaptively distributed in each direction.

12.4.2. Numerical results

The flame structure—after symmetrization—is illustrated in Figures 2 to 7 in the domain of interest [-0.5,0.5]×[0,1] in centimeters. On these figures, we present the distributions of temperature, H_2, H, OH, H_2O_2 mole fractions, and the absolute velocity $\sqrt{u^2+v^2}$.

The temperature distribution in Figure 2 reveals the cold, dark inner zone of the flame. The inner zone is approximately of conical shape in agreement with the theory that assimilates the flame to a hydrodynamic discontinuity [Wil85]. Ignition takes place in a ring above the tube lip, where the flow velocity is substantially lower than the average velocity in the tube. The maximum temperature is around 1730 K and not reached on the axis. According to the Clavin and Williams theory [CW82], this phenomenon can be attributed to the overall Lewis number of hydrogen—which is lower than one—and to the negative stretch at the cone tip. Radial cooling—towards the coflow—of hot gases produced at the flame front also gradually takes place in the flame. As a result, the maximum temperature zone has the shape of a torus.

Figure 3 presents the hydrogen mole fraction distribution. This quantity is 20% in the cold zone, and hydrogen is fully burnt at the conical flame front. On the other hand, as shown in Figure 4, the H atom mole fraction

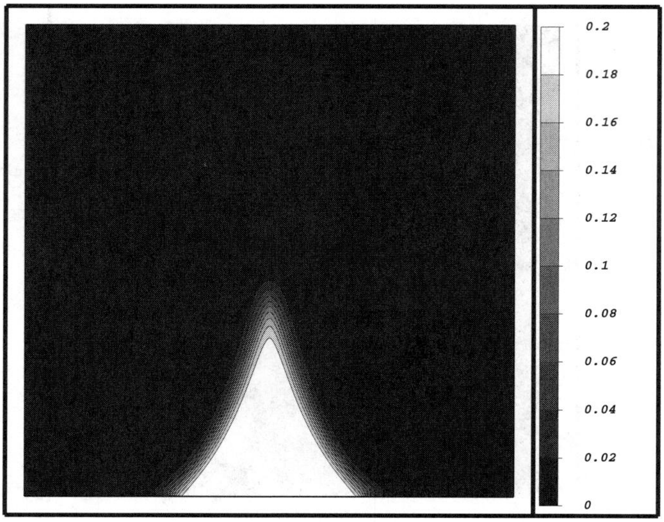

Figure 3. H_2 mole fraction distribution in a hydrogen-air flame.

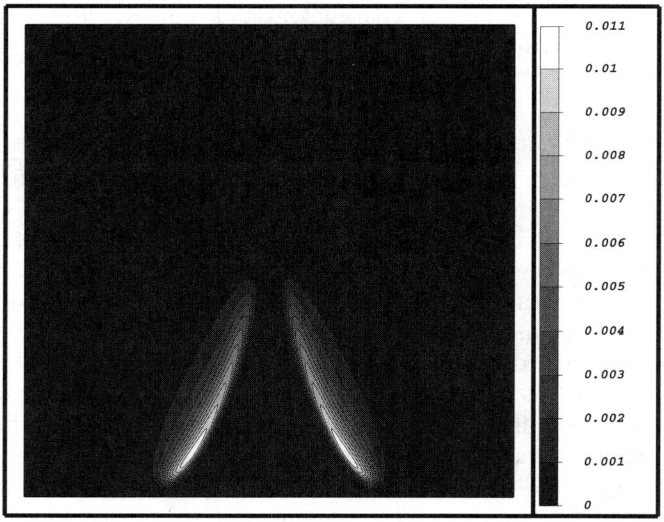

Figure 4. H mole fraction distribution in a hydrogen-air flame.

peaks in the flame front, but is reduced above the ignition ring, at the cone tip, and downstream of the flame, where the temperature is reduced. Similarly, the OH radical isopleths depicted in Figure 5 reveal a maximum

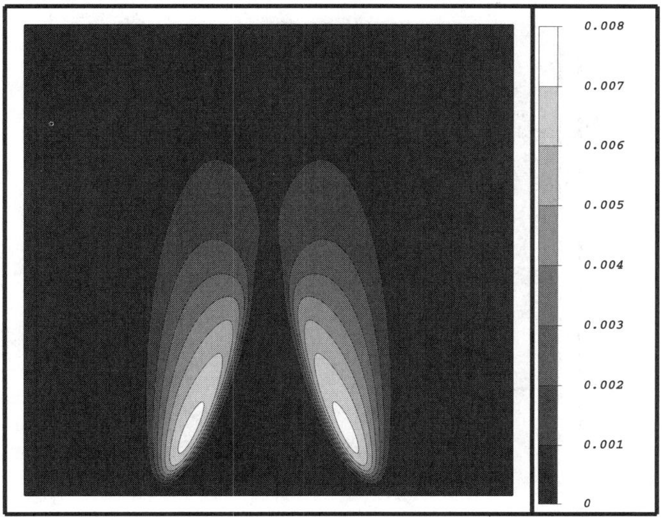

Figure 5. OH mole fraction distribution in a hydrogen-air flame.

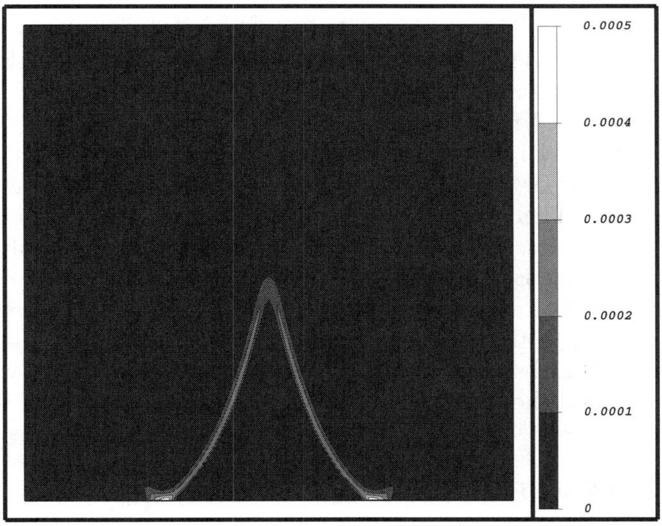

Figure 6. H_2O_2 mole fraction distribution in a hydrogen-air flame.

concentration located behind the flame front that is nonuniform along the cone vertex. The OH radical exhibits a higher maximum where the temperature is larger, but gradually disappears as a result of radial cooling.

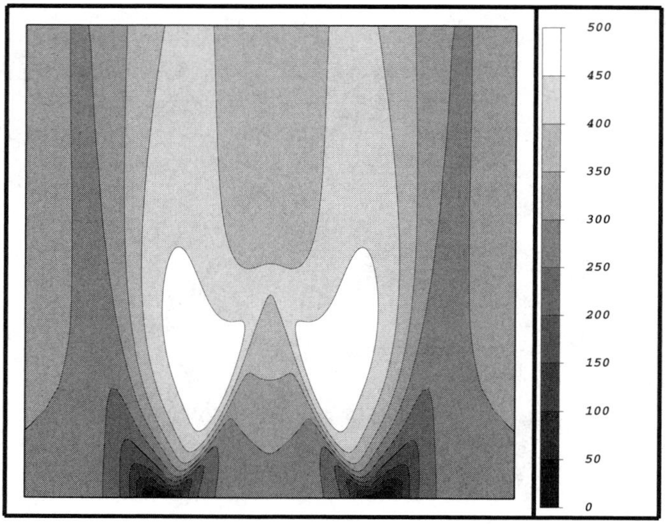

Figure 7. Absolute velocity distribution in a hydrogen-air flame.

The reduction of H and OH at the cone tip can be attributed to the lower temperature, but is also significantly enhanced by thermal diffusion effects, as shown in [EG98b].

As reflected in Figure 6, the H_2O_2 radical is not only present in the flame front, as for one-dimensional flames, but also above the tube lip. On the outer part of the tube, the isopleths for this radical take on the form of a corolla. There is indeed a small amount of hydrogen diffusing in the coflow at the base of the flame and this results in a small diffusion flame.

Finally, the Euclidean norm of the velocity fields $\sqrt{u^2 + v^2}$ is presented in Figure 7. The absolute velocity is lower above the tube lips and reaches its maximum in the flame front as a consequence of gas expansion in the hot zone. The absolute velocity isolines then gradually align vertically due to gravity and the coflow jet.

ACKNOWLEDGMENT

The author is thankful to the Institut du Développement des Ressources Informatiques et Scientifiques (IDRIS, France) for providing the computational resources.

12.5. References

[BPB96] T. Baritaud, T. Poinsot, and Markus Baum, *Direct Numerical Simulation for Turbulent Reactive Flows,* Editions Technip, Paris, (1996).

[BBR99] R. Becker, M. Brack, and R. Rannacher, *Numerical Simulation of Laminar Flames at Low Mach Number by Adaptive Finite Elements,* Comb. Theory Mod., (1999), (in press).

[BS98] B. A. Beth and M. D. Smooke, *Local Rectangular Refinment with Application to Combustion Problems,* Comb. Theory Mod., **2**, (1998), pp. 221–258.

[CW82] P. Clavin and F. Williams, *Effects of Molecular Diffusion and of Thermal Expansion on the Structure and Dynamics of Premixed Flames in Turbulent Flows of Large Scale and Low Intensity,* J. Fluid Mech., **116**, (1982), pp. 251–282.

[Dix84] G. Dixon-Lewis, *Computer Modeling of Combustion Reactions in Flowing Systems with Transport,* in W. C. Gardiner, Ed., *Combustion Chemistry.* Springer-Verlag, New York, (1984), pp. 21–125.

[EG94] A. Ern and V. Giovangigli, *Multicomponent Transport Algorithms,* Lecture Notes in Physics, New Series "Monographs", **m 24**, Springer-Verlag, Berlin, (1994).

[EG96d] A. Ern and V. Giovangigli, *EGlib Server and User's Manual,* http://www.cmap.polytechnique.fr/www.eglib/

[EG98b] Ern A. and V. Giovangigli, *Thermal Diffusion Effects in Hydrogen-Air and Methane-Air Flames,* Comb. Theory Mod., **2**, (1998), pp. 349–372.

[EGS96] A. Ern, V. Giovangigli, and M. Smooke, *Numerical Study of a Three-Dimensional Chemical Vapor Deposition Reactor with Detailed Chemistry,* J. Comp. Phys., **126**, (1996), pp. 21–39.

[EGKS94] A. Ern, V. Giovangigli, D. Keyes, and M. Smooke, *Towards Polyalgorithmic Linear System Solvers for Nonlinear Elliptic Problems,* SIAM J. Sci. Stat. Comp., **15**, (1994), pp. 681–703.

[ES93] A. Ern and M. Smooke, *The Vorticity-Velocity Formulation,* J. Comp. Phys., **105**, (1993), pp. 58–68.

[FP96] J. H. Ferziger and M. Perić, *Computational Methods for Fluid Dynamics,* Springer-Verlag, Berlin, (1996).

[GD88] V. Giovangigli and N. Darabiha, *Vector Computers and Complex Chemistry Combustion,* in C. Brauner and C. Schmidt-Laine, Eds., *Proc. Conference Mathematical Modeling in Combustion and Related Topics,* NATO Adv. Sci. Inst. Ser. E, Martinus Nijhoff Publishers, Dordrecht, **140**, (1988), pp. 491–503.

[GiRa86] V. Girault and P. A. Raviart, *Finite Element Methods for the Navier–Stokes Equations,* Springer-Verlag, Heidelberg, (1986).

[GoRa96] E. Godlewski and P. A. Raviart, *Numerical Approximation of Hyperbolic Systems of Conservation Laws,* Springer-Verlag, New York, (1996).

[HH98] G. Hauke and T. J. R. Hughes, *A Comparative Study of Different Sets of Variables for Solving Compressible and Incompressible Flows,* Comp. Meth. Appl. Mech. Eng., **153**, (1998), pp. 1–44.

[Hug87] T. J. R. Hughes, *Recent Progress in the Development and Understanding of SUPG methods with Special Reference to the Compressible Euler and Navier–Stokes Equations,* in R. H. Gallagher, R. Glowinsky, P. M. Gresho, J. T. Oden, and O. C. Zienkiewicz, Eds., *Finite Elements in Fluids*, John Wiley & Sons, Ltd, Chichester, (1987), pp. 273–287.

[Joh90] C. Johnson, *Adaptive Finite Element Methods for Diffusion and Convection Problems,* Comp. Meth. Appl. Mech. Eng., **82**, (1990), pp. 301–322.

[KMJ80] R. J. Kee, J. A. Miller, and T. H. Jefferson, *Chemkin: A General-Purpose, Problem-Independent, Transportable, Fortran Chemical Kinetics Code Package,* SANDIA National Laboratories Report, SAND80-8003, (1980).

[KRM87] R. J. Kee, F. M. Rupley, and J. A. Miller, *The Chemkin Thermodynamic Data Base,* SANDIA National Laboratories Report, SAND87-8215, (1987).

[KRM89] R. J. Kee, F. M. Rupley, and J. A. Miller, *Chemkin II: A Fortran Chemical Kinetics Package for the Analysis of Gas Phase Chemical Kinetics,* SANDIA National Laboratories Report, SAND89-8009B, (1989).

[KMK96] D. A. Knoll, P. R. McHugh, and D. E. Keyes, *Newton–Krylov Methods for Low–Mach–Number Compressible Combustion,* AIAA J., **34**, (1996), pp. 961–967.

[LP96] B. Lucquin and O. Pironneau, *Introduction au Calcul Scientifique,* Masson, Paris, (1996).

[OB87] E. Oran and J. P. Boris, *Numerical Simulation of Reactive Flows,* Elsevier, New York, (1987).

[PT83] R. Peyret and T. Taylor, *Computational Methods for Fluid Flow,* Springer-Verlag, New York, (1983).

[SS86] Y. Saad and M. H. Schultz, *GMRES: A Generalized Minimal Residual Algorithm for Solving Nonsymmetric Lineae Systems,* SIAM J. Sci. Stat. Comp., **7**, (1986), pp. 856–869.

[Smo82] M. D. Smooke, *Solution of Burner Stabilized Premixed Laminar Flames by Boundary Value Methods,* J. Comp. Phys., **48**, (1982), pp. 72–105.

[WMD96] J. Warnatz, U. Maas, and R. W. Dibble, *Combustion,* Springer-Verlag, Berlin, (1996).

[Wil85] F. A. Williams, *Combustion Theory,* Second ed., The Benjamin/Cummings Publishing Company, Inc., Menlo Park, (1985).

Index